160

THE HOLT SCIENCE PROGRAM

MODERN SCIENCE: Earth, Life and Man by Blanc, Fischler, and Gardner
MODERN SCIENCE: Matter, Energy and Space by Blanc, Fischler, and Gardner
MODERN SCIENCE: Forces, Change and The Universe by Blanc, Fischler, and Gardner
   TEACHER'S EDITION MODERN SCIENCE: Earth, Life and Man
   EXERCISES AND INVESTIGATIONS FOR MODERN SCIENCE: Earth, Life and Man
   TESTS FOR MODERN SCIENCE: Earth, Life and Man

OTHER TITLES IN THE HOLT SCIENCE PROGRAM

MODERN ELEMENTARY SCIENCE (GRADES 1–6), by Fischler, Lowery, and Blanc

SCIENCE 1: OBSERVATION AND EXPERIMENT, by Davis, Burnett, Gross, and Pritchard
SCIENCE 2: EXPERIMENT AND DISCOVERY, by Davis, Burnett, Gross, and Pritchard
SCIENCE 3: DISCOVERY AND PROGRESS, by Davis, Burnett, Gross, and Johnson

MODERN PHYSICAL SCIENCE, by Tracy, Tropp and Friedl
MODERN EARTH SCIENCE, by Ramsey, Burckley, Phillips and Watenpaugh
MODERN LIFE SCIENCE, by Fitzpatrick and Hole
LIVING THINGS, by Fitzpatrick, Bain, and Teter

MODERN HEALTH, by Otto, Julian, and Tether
MODERN SEX EDUCATION, by Julian and Jackson

MODERN BIOLOGY, by Otto and Towle
MODERN PHYSICS, by Williams, Metcalfe, Trinklein and Lefler
MODERN CHEMISTRY, by Metcalfe, Williams, and Castka

SELECTIONS FROM THE HOLT LIBRARY OF SCIENCE

ANTARCTICA, by Eklund and Beckman
COMPUTERS, by Thomas
EXPERIMENTS IN PSYCHOLOGY, by Blough and Blough
FLYING, by Caidin
GLACIERS AND THE ICE AGE, by Schultz
OCEANOGRAPHY, by Yasso
OPTICS, by Gluck
OUR ANIMAL RESOURCES, by Fitzpatrick
OUR PLANT RESOURCES, by Fitzpatrick
OURS IS THE EARTH, by Sollers
PHOTOSYNTHESIS, by Rosenberg
A TRACER EXPERIMENT, by Kamen
VIRUSES, CELLS, AND HOSTS, by Siegel and Beasley
WINGS INTO SPACE, by Caidin

MODERN SCIENCE

Earth, Life and Man

SAM S. BLANC
ABRAHAM S. FISCHLER
OLCOTT GARDNER

HOLT, RINEHART AND WINSTON, INC.
New York • Toronto • London • Sydney

THE AUTHORS OF MODERN SCIENCE: EARTH, LIFE AND MAN

Dr. Sam S. Blanc is an Assistant Professor of Elementary Education at San Diego State College, San Diego, California; formerly, Director of Science and Mathematics, Cajon Valley School District, El Cajon, California.

Dr. Abraham S. Fischler is Dean of the Education Center and Professor of Science Education at Nova University, Fort Lauderdale, Florida; formerly, Professor of Science Education, University of California, Berkeley, California.

Mr. Olcott Gardner is a science teacher at the Jamesville-Dewitt High School, Dewitt, New York.

CONSULTING SCIENTIST OF THE HOLT SCHOOL SCIENCE ADVISORY BOARD

Dr. John H. Pomeroy is Assistant Director, Lunar Sample Program, NASA Headquarters, Washington, D.C.

Copyright © 1971 by
HOLT, RINEHART AND WINSTON, INC.
Parts of book formerly copyrighted © 1967 under the title of **MODERN SCIENCE: Earth, Space and Environment** and copyrighted © 1963 under the title of **MODERN SCIENCE 1** by
Holt, Rinehart and Winston, Inc.
  All Rights Reserved
Printed in the United States of America
ISBN: 0-03-084224-7
2 3 4 5 6 7 8 9 0   0 7 1   9 8 7 6 5 4 3

Preface

The basic knowledge about our world is rapidly increasing as a result of the advances in all scientific fields. This expansion often results in the application of much of this knowledge in our modern society. Therefore, the understanding of basic principles underlying modern science becomes increasingly important for all students.

The junior high school is often the starting point for students who elect scientific careers in later life. However, the content and processes of science can be a challenge to every young person, regardless of his future vocational or professional plans. Science can provide all youth with an appreciation and understanding of his relationship with his environment. The impact of science on the personal, social, and economic lives of all people in our country makes a scientifically literate citizenry essential.

The **MODERN SCIENCE SERIES** presents a modern approach to the teaching of science at the junior high school level with emphasis on fundamental principles and concepts. Many key concepts treated in Holt, Rinehart and Winston's **Modern Elementary Science Program (Grades 1–6)** by Fischler, Lowery and Blanc are studied in greater depth in the **Modern Science Series.** These concepts are brought together in their presentation as part of the four major units in each book.

In **Modern Science: Earth, Life and Man,** the student explores the world in which he lives. With the first unit, the student studies the earth, its composition, and importance in serving man. The student is led to appreciate the preservation of his environment in the second unit. A discussion of the atmosphere and the ocean and the changes that take place in this part of his environment, emphasize for the student the factors that make life possible on our planet. With the third unit, the relationship of man to the living things on the earth, and the interrelationships found among organisms in the balance of life are explored. The last unit concerns the main systems of the human body. In it, the student studies subjects such as the release of energy from food, the structure, functions and care of the body. The four main units in this book then, present the general areas of the earth, the surrounding environment, living things and the human body.

All the concepts and principles treated in each unit relate to the central theme of the unit. They offer the student an opportunity to explore the field in depth and to experience some of the excitement that comes with the discovery of new ideas. The major concepts included in **Modern Science: Earth, Life and Man** are generally not duplicated in **Modern Science: Matter, Energy and Space.** However, when it has been necessary to reintroduce certain concepts in either book, they are used to reinforce some basic understandings or to focus on new experiences of major importance.

The **MODERN SCIENCE SERIES** develops a three-year program in science for grades seven, eight, and nine that makes it possible to teach the topics in reasonable depth, to allow students to engage in actual laboratory experimentation, and to explore individual and group interests. The program includes supplementary matter for each book as follows:

1) Teacher's Edition.
2) Exercises and Investigations for Modern Science.
3) Tests for Modern Science.
4) Key to Tests for Modern Science.

The authors are sincerely grateful to Dr. John H. Pomeroy of the Holt, Rinehart and Winston School Science Advisory Board for his valued contributions to and general evaluation of the **MODERN SCIENCE SERIES.**

Contents

Using Your Book Scientifically — X
Your Exercises and Investigations and Tests — XII
The Ways of Science — XV

MAN EXPLORES THE EARTH — 1 — UNIT ONE

1 The Structure of the Earth — 3
  A What Is the Earth's Composition? — 3
  B What Makes Up the Crust of the Earth? — 8
  C How Are Minerals Identified? — 15

2 The Changes in the Earth — 24
  A What Is the Earth's History? — 24
  B How Have Internal Forces Changed the Earth? — 35
  C How Have External Forces Changed the Earth? — 42

3 The Resources of the Earth — 52
  A How Are Soil Resources Conserved? — 52
  B How Are Water Resources Conserved? — 60
  C How Are Mineral Resources Conserved? — 64

MAN EXPLORES HIS ENVIRONMENT — 77 — UNIT TWO

4 The Changing Atmosphere — 79
  A How Do Atmospheric Conditions Change? — 79
  B How Is Our Earth Heated? — 85
  C What Are the Properties of Air? — 89
  D What Is the Composition of Air? — 95

5 The Changing Oceans — 106
  A How Do Scientists Study the Ocean? — 106
  B What Do We Know About the Ocean? — 115
  C How Does the Ocean Affect Us? — 123

6 The Changing Weather — 134
  A What Causes Winds? — 134
  B What Causes Clouds? — 142
  C Why Does It Storm? — 145
  D How Is Weather Forecast? — 152
  E What Determines the Climate? — 157

UNIT THREE — MAN EXPLORES LIVING ORGANISMS — 171

7 The Nature of Life — 173

- A What Is a Living Organism? — 173
- B How Do Organisms Produce and Consume Food? — 180
- C How Do Living Organisms Reproduce? — 186
- D How Are Organisms Classified? — 193

8 The Relationships of Life — 203

- A How Do Organisms Affect Each Other? — 203
- B What Forms of Life Are Found on Land? — 211
- C What Forms of Life Are Found in Inland Waters? — 221
- D What Forms of Life Are Found in the Oceans? — 229

9 The Diversity of Life — 238

- A What Are the Main Plant Groups? — 238
- B What Are the Characteristics of Higher Plants? — 245
- C What Are the Characteristics of Single-celled Animals? — 251
- D What Are the Characteristics of Invertebrate Animals — 255
- E What Are the Characteristics of Vertebrate Animals? — 266

10 The Continuity of Life — 281

- A How Are Characteristics in Living Organisms Transmitted? — 281
- B Why Have Living Organisms Changed? — 294
- C How Are Living Organisms Conserved? — 300

UNIT FOUR — MAN EXPLORES THE HUMAN BODY — 315

11 Energy for the Body — 317

- A Why Does the Body Need Food? — 317
- B What Are the Main Classes of Nutrients? — 323
- C How Is Our Food Digested? — 333
- D How Is Energy Released from Food? — 338

12 Functions of the Body — 345

- A How Do Cells Function? — 345
- B How Are Nutrients Transported? — 352
- C How Is Movement Controlled? — 365

13 Growth of the Body — 374
- A What Changes Occur in the Body? — 374
- B What Is Inheritance? — 379
- C What Are the Chemical Regulators? — 383
- D How Does the Body React? — 387

14 Health of the Body — 397
- A What Causes Disease? — 397
- B How Do Microorganisms Infect Us? — 403
- C What Are Our Natural Defenses? — 409
- D What Are Organic Diseases? — 414
- E What Are Habit-Forming Drugs? — 418

Words in Science and Key to Pronunciation — 428
Index — 441
Acknowledgments — 449

Using Your Book Scientifically

As the study of science requires the activity of the mind in addition to the necessary memorizing of facts, this book is planned to help you discover *scientific principles* and apply them. You must, of course, know facts before you can reason about them. This book provides the tools for understanding and applying this knowledge. Scientific information is the sum of present knowledge about the world and you, which has been brought together and organized. You, the student, absorb this knowledge by *reading, remembering, and recording each fact.* This is a process which must take place before you can relate or connect one idea or fact with the next.

There will be many new words in your study of science. *Each scientific term has an exact meaning.* Be sure that you understand what each new word means. Try to make these scientific terms part of your vocabulary. To communicate with others in science, you must know the words if you are to express your ideas accurately. To help you in this way, the important words in the text have been italicized, pronounced, and explained. The main terms, with a *Key to Pronunciation*, are listed in *Words in Science* at the back of the book.

After you understand the meanings of the words, work at learning the meaning of each sentence, then each paragraph. Then, relate what is in each paragraph to the topic you are studying. At the end of each section, try to tell yourself what the book said without looking at the text. At the end of each section, and without marking the book, try to answer the questions in *Review*. Remember that you are learning new facts to help you solve problems and to understand scientific principles. Also, try to find ways of using these facts and principles in understanding events in your daily life.

Do as many of the *Student Activities* in the book as you can. The activities will help you to better understand a statement of fact, and they will help you solve the problems presented. The information you acquire from all the activities should help you to form correct conclusions about many scientific problems. In addition, you can explain in your own words how some of the scientific ideas you have learned are applied in modern science.

Most of the activities can be done in one class period or less. Some of them may be carried out individually or with friends at home. As you handle the equipment and chemicals called for by the activities, try to foresee accidents. Remember that *the safe way to work is the best way*.

Sample Problems are types of problems that occur inside some chapters. They help you to use mathematics in understanding some scientific ideas explained in these chapters.

Proper use of the textbook should include careful examination of the many pictures and drawings within the text. Doing this will help you to understand more clearly what you have read. Looking at these drawings and photographs and relating them to the discussion develop your ability to observe and make connections between related things in your mind. In addition, the pictures and drawings show how the materials for many activities are arranged.

Have you ever read a page in a book, and when you finished, could not remember what it was all about? This happens unless you have learned to concentrate. That is, you must not be thinking of other things while you are studying your assignment. To repeat, reading in science is to give you facts to help solve problems. To help you do this better, at the end of each section is a *Review* in the form of five questions. This Review is designed to help you find out if you are understanding and remembering the main points.

There are questions and problems at the end of each chapter that will help you in reviewing. One section is called *Thinking With Science*. Each chapter lists some projects called *Research in Science*. You can do these projects on your own or with the help of your teacher as a club activity. Perhaps you can think of an original science project of your own. Another section, *Mathematics In Science,* accustoms you to a main tool of science.

Each unit starts with a brief history and pictures of some of the famous scientists who helped develop the principles and laws on which the unit is based. At the close of the unit, under the heading *Readings In Science,* you will find a list of interesting, up-to-date books to fill out your knowledge of what you have been studying.

Science is an adventure. It is really "seeing" with all your senses. It can provide the thrill of discovering by yourself the puzzle of how and why things happen. Read thoroughly, observe carefully, and study hard. Above all, *enjoy* learning science.

EXERCISES AND INVESTIGATIONS and TESTS

Accompanying the textbook is *Exercises and Investigations for Modern Science: Earth, Life and Man.* In addition to learning basic scientific facts and principles, a student needs to develop skills in scientific thinking. The questions and activities in *Exercises and Investigations* are designed to review the material that you are studying and to set it in your mind. Working out the problems is the reinforcement of your classroom lessons and your textbook reading. *Exercises and Investigations* is organized according to the units and chapters in **Modern Science: Earth, Life and Man.** It includes an EXERCISE and two or more INVESTIGATIONS to correspond with each chapter. Included is a Practice Test for each unit.

With each EXERCISE, you are led to think through the development of the material, to apply your knowledge in simple problem situations, and to relate the facts and principles in each section by means of a review of terms and a general statement at the end.

The INVESTIGATIONS are of two types: one which requires no laboratory materials and the other which does. Not all science classrooms have all the equipment and facilities that are needed in a laboratory course. For this reason, INVESTIGATIONS have been included that present an Experimental Situation not needing laboratory facilities for its solution. Each INVESTIGATION is described in detail and presents a group of data for you, the student, to use. You are then asked to apply the results of this experimental situation by means of a series of questions in the Interpretation. To check further your understanding of the Problem and its solution, you then make an Application of the basic principle in a common situation.

When common laboratory materials are available, and when you are able to carry on independent laboratory study, the INVESTIGATIONS which require such materials will provide you with a challenge. These INVESTIGATIONS present a problem, a list of materials, and a simplified procedure. At the completion of the laboratory work, you answer a series of questions in the Interpretation based on your observations. An Application is included that relates and connects your knowledge.

Through the use of this workbook to extend your science experiences, you will show increased skills in your understanding of the facts and principles of science. In addition, the activities devoted to developing methods of investigation and thinking should result in a more advanced understanding of the methods of science: (1) identification of a problem, (2) skills in observation, (3) experimental procedures, (4) interpretating data, and (5) drawing conclusions.

The *Tests for Modern Science: Earth, Life and Man* follow the organization of your textbook. There is a test for each chapter, one for each of the four units, and a final test of the textbook. Each chapter test covers the main facts and principles included in the chapter. The first two parts of a chapter test contain short-answer type questions designed to test your recall and understanding of the factual knowledge. The third part of a chapter test is designed to challenge you to think, draw conclusions, see relationships, and show a mathematical understanding of science.

The unit tests are broader in scope and emphasize applications of science principles as well as your recall of facts. The final test measures your knowledge of (1) the general fields of science, laboratory materials, units of measurement, etc., (2) the implications of the major laws of theories, and (3) the underlying principles of science.

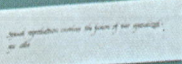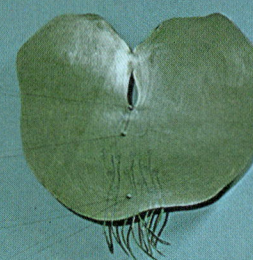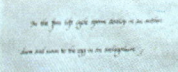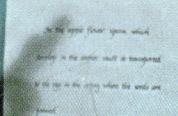

xxii / THE WAYS OF SCIENCE

Inferences depend upon the amount of knowledge a scientist has and the *insight* he brings to the problem. Insight is the degree of understanding a scientist possesses. One scientist may "see" more deeply than others into a particular problem. For example, it was known for many years that molds prevented bacterial growth. It remained for *Sir Alexander Fleming (1881–1955)*, a British scientist, to *see* the possibility of producing a drug from molds to fight body diseases caused by bacteria. The result of his efforts was the wonder drug, *penicillin* (Fig. B-4).

Scientists develop theories to explain the world around them. When the scientist has uncovered a group of facts about an occurrence in nature, he tries to find a relationship or generalization among them. A theory, which is an explanation of this grouping of facts, is proposed. The theory or explanation is a mental picture or model created by the scientist to fit the facts. The model will be further tested by future observations. Additional facts may help to strengthen the theory or to destroy the theory. New facts may destroy an established theory, too (Fig. B-5).

You have probably learned that an atom is made up of a nucleus and a number of particles moving around a nucleus. But have you ever seen an atom? Has any scientist seen an atom? No! The picture of an atom which you have in your mind is a *mental model* that fits only the known facts about atomic behavior. As such, it is a useful scientific theory until a better one replaces it (Fig. B-6, page XXIII).

Sir Isaac Newton (1642–1727), based upon his observations of objects attracted to the earth, developed the important *law of universal gravitation.* This natural law explains the attraction among all physical objects and includes a description of the motion of the planets around the sun. *Gregor Mendel (1822–1884)*, an Austrian monk and scientist, discovered the *laws of heredity* from facts uncovered in a careful study of pea plants. *Albert Einstein (1879–1955)* was born in Germany; he later became an American citizen. One of the truly great scientists, he developed the *theory of relativity* which became the modern basis for understanding the behavior and relationship of matter and energy.

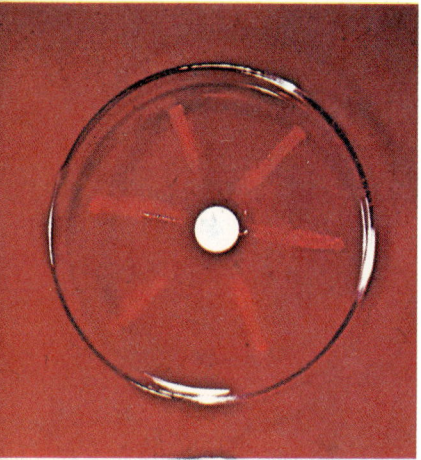

B-4 The bacteria shown here as streaks cannot grow. The Petri dish has been soaked with fluids produced by molds.

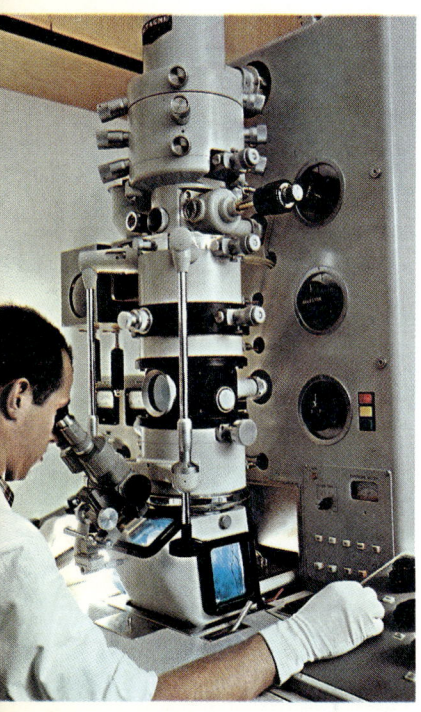

B-5 The electron microscope helps scientists study the smallest living things. How can this help him form new theories?

THE WAYS OF SCIENCE / xxi

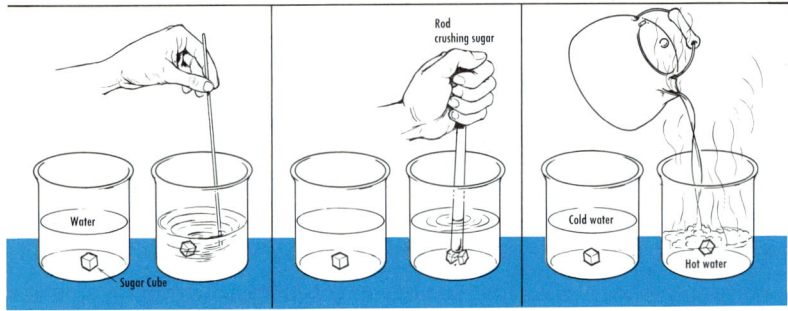

B-3 Why does the drawing illustrate an example of a controlled experiment?

You have probably noticed that an automobile standing outside in the cold for some time develops frost on the inside windows. What hypotheses can you make to account for the frost? At this point it is necessary to determine which of your hypotheses is correct and then to design experiments.

A scientist designs experiments to test variables in the problem. Unless you plan your experiments carefully, the results may not be clear. A result may be due to one or several different *variables*. Variables are the factors or conditions that can change in an experiment. In a *controlled experiment* only one variable factor is allowed to change. The other variables are controlled.

For example, how can you make a solid dissolve faster in water? You might say, "Stir it up," or "Crush it into small pieces," or "Heat the water." What you have done, then, is to point out three variables: (1) movement of the water, (2) size of the pieces, and (3) temperature of the water. Remember: In designing a worthwhile experiment, you must allow *only one factor to change at a time* (Fig. B-3).

A scientist will repeat each trial more than once to check his results. He does this to eliminate possible sources of errors. Arriving at the same results gives the scientist confidence. Furthermore, it is necessary that he organize his findings in his investigations. Others reading his published results can then investigate the research on their own.

A scientist studies his results to form inferences. *Inferences* are logical conclusions drawn from the findings of the research. It is possible for two scientists studying the same results to form inferences that are entirely different.

Your ability to make accurate observations depends upon the use of all your senses and your past experience with what you are studying. Answer the following question with each of your five senses: What differences have you observed between a morning in the fall and a morning in the spring?

The scientist usually develops a certain plan of attacking a problem. Although there is no one scientific way to solve a problem, the following might summarize the steps involved in the solution of many problems.
1. Stating the problem clearly.
2. Developing hypotheses (intelligent guesses).
3. Experimenting to test each hypothesis.
4. Testing several times to verify the results.
5. Drawing conclusions from the results.

A scientist must state the problem clearly. Stating the problem is often difficult. Usually, the difficulty lies in trying to narrow down the problem area to a small, specific problem.

A scientist tries to find a logical explanation for the problem. Once a problem is carefully stated, more study is done to develop reasonable solutions. Each possible solution that is tried is an intelligent guess or hypothesis. The scientist may test many *hypotheses* (plural for hypothesis) to learn which is correct (Fig. B-2). Though several hypotheses prove to be incorrect, each trial adds some knowledge.

B-2 Why do scientists test their hypotheses?

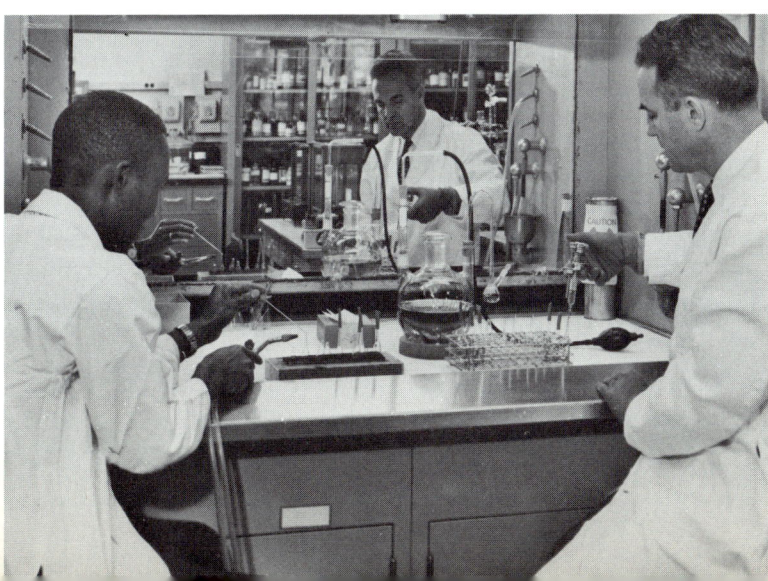

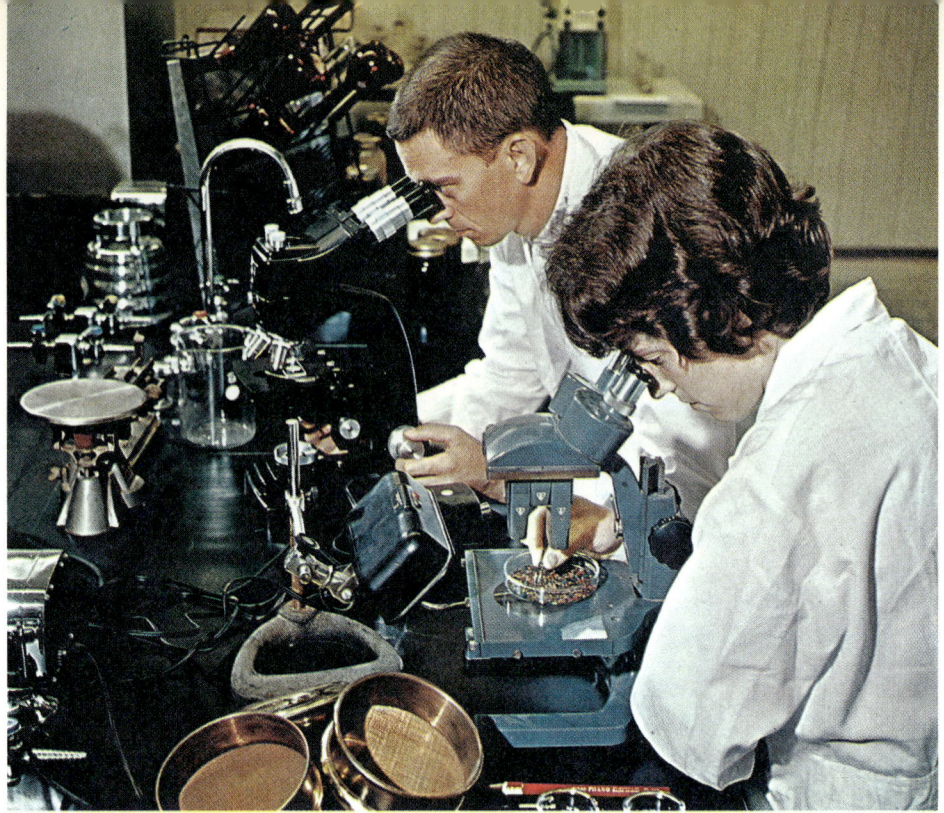

B-1 What is the first step in scientific study?

as the next step in the investigation. He attempts to gain as much background information as possible. This helps him to develop an intelligent guess or *hypothesis* (hye-PAH-thih-sihs) explaining the problem. After formulating an hypothesis, he experiments to determine whether or not it is correct. The results of his experiment may lead to an experimental *fact* or *truth*. He will repeat the experiments to check his results. He then publishes his findings to inform other scientists and to allow them to check for errors. This system of checking requires that the scientist be extremely accurate and honest in writing his experimental results.

After the scientist gathers enough experimental facts, he develops a *generalization*, a broad statement based on existing facts. Many generalizations related to a particular happening help him to state a *theory* which attempts to answer the question "How?" or "Why?" A theory may be changed or even abandoned after further observation and information. Can you think of theories that have changed?

xxiv / THE WAYS OF SCIENCE

accurate measurement. Scientists like to record *quantitative* (KWAHN-tih-*tay*-tihv) *differences* because they reveal more exact information (Fig. C-1). The question "How much?" leads you to answer in a quantitative way. Do the following activity to see how you can measure differences in temperature.

RECORD

Pour some water into a beaker. Bring the water to a boil and record the temperature. Add one teaspoonful of sugar to the water and bring to a boil again. Record the temperature. Does sugar, dissolved in water, affect the temperature at which it boils?

Pour some water into another beaker. Add crushed ice. Allow the mixture of ice and water to stand until the reading on your thermometer does not change. Record your data. Now, dissolve one teaspoonful of salt in the water. Stir it thoroughly. Record the temperature of the water. Is there a difference in the two readings? Could you use this test to determine whether water contains dissolved material? Check your data with your classmates. Do they all agree with your answer? If not, how can you account for differences?

Scientists use a common system of measurement. If three people used a different system in measuring a board, for example, would each one understand the measurement of the other? Of course not! A unit of measure must first be agreed upon by all. It can be anything you choose. What is important is that the unit chosen is used for future measurements.

The English system of measurement developed from units derived from the width of a man's thumb (inch), the length of a man's stride (yard), or how much grain a man held in his two cupped hands (pint). Why do you think this system was not accurate? Today *length* (inch), *volume* (quart), and *mass* (pounds), are based on standard units.

The metric system is used in science throughout the world. The metric system of measurement was developed in France in the eighteenth century. It is based upon a unit of length called the *meter* (MEE-tuhr), a unit of volume called the *liter* (LEE-tuhr), and a unit of mass, the *kilogram*.

C-1 What are these students measuring?

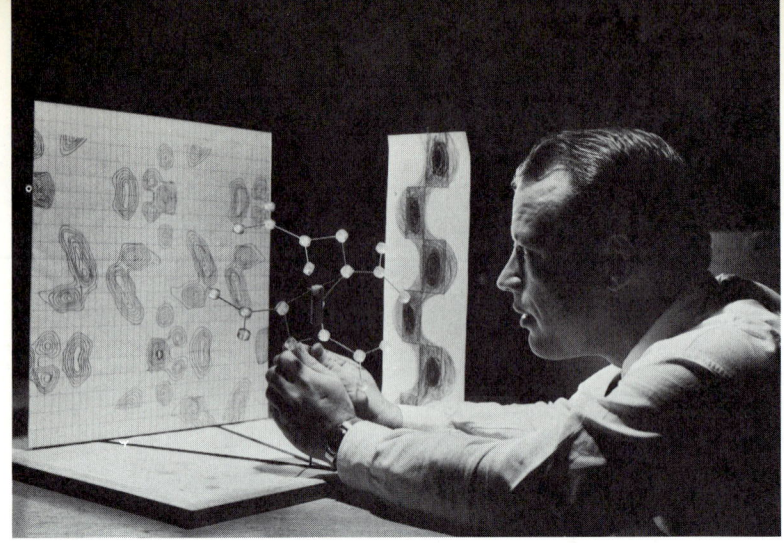

B-6 How do scientists make use of models such as the one shown?

REVIEW

1. List five steps that might be employed by a scientist to attack a problem.
2. Briefly explain each step above.
3. (a) In an experiment, what is meant by the control, (b) the variable?
4. (a) What is a theory? (b) How do mental models help a scientist?
5. What is meant by the "insight" of a scientist?

C/How Do Scientists Use Measurement?

Scientists require accurate information. If you record your observations as a candle burned, what type of data would you collect? Probably you would note *qualitative* (KWAL-ih-*tay*-tihv) *differences* which show a change in *color, shape, size,* or *structure.* For example: the candle became shorter; the flame was made up of three color areas. You might attempt to describe the colors. Is this accurate enough? Does everyone see colors or estimate sizes alike?

Another way of recording the information is to find out by *accurate measurement* how much smaller (decrease in length) the candle became over a given period. By using an instrument measuring the temperature and color relationships of a candle flame, you could record still another kind of

xviii / THE WAYS OF SCIENCE

Science is a way of solving problems. For centuries, early scientists attempted to explain what they saw without testing their explanations. They thought the sun moved around the earth. ("Just look, the sun rises in the east and sets in the west.") Others thought the earth was flat. ("Just look at the earth from wherever you are. Isn't it flat?") Still others thought that flies came from decaying meat. ("Just look. Every time I let meat stand, flies appear from it.")

If you do not accept such untested "common sense" explanations, then you are faced with a problem. You must prove by experimentation whether or not your own explanation is correct. As you proceed through the study of science, you should continually ask yourself three questions:

1. How do I know?
2. How well do I know?
3. How can I find out?

REVIEW

1. How can everything in nature be explained according to the scientist?
2. (*a*) What is the difference between pure science and applied science? (*b*) Give three examples of new technological developments.
3. How does a superstition differ from a scientific conclusion?
4. What three questions should a science student ask himself as he studies his lesson?
5. Define four areas of science.

B/How Do Scientists Solve Problems?

Accurate observation is the first step in scientific study. Usually, a scientist observes a particular occurrence and begins to wonder *why* and *how* it happens. He is curious by nature (Fig. B-1). Once a scientist becomes interested, he spends time reading books, scientific journals, and magazines

Science is divided into major fields of study. Scientific knowledge is further organized into special fields. For the scientist this arrangement is very useful, and most scientists specialize in one field. The field of *biology* studies living organisms and their activities. *Chemistry* studies the composition of matter and the changes it undergoes. The field of *physics* is the study of motion, energy, and forces affecting matter. Some other major fields are *astronomy*, the study of bodies in space; *geology*, the study of the earth; *botany* (a division of biology), the study of plants; *zoology* (a division of biology), the study of animals. Can you name and describe other basic fields of science?

The organized knowledge of science becomes a written record. Over the centuries much of the organized knowledge in science has been preserved in some way. Observations and findings, as they are uncovered, are written down to appear usually in scientific journals or other publications. There are thousands of scientific journals in our country alone. Accepted findings will, in time, appear in textbooks for use in schools and universities.

Scientific knowledge grows and changes rapidly. More scientists are engaged in research today than ever before. By the time you reach high school, some of the theories which you learned may no longer be held. Do you think the modern jet is a result of new advances in knowledge (Fig. A-3)?

A-3 How is today's modern jet an application of new scientific knowledge?

A-1 How does an electric vehicle, an example of applied science, fight air pollution?

people and all the wonders of nature. He arranges this knowledge. An applied scientist invents or makes practical use of this knowledge (Fig. A-1). Some of the practical benefits have been in better foods, improved health, rapid transportation and communication, and control over our environment. Some people call these organized groups *pure science* and *technology* (tek-NAH-lo-jee). It is important to understand the difference between the two.

Science is the enemy of superstitions. *Superstitions* are beliefs which have no scientific basis. That is, they are beliefs that are untested by experimentation. They begin from curiosity and observation, but no experimentation is done to test whether the conclusions reached are correct or not. Many of the common superstitions of today are based on hearsay and opinions, scientifically untested. How could you disprove, by experiment, the superstition about the dangers of a black cat crossing a person's path? (Fig. A-2).

A-2 Are superstitions based on scientific facts? Explain.

The Ways of Science

A/What Is Science?

Science is the study of the world and the universe. As scientists observe and study this truly amazing universe, they notice that the universe presents regular patterns of behavior. For example, scientists notice that objects, when released, fall to the earth. These observed facts are summarized into a *natural law*. Laws of the universe or nature are really man-made summaries or descriptions of what happens around us. All the living and nonliving objects, making up our environment, act according to natural laws. The accurate and orderly arrangement of this knowledge may be thought of as *science*. Scientists then try to develop explanations or theories that account for the way everything behaves. Scientists that are investigating nature are doing basic research.

Science is organized knowledge. We profit from the accumulated knowledge of thousands of years because this knowledge is well organized. One type of organization places man's scientific knowledge into two groups: *pure science* and *applied science*. A pure scientist uncovers scientific knowledge about the stars, planets, oceans, hills, rocks, plants, animals,

THE WAYS OF SCIENCE / xxv

Until recently, the standard meter was the distance marked on a metal bar kept at the International Bureau of Weights and Measures near Paris, France. The standard kilogram of mass is the mass of a metal cylinder also kept at the International Bureau of Weights and Measures near Paris, France (Fig. C-2). Copies of these standards are used by all countries.

So that every scientist could more easily prepare a standard from which to compare his measurements, a new international definition of the meter was established in 1960. *The meter is now measured as 1,650,763.73 times the wavelength of the orange-red light emitted by atoms of the element krypton-86 when it is heated and glows.* This is the same length as the standard marked on the metal bar. Any properly equipped laboratory anywhere in the world can reproduce this standard with great accuracy (Fig. C-3).

C-2   The standard kilogram. Why is a standard unit of mass necessary?

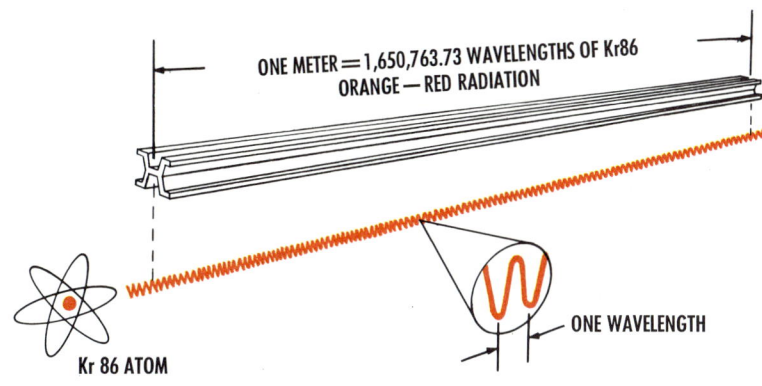

C-3   What is the new international definition of the meter?

The divisions of the metric system are in units of ten. Metric units are simpler to use than the English units. For example, when a *meter* is divided into 10 equal parts, each part is a *decimeter* (DES-ih-*mee*-tuhr) (*deci* means tenth). When a meter is divided into 100 equal parts, each part is a *centimeter* (CEN-tih-*mee*-tuhr) (*centi* means hundredth). Other prefixes and their meanings are given in the table (page XXVI).

The amount of space an object occupies is its volume. A cubic box with inside dimensions of 10 cm on each side has a volume of 1000 cubic centimeters (length × width × thick-

xxvi / THE WAYS OF SCIENCE

ness). This box can hold 1 liter of water (at 4°C). One liter, then, is equal to 1000 cubic centimeters or 1000 milliliters (Fig. C-4). What would ten liters be called? 1000 liters? 1/10 of a liter?

Table of Metric Measures

PREFIXES

$milli = \frac{1}{1000}$ $kilo = 1000$
$centi = \frac{1}{100}$ $hecto = 100$
$deci = \frac{1}{10}$ $deca = 10$

LENGTH

10 millimeters (mm) = 1 centimeter (cm)
100 centimeters (cm) = 1 meter (m)
1000 meters (m) = 1 kilometer (km)

MASS

1000 milligrams (mg) = 1 gram (g)
1000 grams (g) = 1 kilogram (kg)

VOLUME

1000 milliliters (ml) = 1 liter (l)
1000 liters (l) = 1 kiloliter (kl)

One liter of water at 4°C has a mass of 1000 grams or 1 kilogram. In other words, 1000 cubic centimeters of water equals a standard mass of 1 kilogram. *Therefore, 1 cubic centimeter or 1 milliliter of water has a mass of 1 gram.* How many grams are there in a kilogram? a decigram? a hectogram?

The table below will show you some of the relationships between the English and metric systems of measurement.

Table of Metric-English Equivalents

1 mile = 1.6 kilometers
1 inch = 2.54 centimeters
1 ounce = 28.3 grams
1 meter = 39.37 inches
1 liter = 1.06 quarts
1 kilogram = 2.2 pounds

English measures can be changed into metric units. With simple arithmetic, you can change measurements in English units into metric units. The activity below will give you some practice.

COMPARE

Compare the metric system and the English system of measurements. Measure the length of your book to the nearest millimeter

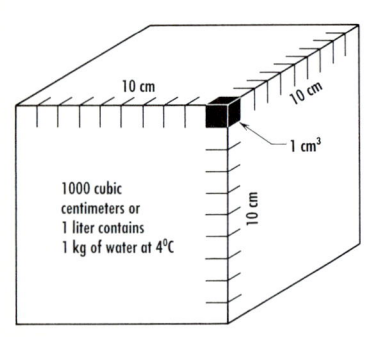

C-4 The diagram shows a box containing 1000 cubic centimeters. Notice the size of one centimeter.

with a metric ruler and the nearest sixteenth of an inch with an inch-foot ruler. Using decimal figures, divide the number of inches into the number of centimeters.

How many centimeters equal one inch? Using a laboratory balance, find the mass of a block of wood or metal in grams and then in ounces. Then, divide the number of ounces into the number of grams. How many grams equal one ounce?

As you can see, English and metric measures can be changed from one into the other. Actually, all that changes are the units used in measuring. In other words, the two systems are equivalent or equal to each other.

The metric system is a decimal system based on units that are easily divided or multiplied by tens, hundreds, or thousands. Each unit of length, volume, or mass can be changed into larger or smaller units simply by moving the decimal point.

SAMPLE PROBLEMS

A. How many centimeters are there in 5 meters?
   Solution:
   1. Move the decimal point 2 places (hundreds) to the right.
   2. 5.00 m = 500 cm
B. How many liters are there in 2000 milliliters?
   Solution:
   1. Move the decimal point 3 places (thousands) to the left.
   2. 2000 ml = 2.0 liters
C. How many grams are there in 10 kilograms?
   Solution:
   1. Move the decimal point 3 places (thousands) to the right.
   2. 10 kg = 10,000 g

Two common scales are used to measure temperature. The temperature at which a substance boils is called the *boiling point* of the substance. Similarly, the temperature at which a substance freezes is called the *freezing point* of the substance. Most thermometers use the boiling point and freezing point xxviii / THE WAYS OF SCIENCE

of water as a standard in determining the fixed points on their scales. The *Celsius* (SEL-see-uhs) *scale* (C), also called the *centigrade* (SEN-tih-*grayd*) *scale*, is based on the freezing point of water at 0° (degrees). This compares to 32° on the more commonly used *Fahrenheit scale* (F). The boiling point of water on the Celsius scale is 100°, but this point on the Fahrenheit scale is 212° (Fig. C-5). The following activity will show you how the scale on a thermometer is developed.

CALIBRATE

Use three beakers and fill each one about halfway with water. To the first beaker, add several pieces of ice. The second beaker is left at room temperature. The water in the third beaker is brought to a boil. Using a thermometer marked in both Fahrenheit and Celsius degrees (or two thermometers, one Fahrenheit and one Celsius), measure the temperature in each beaker. Record the temperatures in the table below. **Do not write in your book.**

Beaker	Temperature °F	Temperature °C
Ice water		
Room temperature		
Boiling water		

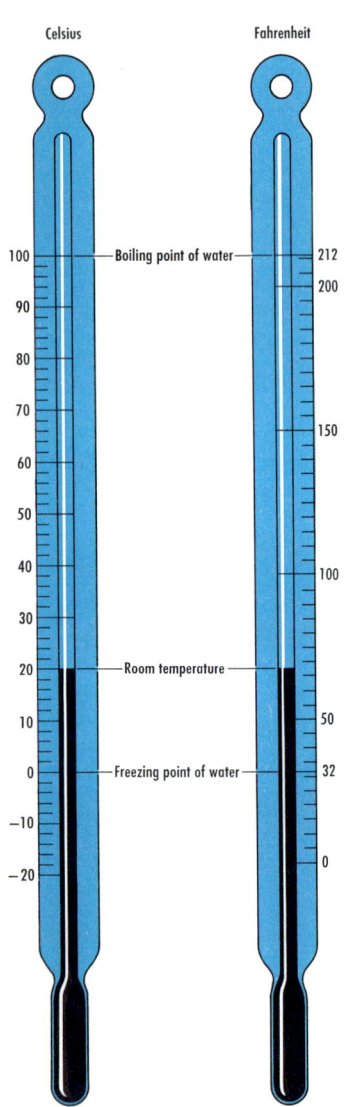

C-5 Notice the freezing and boiling point on each thermometer. How do they compare?

Write the formula for changing a Fahrenheit reading to Celsius. Write the formula for changing a Celsius reading to Fahrenheit. Are these equivalent?

As you can see, Celsius degrees are larger than those used on the Fahrenheit scale. One hundred steps (degrees) on the Celsius scale equal 180 steps on the Fahrenheit scale, or 5° on the C scale = 9° on the F scale (Fig. C-5). You should note that the Celsius system of temperature measurement is also a decimal system with 100 degrees, or divisions, between the freezing point and boiling point of water.

SAMPLE PROBLEMS

A. What is the temperature on the Celsius scale that corresponds to 68°F?

Solution:
1. Subtract 32° from 68° = 36°
2. Multiply 36° by 5/9 (or .556) = 20°
3. 68°F = 20°C

B. What is the temperature on the Fahrenheit scale that corresponds to 30°C?

Solution:
1. Multiply 30° by 9/5 or 1.8 = 54°
2. Add 32° to 54° = 86°
3. 30°C = 86°F

REVIEW

1. What is the difference between qualitative and quantitative observations?
2. Why must a unit of measurement first be agreed upon?
3. (*a*) What are the metric units for measuring length, mass, and volume? (*b*) What are the English units for measuring length, mass, and volume?
4. What is longer, a yard or a meter? by how much?
5. What two fixed points are used in determining the scale on an ordinary thermometer?

unit 1

Man Explores The Earth

Man has always been curious about the earth on which he finds himself. What makes up the inside of the earth? How was the earth formed? What is on the bottom of the oceans? To answer some of these questions, primitive man invented stories and myths because he believed that the earth had always been the same.

The ancient Greek philosopher-scientists, however, observed that changes were taking place. For example, *Herodotus (484–425 B.C.)* noted that the Nile delta was formed by the river which had carried materials washed away from regions farther upstream. The early investigators also studied rocks and fossils to determine what changes had taken place in the earth's history. For example, Greek scientists learned that Vesuvius was a "sleeping" volcano that one day would erupt, even though no human record of an eruption from that mountain was available.

2 / UNIT 1   MAN EXPLORES THE EARTH

Herodotus

They identified land masses that had risen and others that had sunk. They suggested hypotheses for the explanation of these occurrences.

Nicholas Steno (1638–1687) of Denmark made important discoveries regarding crystal forms of rocks which revealed the past history of the earth. For example, he discovered that crystals follow a definite growth pattern and that large areas of present lands had once been sea areas.

Geology, the study of the earth's history, structure, and composition was placed on a scientific foundation in the eighteenth century with the works of *James Hutton (1726–1796)*, a Scottish geologist. In his book, *Theory of the Earth*, published in 1795, Hutton proposed how we can account for the earth's geological features. He assumed that the forces now at work in building up and tearing down land features have not changed very much over vast periods of time.

Cuvier

Early in the nineteenth century, *George Cuvier (1769–1832)*, a French geologist, and *William Smith (1769–1839)*, an English geologist, carried on studies in their countries in which they mapped out the patterns of different rock layers. Then, by comparing these patterns, they were able to make the first geologic maps, showing the kinds of rock layers found in different regions. Cuvier also noted that the fossils found in rock layers closest to the surface were similar to living plants and animals, but that at greater depths the fossils differed more and more from present forms. This observation led to the first method of historical dating of various layers of rock. It became the basis for the science of *paleontology*, the study of ancient life as revealed by fossils found in rocks.

Hutton

Among the many scientists interested in the geology of the New World, *James Dana (1813–1895)*, an American geologist, was the most prominent. Revisions of his book, *Systems of Mineralogy*, first published in 1837, are still used today. The book describes the locations, shapes, properties, and formulas for thousands of different minerals.

In their studies, geologists go back in time to the earth's early development and try to find out what happened at each stage. They also study plant and animal forms that lived on the earth during these stages. Geology is a science that requires knowledge of biology, chemistry, and physics. In this unit, we shall examine some of the evidence to explain what happened in the geologic history of the world.

chapter 1
The Structure of the Earth

A/What Is the Earth's Composition?

The earth is a huge mass of rock, water, and gases. The general shape of the earth is that of a flattened sphere. It is not perfectly round but slightly flattened at the poles. Artificial satellites have confirmed that the earth is 7927 miles in diameter at the equator and about 28 miles less in diameter at the poles. This is a small difference amounting to only one-third of one percent when compared to the earth's total size.

The surface of the earth is fairly regular even though there are mountains rising high into the air and deep trenches present in the ocean floor. Mt. Everest, a mountain in southern Asia, is the highest mountain in the world. It rises about five and a half miles above sea level. The deepest part of the oceans that is presently known to man is about seven miles. Therefore, the greatest known distance from the highest to the lowest points on the earth's surface is about 12½ miles. This is less than half the difference between the earth's equatorial and polar diameters. In other words, the irregularities on the earth's surface are very small in comparison with its size.

The solid earth is made up of four layers. The solid part of the earth is called the *lithosphere* (LITH-uh-*sfeer*). Actually, we know more about the water and gases covering the earth than we know about the solid earth itself. We can see the surface of the earth, and mines and deep wells tell us a little of what is under the surface. However, the deepest wells extend only about five miles below the surface. Thus, compared to the almost 8000-mile diameter of the earth, this is only a scratch on its surface. What is below this five-mile depth? No one knows for sure, but scientists believe the earth is made up of several layers (Fig. 1-1).

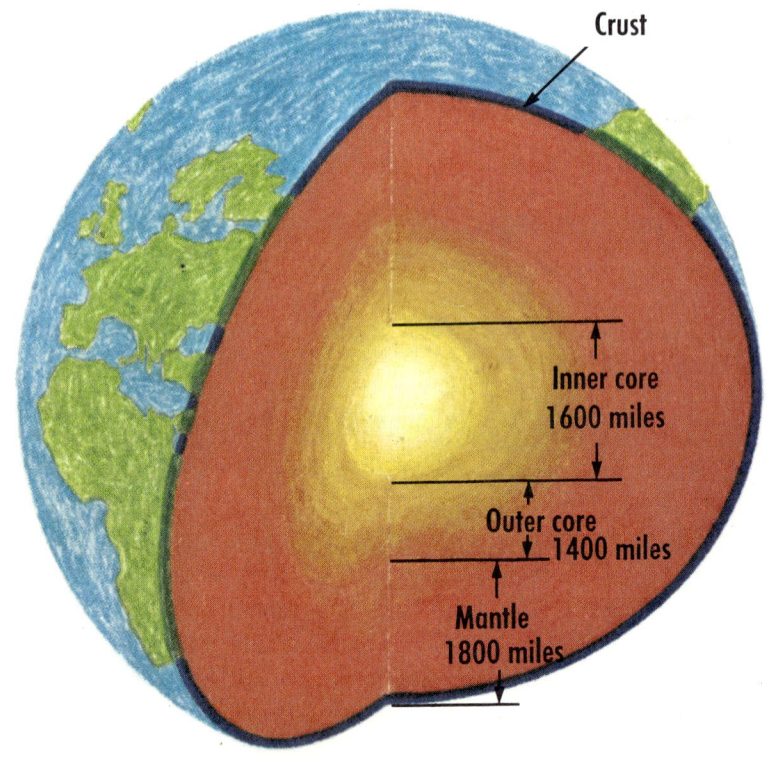

1-1 How many layers make up the earth?

The outermost layer, the part we see on the surface of the earth, is the *crust*. This is believed to be about 7 to 30 miles thick in land areas and thinner under the oceans. Geologists have found that the crust is about 4 to 10 miles thick under most of the oceans. The rocks of the crust actually form a thin shell which covers the surface of the globe.

The *mantle* (MAN-tuhl), which is about 1800 miles thick, lies underneath the crust. It is thought to be composed of heavier rock than the material making up the crust. Because of the great pressure and heat in this layer, the mantle is not quite a solid and not quite a liquid. In other words, the material making up the mantle is thought to be in a plastic state. The land masses making up the crust are thought to "float" on this plastic layer.

The next layer is the *outer core,* nearly 1400 miles thick. It is believed to be composed mostly of iron and nickel in a molten state at a very high temperature. Evidence indicates that the outer core is about twice as dense as the material in the mantle.

The innermost layer, the *inner core,* extends about 800 miles to the earth's center. It also is probably composed of iron and nickel. Because the pressure is nearly 60,000,000 pounds per square inch, the inner core is more like a true solid. The material making up the inner core is thought to be about three or four times as dense as the material making up the mantle. You might think of the earth as being constructed much like a baseball with a two-layered core: a thick layer around the core (mantle), and a thin skin on the surface (crust).

By studying earthquakes, which will be discussed in a later chapter, scientists have learned much about the inner structure of the earth.

The interior of the earth is very hot. As we drill into the crust, we find that the temperature increases about 1°C for every 200 feet, or about 27°C for each mile in depth. How hot do you suppose it gets in mines that are two miles below the surface? *Geologists* (jee-AHL-uh-jists), scientists who study the structure of the earth, believe that the bottom of the earth's crust has a temperature of about 500°C and that the temperature of the lower mantle layers is about 2500°C. The inner core of the earth is thought to have a temperature as high as 3000°C.

You may know that hot springs or *geysers* (GYE-zurz) result when underground water seeps into cracks in the crust, called *fissures* (FIHSH-uhrz). These may be thousands of feet deep. The heat at this depth causes the water to boil and expand. As some of the water at the top of the fissure is forced out, the pressure on the heated water is reduced. Then, as can be seen in Fig. 1-2 (page 7), the water in the fissure explodes into steam, blowing hot water and steam above the surface. One of the best known geysers in this country is Old Faithful in Yellowstone National Park. It erupts on the average of every 65 minutes, throwing water and steam to a height of about 150 feet. Most geysers are located in Yellowstone National Park, New Zealand, Iceland and Japan. You can duplicate this effect in the next activity.

DEMONSTRATE

Fill the flask approximately one-third full of water. Place one end of the S-shaped tube through the hole in the stopper. Then attach a small piece of rubber tubing to the other end of the tube. Now attach a funnel to the top of the rubber tubing, as shown. Heat the water gently until it comes to a boil.

What do you observe? Can you explain your results? How is this similar to a geyser?

Water occupies about three-quarters of the earth's surface. The *hydrosphere* (HYE-druh-*sfeer*) is the water layer covering the earth. Water and air have been the most important factors in changing the surface of the earth throughout its history. Most of the earth's water is in the oceans, but rivers and lakes are also a part of the hydrosphere. How much water is there on the earth? Scientists have estimated that if all the mountains were leveled off and the sea bottom were raised, there would be enough water to cover the earth to a depth of a mile and a half.

Among the great oceans of the earth (Fig. 1-3), the Pacific Ocean is the largest and the deepest. It covers 64 million square miles. In places it extends to a depth of almost seven

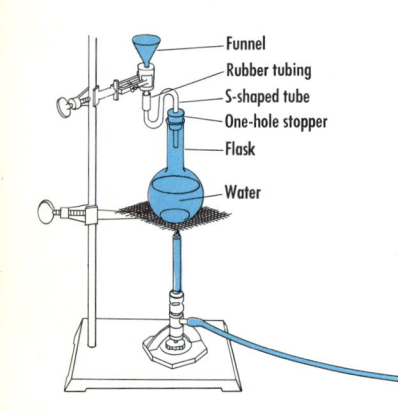

CHAPTER 1 THE STRUCTURE OF THE EARTH / 7

miles. All the continents of the world could be put in the Pacific Ocean with room to spare. The Atlantic Ocean is the second largest with an area of 31 million square miles. The Indian Ocean, which covers an area of over 28 million square miles, is the only entirely warm-water ocean on earth. The Arctic Ocean is the smallest, with an area of about 5½ million square miles. Arctic Ocean explorations with nuclear submarines have shown that in certain places it is over 17,000 feet deep. Scientists estimate that the *average* depth of the oceans is about 12,500 feet.

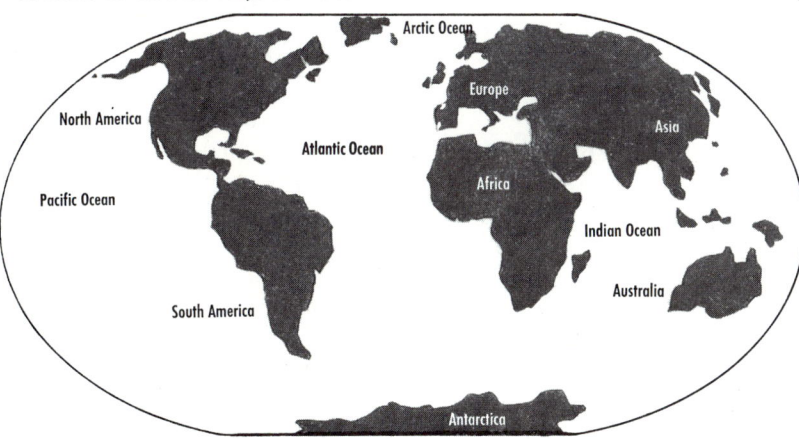

1-3 Why is the exploration of the hydrosphere so important to man?

1-2 What causes water to shoot out of this geyser?

The atmosphere surrounds the lithosphere and the hydrosphere. The *atmosphere* is made up of gases surrounding the solid and liquid parts of the earth. In our study of the earth, we will be concerned mainly with only a few of these gases. Oxygen is important because it is combined with many other elements to form a large percentage of the rocks and minerals in the earth's crust. Carbon dioxide, in water solution, also has helped to form many rocks and minerals. Dry carbon dioxide gas has no effect on rocks.

As we shall soon see, the water vapor in the earth's atmosphere is responsible for many geological changes. When this water vapor condenses as rain or snow, or freezes to form ice, it becomes one of the most important agents by which the earth's surface has been changed over many, many millions of years.

REVIEW

1. What is the earth's diameter at the equator?
2. Name the four main layers of the lithosphere.
3. How much of the earth's surface is covered by the hydrosphere?
4. Name the earth's four major oceans.
5. Which gases in the atmosphere are of great geological importance?

B/What Makes Up the Crust of the Earth?

The crust is composed of rocks and minerals. To understand the structure of the earth, we must first gain some knowledge of the materials in it. Scientists, for the most part, have only been able to explore the earth's surface. However, where there has been a great deal of *erosion* (ee-RO-zhun), a wearing away of the surface, they can study exposed rock layers that were buried deep in the crust (see Fig. 1-4).

Rock layers are of three main kinds. *Igneous* (IHG-nee-uhs) *rock* was formed directly from molten material and means "firemade." These are the original rocks of the earth from which all other kinds were eventually formed. Igneous rock

1-4 How were these rock layers in Bryce Canyon National Park (Utah) formed?

CHAPTER 1 THE STRUCTURE OF THE EARTH / 9

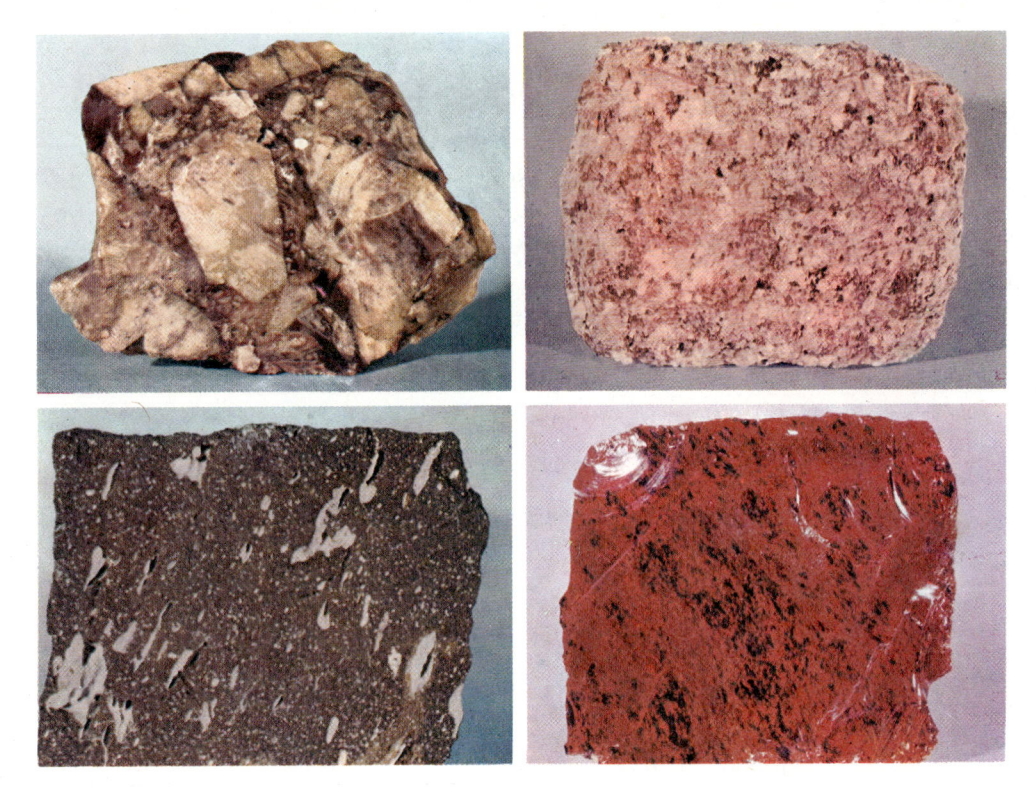

1-5 Igneous rocks: rhyolite breccia (top left), pink granite (top right), basalt (bottom left), and obsidian (bottom right). Which ones were formed deep in the earth?

makes up about 95 percent of the earth's crust. It was formed long ago by the cooling of the hot molten rock called *magma* (MAG-muh), deep in the earth, or on the surface.

If the rock was formed when the magma cooled below the earth's surface, the crystals in the rock are large, as in *granite*. If the magma cooled on the surface of the earth, the rocks are made of fine crystals or are glassy, like *obsidian*. Other examples of igneous rocks are seen in Fig. 1-5.

Sedimentary (sed-ih-MEN-tuh-ree) rocks were formed from sediment, such as mud, sand, and gravel. These rocks were carried by rivers into the seas and deposited in underwater layers. Most of the rocks we see on the surface of the earth are sedimentary; yet, these make up less than five percent of the total rocks in the earth's crust. Since sedimentary rock

10 / UNIT 1 MAN EXPLORES THE EARTH

contains fossils, it provides geologists with much information about the more recent history of the earth. As these layers of sediment were built up, great pressure and chemical changes cemented them together. Often, shells of ancient sea animals can be found mixed in with the mineral *calcite* (KAL-site) making up *limestone,* another sedimentary rock. Some other examples of sedimentary rocks are *shale,* made from clay, and *conglomerate* (kahn-GLAH-mer-uht), made from gravel (Fig. 1-6). To show how sedimentary rock layers are formed, do the next activity.

OBSERVE

Fill an empty milk carton about a third full of coarse gravel, sand, and clay. Fill the carton with water. Add about a tablespoon of Portland cement. Close the top of the carton and shake to mix the materials. Let it stand for several hours. Carefully pour off the water in the carton, so as not to disturb the layers. Let the carton stand for several days until the cement hardens. Remove the sides of the carton. Which materials settled to the bottom? Can you see layers in the "rock"? The hardened material represents sedimentary rock.

Metamorphic (met-uh-MOR-fihk) *rock* means rock "changed in form." These metamorphic rocks were originally either igneous or sedimentary, but were changed by high temperature, great pressure, and other conditions in the earth. Because metamorphic rock is formed under great pressure, it is usually much denser and harder than the original materials from which it was formed. Have you ever noticed the *marble* used on the walls of many public buildings? Marble was formed from limestone, a sedimentary rock, by great heat and pressure beneath the earth's surface. Other examples of metamorphic rocks are *slate,* formed from shale; *quartzite* (KWARTS-ite), formed from sandstone; and *gneiss* (NYES), formed from granite (see Fig. 1-7).

Useful minerals are found in the crust. The rocks making up the crust are composed of over 2000 different minerals. *Minerals* are single elements or compounds found naturally in

1-6 Sedimentary rocks (top to bottom): conglomerate, sandstone, shale, limestone. Each was formed from what materials?

1-7 Common metamorphic rocks: mica schist (top left), marble (top right), slate (bottom left), gneiss (bottom right). What processes formed them?

the earth's crust. As you probably know, minerals are made up of definite combinations of the same chemical elements of which everything else on the earth is composed. Only eight elements make up about 98 percent of the earth's crust, as shown below.

Main Elements Making Up the Crust			
Oxygen	46.60%	Calcium	3.63%
Silicon	27.72%	Sodium	2.83%
Aluminum	8.13%	Potassium	2.59%
Iron	5.00%	Magnesium	2.09%

During the formation of the crust, many metals were separated from the magma to form *ores*. In a few places, pure metals, like gold and copper, are found in outcrops of rocks

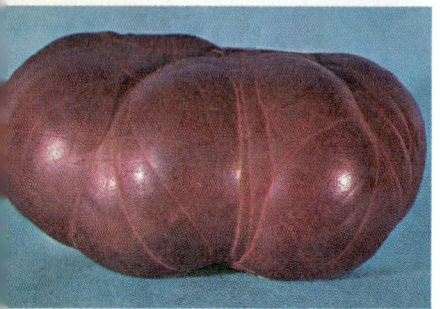

12 / UNIT 1 MAN EXPLORES THE EARTH

or in veins in the rock. Most metals, however, are found in chemical combinations with other elements and with impurities from which they must be separated. Some common ores we use are *oxides* (the metal combined with oxygen) like iron oxide, *carbonates* (the metal combined with carbon and oxygen) like copper carbonate, and *sulfides* (the metal combined with sulfur) like zinc sulfide. (See Fig. 1-8.)

Gems are precious minerals. Gems are minerals that are prized for their color, luster, and hardness. Interestingly enough, gems are really simple crystalline forms of common minerals. For example, diamonds are composed of the common element carbon, the same as coal or graphite. But the crystal structure of diamonds is such that they can be cut to refract or bend light rays so that they sparkle. The difficulty of obtaining the gems and the demand for them determines their value. Gems are also valued for size, color, hardness, and lack of flaws.

Most gems are classified as *precious* or *semiprecious*, depending on their value. Diamonds, rubies, emeralds, and sapphires are precious gems. Topaz and tourmaline are considered semiprecious gems (see Figs. 1-9 and 1-10).

A number of precious minerals have been man-made by various laboratory processes. Tiny diamonds, suitable for industrial use rather than for jewelry, have been produced by heating *graphite* and other substances under great pressure. Man-made rubies, sapphires, and emeralds have also been produced by heating powdered *alumina* (aluminum oxide) under high pressure.

Nonmetallic minerals also have great commercial value. Since many nonmetals dissolve readily in water, they may be carried long distances from their original sources. They are then deposited in sedimentary layers in lakes and oceans. As geologic changes took place, ancient lakes and seas dried up and nonmetallic deposits, such as salt beds and limestone formations, were left exposed in many regions.

An example of a large salt bed is found around the edges of the Great Salt Lake in Utah. Many valuable nonmetallic minerals are obtained from deposits of this kind, like *borax* and *gypsum* (JIHP-suhm), used in making plaster.

1-8 Common metallic ores (top to bottom): sphalerite, cinnabar, hematite, and azurite. How do ores differ from common rocks?

1-9 Precious minerals before being cut into gems: opal (top left), beryl (top right), garnet (bottom left), ruby (bottom right).

When we think of useful substances obtained from the earth, we usually think of such metals as iron, aluminum, copper, zinc or tin. However, nonmetallic minerals that are mined in the United States have over three times the value of the metallic minerals. Could we get along without salt, asbestos, sand, or limestone? Study this question at this time in the next activity.

EXAMINE

Make a collection of nonmetallic substances that are obtained from the earth. Fasten these samples on a large board. Make a chart showing the use of each. Look in encyclopedias in your library at home or at school to find out the value of each of these mined in the United States every year. Find out in which states each nonmetallic substance is plentiful.

1-10 Tourmaline (top), topaz (bottom) are semiprecious minerals.

Fuels are found in the crust. You may know that fuels were formed from decayed plants in the earth many millions of years ago. One of the most important of these fuels is *coal*. All of the present coal deposits were formed in the same way. However, the beds from which we get *anthracite* (AN-thruh-site) or hard coal, were under greater pressure than others. The coal deposits that are found in Eastern Pennsylvania are the main source of anthracite coal in the United States. *Bituminous* (bye-TYOO-mih-nuhs), or soft coal, is found in regions where the beds have not been under as much heat and pressure as the hard-coal beds. West Virginia leads all the other states in the production of bituminous coal. Deposits, however, are found throughout the eastern, central, and southern states. Compare how hard and soft coal burn in the next activity.

COMPARE

Obtain some samples of hard and soft coal. Compare their appearance. Scratch each piece with a nail. Which kind is harder? Break coal into small pieces and place a few pieces of each kind in a separate evaporating dish. Use a Bunsen burner to light each sample. Are the colors of the flames different? Which type of coal seems to produce more heat? Which type of coal makes the most smoke?

Petroleum, another important fuel, is found in sedimentary rock layers, like sandstone or shale. The crude oil collects in pores, or tiny spaces, in this rock. It is kept from escaping by layers of nonporous rock into which the oil cannot pass. Geologists speak of these rock layers containing oil deposits as oil pools. Since petroleum is known to have been formed from countless prehistoric forms of sea life, salt water is usually found underneath an oil pool. Some oil pools are still under water, and some oil wells have been drilled offshore along the coasts of California, Texas, and Louisiana (Fig. 1-11).

Offshore oil wells sometimes present serious problems. Many birds and fish were destroyed from leaking oil wells off California and Louisiana in 1969.

CHAPTER 1 THE STRUCTURE OF THE EARTH / 15

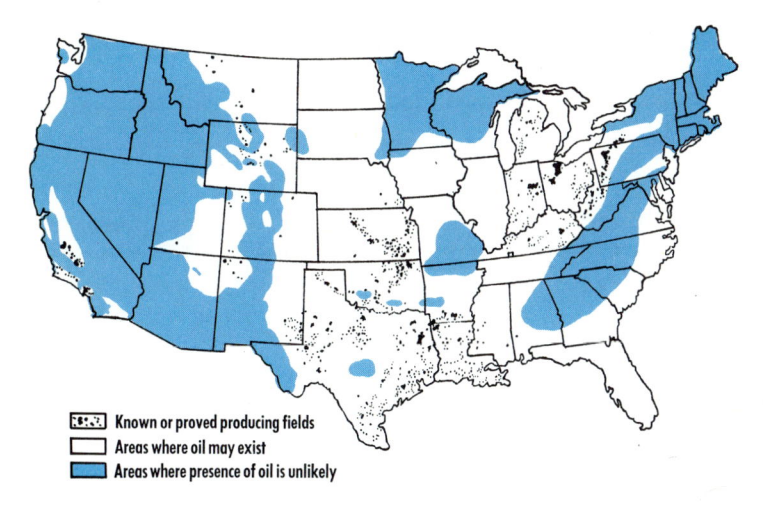

1-11 The map shows oil deposits in the continental United States. Why do scientists believe oil deposits once were under ancient seas?

REVIEW

1. Name the three main kinds of rock making up the crust.
2. List the eight main elements making up the crust.
3. What chemical compounds make up some common ores?
4. Give some examples of: (*a*) precious gems; (*b*) semi-precious gems.
5. Name two important fuels found in the earth.

C/How Are Minerals Identified?

Physical properties of minerals are used for identification. As you have learned previously, some rocks consist of only one mineral, but most rocks are made up of mixtures of minerals. For example, limestone usually consists of calcite, but granite (Fig. 1-5) is always made up of quartz, feldspar, and other minerals. Only about 50 minerals are commonly found in the rocks of the earth.

Physical properties which geologists use to identify many minerals are *luster, color, streak, crystal shape, cleavage* and

16 / UNIT 1 MAN EXPLORES THE EARTH

fracture, and *harness.* Luster is the shine of the minerals which helps in identifying metal with a characteristic luster. A piece of iron pyrite, for example, has a brassy shine. We would say its luster is metallic. A piece of limestone, which has no sparkle or shine, is said to have a nonmetallic luster.

The true color of a mineral can best be seen in a freshly broken piece because the action of air and water usually changes the natural color on the outside of the sample (see Figs. 1-12 A and B). Colors of some of the more common minerals are listed in the table below.

COLORS OF COMMON MINERALS

Minerals	Composition	Color
Cuprite	copper oxide	red
Graphite	carbon	black or gray
Gypsum	calcium sulfate	white
Jasper	silicon dioxide	red
Malachite	copper carbonate	green
Olivine	magnesium-iron silicate	yellow-green
Pyrite	iron disulfide	brassy
Quartz	silicon dioxide	white or pink

In addition to the color of the mineral itself, the color of its fine powder is a help in identification. This method calls for rubbing a piece of the mineral on a piece of unglazed porcelain to form a streak. You can see this by doing the following activity.

COMPARE

Use pieces of unglazed porcelain, or the back of a white ceramic tile used in kitchens. Rub samples of different minerals on them to produce typical streaks. Do all minerals leave a streak of fine powder on the porcelain? Compare the colors of the streaks with those described in the tables to see if you can identify your samples.

1-12A Compare these minerals using the tables. (Top to bottom) quartz, biotite, sulfur, malachite.

CHAPTER 1 THE STRUCTURE OF THE EARTH / 17

Mineral	Composition	Color
Azurite	copper carbonate	light blue
Cuprite	copper oxide	red
Graphite	carbon	black
Hematite	iron oxide	red-brown
Malachite	copper carbonate	light green
Kaolin	magnesium silicate	white
Pyrite	iron disulfide	greenish or brownish black

Most minerals have definite crystal structures. The types of crystal structure can usually be identified by studying the surface of a freshly broken piece of the mineral with a hand magnifier. Many crystals have interesting shapes in the form of cubes, needles, and prisms as you can see in the following activity.

OBSERVE

Collect various minerals, like mica, quartz, and feldspar. Place these in a strong paper bag and tap the minerals with a small hammer. Observe the various broken pieces for fracture and cleavage. Did they all break in the same manner? Can you tell the difference in their crystal structures by the way each mineral splits? Was the break smooth or rough? Use a hand lens to find the shape of the crystals.

Most crystals are arranged in definite patterns. This is because the atoms and molecules of which they are composed assume regular geometrical shapes in forming the crystal.

Minerals form crystals in the basic forms or systems as shown in Fig. 1-13 (page 18). Scientists find that most mineral samples show more complex crystal structures which are combined forms of these basic systems. Therefore, it will not always be easy to definitely identify the crystal form of a given mineral from this brief discussion. If you are interested, you may consult more advanced references on *mineralogy*. You can see how crystals form in the following activity.

1-12B Compare these minerals using the tables. (Top to bottom) cuprite, jasper, graphite, gypsum.

18 / UNIT 1 MAN EXPLORES THE EARTH

EXAMINE

Put a drop of silver nitrate on a clean slide and lay a single strand of fine copper wire in the drop. **(Caution: Silver nitrate is poisonous. Do not get it on your hands.)** Do not put a cover glass on the slide. What do you see when you observe with a microscope or a microprojector? Use a desk lamp to warm the slide. Draw the crystals that form along the wire. Put a drop of sodium chloride or ammonium chloride solution on another slide. Note the crystals that form under a microscope or microprojector as the water evaporates. If the crystal growth is too slow, use a more concentrated solution. If it is too fast, use a more dilute solution. Draw the crystals. Compare your drawings with Fig. 1-13.

1-13 Can you identify the shape of these crystals?

The hardness of minerals can be used in identification. Minerals vary in hardness from *talc*, which is so soft that it can easily be scratched with your fingernail, to *diamond*, which is the hardest of all natural minerals. Since the hardness of a mineral cannot be determined by simply looking at it, a way of comparing this property is needed. Scientists have selected the ten minerals listed in the table. They are used as standards in comparing the hardness of minerals. If a sample can scratch quartz but not topaz, then its hardness is between 7 and 8.

Hardness Scale of Minerals	
1. Talc	6. Orthoclase
2. Gypsum	7. Quartz
3. Calcite	8. Topaz
4. Fluorite	9. Corundum
5. Apatite	10. Diamond

However, to test the approximate hardness of unknown minerals when you are on a field trip, the following method can be used. If the sample can be scratched with your fingernail, it has a hardness of less than 2.5 on the scale. If it can be scratched with a copper penny, it has a hardness of less than 3. If it can be scratched with a knife blade or a piece of glass, it has a hardness of about 5.5. If only a steel file will scratch the sample, then the hardness is about 6.5 or more. Compare the hardness of different minerals in the following activity.

RECORD

Collect samples of different minerals in your area. Wash the various pieces of mineral in water. Scrub them carefully with a small brush to remove any soil. Put each sample in a strong paper bag and break the piece with a hammer. Note the luster and color of the broken surface of each sample. Examine the surface with a hand magnifier to find out the crystal shape. Determine the approximate *hardness* of each sample as described above. Copy the following table in your notebook and record your observations. **Do not write in your book.**

Kind of mineral	Luster	Color	Crystal shape	Hardness

Chemical properties are used to identify minerals. One of the most common chemical tests is the one for *calcite* (calcium carbonate), the mineral making up limestone and marble. If a drop of cold, dilute hydrochloric acid or vinegar is placed on a rock sample containing calcite, the acid begins to bubble because a gas is given off. In this chemical reaction, the acid acts on the calcite. The products are calcium chloride, water, and carbon dioxide gas.

Some metallic minerals can also be identified by means of *flame tests*. A pinch of the powdered mineral, moistened with acid, is held on a clean platinum wire in a flame. The color change of the flame indicates the presence of certain metals. You can easily show the yellow color that is produced by sodium in a flame. Sprinkle a few grains of table salt (sodium chloride) into the blue flame of a Bunsen burner. The various colors that we see in fireworks displays are produced by adding different minerals to the mixture that is oxidized.

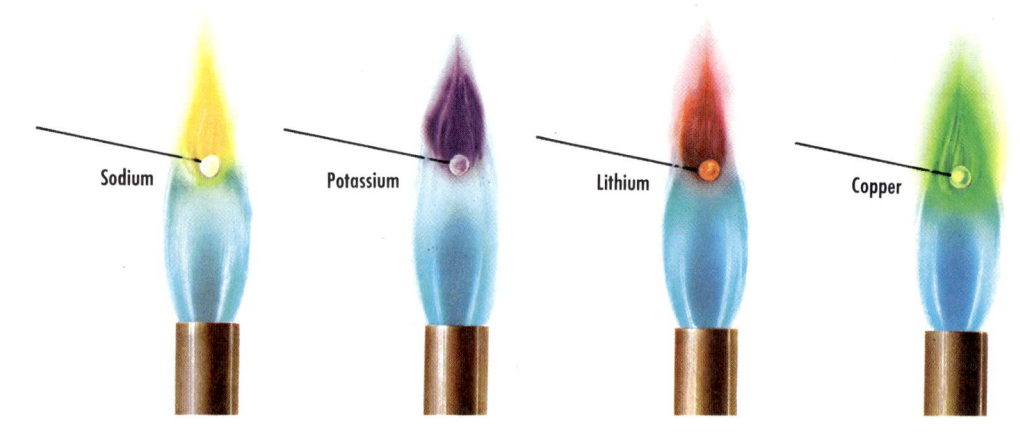

1-14 Flame tests are used to identify metals. Beads of compounds containing minerals show characteristic colors when heated.

The identifying colors of common flame tests are described in the table (also see Fig. 1-14).

COMMON FLAME TEST COLORS	
Metal	Color
Barium	pale green
Calcium	orange
Copper	emerald green
Lithium	crimson
Potassium	violet
Sodium	yellow
Strontium	crimson

REVIEW

1. Name four physical properties used to identify minerals.
2. What is the color of the streak produced by: (*a*) azurite, (*b*) hematite, (*c*) malachite?
3. Why are mineral crystals arranged in definite patterns?
4. In the hardness scale of minerals, which is: (*a*) hardest, (*b*) softest?
5. What colors are produced in the flame test for the metallic minerals of calcium, potassium, and sodium?

THINKING WITH SCIENCE

A. *On a separate sheet of paper write the numbers 1 to 15. After the number of each question, write the letter of the term that correctly completes the statement.* **Do not write in your book.**

1. The layer of plastic rock underneath the crust is the: (*a*) outer core, (*b*) inner core, (*c*) mantle.

2. The gaseous envelope surrounding the earth is called the: (*a*) atmosphere, (*b*) lithosphere, (*c*) hydrosphere.

3. The color of the flame test for minerals containing sodium is: (*a*) red, (*b*) blue, (*c*) yellow.

4. The most common element in the crust of the earth is (*a*) iron, (*b*) oxygen, (*c*) aluminum.

5. The study of the history, structure, and composition of the earth is known as: (*a*) biology, (*b*) geology, (*c*) meteorology.

6. The original rocks of the earth formed directly from molten minerals are known as: (*a*) sedimentary, (*b*) metamorphic, (*c*) igneous.

7. The rocky layers making up the solid earth are known as the: (*a*) lithosphere, (*b*) hydrosphere, (*c*) atmosphere.

8. The rocks that are made from combined layers of mud, sand and gravel are called: (*a*) igneous, (*b*) metamorphic, (*c*) sedimentary.

9. The layer of hot, molten nickel and iron that is found underneath the mantle is the: (*a*) inner core, (*b*) outer core, (*c*) crust.

10. The layer of water covering the earth is called the: (*a*) atmosphere, (*b*) hydrosphere, (*c*) lithosphere.

11. The earth's shape can be compared to that of a (an): (a) egg, (b) ball, (c) flattened sphere.

12. Cracks in the earth's crust are called: (a) magmas, (b) fissures, (c) ores.

13. The largest and deepest of the oceans is the: (a) Atlantic, (b) Indian, (c) Pacific.

14. An example of a semiprecious stone is: (a) ruby, (b) garnet, (c) sapphire.

15. An orange flame test indicates the presence of: (a) copper, (b) sodium, (c) calcium.

B. *Write the answers to the following in your notebook. Be sure to use complete sentences and correct spelling and grammar.*

1. Describe the earth's shape as it might be seen from a spaceship.

2. Describe the composition of the four main parts of the lithosphere.

3. Explain why a geyser erupts regularly in a cloud of steam and hot water.

4. Compare the characteristics of the four main oceans of the earth.

5. Compare the ways in which the three main rock layers of the crust were formed.

6. On what basis are gems valued as precious minerals?

7. In what kind of geologic formations are oil-bearing rocks usually found?

CHAPTER 1 THE STRUCTURE OF THE EARTH / 23

8. How are mineral streaks used in the identification of some minerals?

9. Describe how flame tests are used by chemists to identify certain minerals.

10. Explain how the approximate hardness of minerals can be determined on a field trip.

RESEARCH IN SCIENCE

1. Use modeling clay to make a model showing a cross-section structure of the lithosphere.

2. Arrange a field trip with some classmates to study rock formations in your area.

3. Make a rock collection and identify the different kinds of rocks you have collected.

4. Report on the appearance of certain minerals under ultraviolet light.

5. Make an exhibit of various metallic and nonmetallic minerals to show your class.

6. Write a report on how oil is recovered from oil pools in the ground.

chapter 2
The Changes in the Earth

A/What Is the Earth's History?

Several theories explain how the earth was formed. Who made the *matter* that makes up the universe, of which our earth is a part, and why, scientists do not know. In fact, scientists do not attempt to answer questions like these. To answer "who" made the earth or "why," you must turn to other areas of knowledge. Many people have found satisfactory answers to these questions in religion and philosophy. As we pointed out in the Ways of Science (pages xv and xvi), scientists investigate all living and nonliving substances. They try to develop theories and explanations to account for the way everything behaves. The methods used most frequently are observing, measuring, and experimenting.

The theory, that is most widely accepted, states that in the beginning, the earth was a great mass of dust and gases spinning in space around the sun.

As you know, all objects have an attraction or gravitational pull on each other. This force, acting on the mass of matter that was to become the earth, gradually pulled the dust and gases into a huge ball. The pulling together or contraction is thought to have heated this mass. This resulted in pressures and strains within the forming earth. This caused movements of large masses of molten materials. The next activity illustrates how the force of contraction acts on a hot, molten substance as it cools.

OBSERVE

Melt a large piece of paraffin wax by heating it in a Pyrex beaker in a pan of hot water. **(Caution: Do not heat wax over an open flame.)** Place the breaker in a pan of crushed ice or cold water. Note where the paraffin hardens first. What effect is produced? What formations might have been produced in such a process?

Molten rock formed the water and air. According to one theory, as millions of years passed, the outer layer of molten materials gradually cooled enough so that a crust began to form on the earth. However, the pressure of the magma below the surface produced many tremendous eruptions of fiery rocks and hot gases (Fig. 2-1). These masses of boiling rock are believed to be the main water and air sources that cover the earth today. Rocks can hold a large amount of gas in the form of gas molecules locked in among the rock molecules. When these rocky materials are heated to a high temperature, many of these gases are released.

Over the millions of years in which the earth gradually cooled, eruption after eruption shot gaseous clouds into the space around the earth. Finally, a gaseous atmosphere began to form. The heat from the earth and the sun must have made this atmosphere a boiling, gaseous blanket.

As further cooling took place, the water vapor condensed and fell as rain. The surface was still very hot, however, and the water probably turned to steam, rising up as water vapor into the atmosphere. In time, as the earth's surface cooled, this process slowed down until finally most of the water remained on the earth to form the oceans. The atmosphere, cleared of the water vapor, smoke and dust, gradually changed into a blanket of clear air. This allowed the sun's energy to pass through, producing chemical changes in the oceans. Many scientists agree that such events probably produced a suitable environment for the appearance of the first forms of ocean life.

2-1 What materials do volcanic eruptions add to the atmosphere?

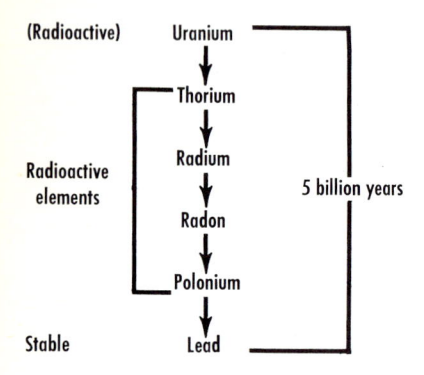

2-2 Stable lead is the end product of radioactive uranium decay. Given one gram of uranium, how much remains after five billion years?

The exact age of the earth is not known. Scientists use several methods to estimate the earth's age. They search for clues and then form hypotheses to explain the facts they have uncovered. For example, geologists make careful studies of the rate at which canyons are cut in solid rock by rivers. The length of time it takes for oceans to become as salty as they are is also determined. Since the conditions under which these processes take place vary greatly, the results obtained from these studies are not very accurate.

Scientists have found a radioactive "clock" that gives a more dependable clue to the earth's actual age. As you probably know, uranium is a naturally *radioactive element*. Its atoms break down or decay into atoms of lighter elements. As this process takes place, the uranium gradually changes into a form of lead. Such changes in radioactive elements take place continually. The decay rate depends on how fast the particular radioactive element decomposes. The time it takes for half of a given amount of radioactive element to decay is called the element's *half-life* (Fig. 2-2).

By measuring the amounts of uranium and lead in a particular rock sample, scientists can estimate how long it took for the change to occur. Using the *uranium-decay* method, geologists have been able to estimate the age of rocks in various parts of the world. Rocks in the Black Hills of South Dakota are about one and a half billion years old. Rocks from parts of western Minnesota are over three billion years old.

Unfortunately, the uranium-decay method of dating rocks has some disadvantages. Many rocks do not contain uranium. In those that do, the uranium-decay rate is so slow that the results are not accurate unless the rock is over a billion years old. A newer method of measuring the age of rocks uses the half-life time in the decay of *radioactive potassium* which changes into *argon gas*. Since potassium is a common element and its half-life is much shorter than that of uranium, the potassium-argon method can be used to find the age of many more rocks.

Fossils give further clues to the age of rocks. As sediment was deposited many millions of years ago to form rock layers, remains of plants and animals were buried. The soft parts of

2-3 What do fossils of plants (top) and animals (below) tell scientists?

these living things soon decayed. The hard parts, like bones, shells, or wood, gradually were replaced by stone or made impressions. These are found as a part of the rock layers. These remains are called *fossils*. (See Fig. 2-3).

Paleontologists (*pay*-lee-ahn-TAHL-o-jihsts) are scientists who study fossil remains of ancient forms of life. They have found the same kinds of fossils in similar rocks layers in different parts of the world. This suggests that all these rock layers must have been deposited during the same time that these forms of life were living.

Carbon dating gives clues to fossil ages. A radioactive form of carbon, called *carbon-14* is found naturally in small amounts in carbon dioxide in the air. It has been used to determine the ages of some fossils. The carbon taken in by all living things has some carbon-14 atoms mixed in with the ordinary carbon atoms. When a living thing dies, the intake of all carbon atoms stop. The carbon-14 atoms in the tissues, however, continue to break down for thousands of years.

Neutrons bombard nitrogen atoms found in the atmosphere, resulting in the formation of protons and radioactive C-14 atoms. The C-14 atoms join the molecules of carbon dioxide that plants need for their growth.

Carbon-14

Living material dies and carbon-14 begins to disintegrate

5568 years later ½ of carbon-14 remains

11,136 years later ¼ carbon-14 remains

16,704 years later ⅛ of carbon-14 remains

2-4 How does carbon-14 measure the age of fossils?

For example, by comparing the amount of carbon-14 in a piece of fossilized wood with the amount of carbon-14 in a similar piece of living wood, the approximate age of the fossil sample can be found. In other words, since the half-life of carbon-14 is 5568 years, a piece of fossilized wood with half as much carbon-14 as a living piece of wood, must have come from a tree that died at least 5568 years ago. In a similar way, a fossil with only one-fourth as much carbon-14 as a similar present-day organism must have been alive as far back as twice the half-life time of carbon-14, or 11,136 years ago. The age of fossils up to 50,000 years old can be determined quite accurately by carbon dating (Fig. 2-4).

Geologic time is divided into five main eras. From a geological viewpoint, the earth is continually changing. However, these changes take place very slowly compared to the human lifespan. According to theories of rock layer formation, the oldest rocks are in the bottom layers (Fig. 2-5). These ancient layers have been studied where the earth's crust has been exposed or in deep canyons cut into the earth. Using this evidence, geologists have developed a chart showing the *eras*, the main time periods, during which the various rock layers were deposited. As shown, the chart starts with the most ancient time period at the bottom and describes the eras up to the present. This chart, showing the various eras, their approximate times, and the forms of life found during each period, is known as a *geologic timetable*. (See Fig. 2-6 on page 30.)

GEOLOGIC TIMETABLE

CENOZOIC ERA *(Age of Mammals)* Began 60 million years ago	Man dominant. Modern mammals. (10,000 years ago)
	Cave man. Ice ages. Extinction of mammoths. Rise of modern horse. Man uses fire. Saber tooth tigers. Whales. Primitive horse.

CHAPTER 2 THE CHANGES IN THE EARTH / 29

MESOZOIC ERA *(Age of Reptiles)* Began 200 million years ago	Dinosaurs become extinct. Flowering plants. True trees. Modern insects. True birds. Giant dinosaurs. Birds with teeth. Turtles. Flying reptiles.
PALEOZOIC ERA *(Age of Fishes)* Began 550 millions years ago	Rise of insects, spiders, and primitive reptiles. Extinction of trilobites. Glacial period. Spore-bearing plants. Sharks. Large amphibians. First air-breathing spiders and scorpions. Coal formed. Tree ferns and dense vegetation. Lung fish. Primitive amphibians. Fish and invertebrates dominant. First land plants. Primitive sharks. First air-breathing animals. Corals. Clams. Starfish. Seaweeds.
PROTEROZOIC ERA *(Age of Invertebrates)* Began 1.2 billion years ago	Development of marine invertebrates. Single-celled primitive animals and plants. No fossil records.
ARCHEOZOIC ERA and AZOIC ERA *(Dawn of Life)* Began 2.1 billion years ago	Some early marine invertebrates. Single-celled primitive life. Seas cover some land areas. Waterless earth. No life. No fossil records.

2-5 What do exposed layers of rock (Grand Canyon) tell scientists?

We probably think of the earliest human records as being very ancient. Yet, according to the geologic timetable, the first appearance of man is a recent event. In all the time that has passed since the oldest civilizations existed, the earth has changed very little. For hundreds of millions of years, before man appeared on the earth, however, its surface was subjected to severe change.

2-6 Examples of plant and animal life in some of the geologic eras.

How can we tell that these changes were taking place? Consider the Grand Canyon in Arizona. Scientists have estimated that the Colorado River is cutting away the rock at the rate of one foot in 1000 years. Since the canyon is 6000 feet deep at present, it must have taken about six million years to form the Grand Canyon. This seems like a long time, but the Grand Canyon is really a fairly recent geologic development. The evidence in the rocks shows that long before the canyon was formed in that area, mountain ranges were worn down to a level plain and the land covered by a sea for millions of years. Then, the whole area was raised high above sea level and the river began cutting away the rock layers to form the canyon.

The Archeozoic Era occurred before the first real evidence of life. As may be expected, the record of an era that began over two billion years ago is very incomplete. Rocks containing the records of the *Archeozoic* (*ahr*-kee-uh-zo-ihk) *Era* are found only in a few places, one being a large outcropping in a remote part of Canada. Some scientists think that primitive single-celled forms of life may have existed, but there is no direct evidence of life that far back.

The Proterozoic Era had very primitive forms of life. The *Proterozoic* (*pro*-te-ro-zo-ihk) *Era* began over a billion years ago and lasted about 650 million years. Again, no real evidence of living things has been found in rocks of this era. However, thick layers of limestone, graphite, and slate are found. Since these rocks contain carbon, an element always found in all living things, it is believed that the ancient seas must have supported many simple plants and animals. Then, as they died, their bodies fell to the sea bottom. In time, these remains became the carbon-containing deposits. The North American continent is believed to have been a larger land mass at that time than at present.

The Paleozoic Era was the time of ancient life. The *Paleozoic* (*pay*-lee-uh-zo-ihk) *Era* lasted about 360 million years, beginning about 550 million years ago and ending 190 million years ago. It was a time during which there were many changes in the earth's surface, as seen in Fig. 2-7 (page 33). One common fossil of this era was a small, lobster-like animal called a *trilobite* (TRYE-luh-*bite*), the earliest shell-covered animal known. During this time, the rich coal fields of the present were formed from ancient forests. Toward the end of the era, the mountains of Vermont and New Hampshire were formed, and the Appalachian Mountains rose as a tremendous range over 25,000 feet high.

The Mesozoic Era was a time of change between ancient and modern life. The *Mesozoic* (*mez*-uh-zo-ihk) *Era*, which lasted about 135 million years, produced the first dry-land forms of life. Giant reptiles, known as *dinosaurs,* lived then. Many smaller reptiles, some adapted for flight in air, appeared. The common fossils found in rocks of this period are coiled shells of *ammonites* (AM-uh-*nites*), a form of shellfish. During much of this period, the central part of North America was covered by a shallow sea, and near the end of the era, the Rocky Mountains were formed. As the Mesozoic Era ended, the North American continent had about the same shape as it has today (see Fig. 2-8).

The Cenozoic Era is the time of modern life. The era in which we are living is known as the *Cenozoic* (*see*-nuh-zo-ihk) *Era*. It began about 60 million years ago. During this era, familiar mammals appeared on the earth. Many of our petroleum and natural-gas deposits were formed during its early part. It is an era, also, in which great mountain ranges were formed, like the Sierra Nevadas in California and the Alps in Europe. The main continents and oceans appeared very much as they do today (see Fig. 2-9).

Many violent volcanic eruptions took place during the Cenozoic Era. Large areas in the states of Washington, Oregon, Idaho and northern California were covered by thick lava. Volcanic cones from these eruptions are common in those areas.

From the evidence found, scientists believe that at different times, large parts of North America and Europe were covered with great ice sheets. During these times, known as the *Ice Ages,* at least four different ice sheets covered parts of North America. Starting over a million years ago and lasting until only a few thousand years ago, the ice sheets formed, melted, and then formed again. The causes of the Ice Ages are not

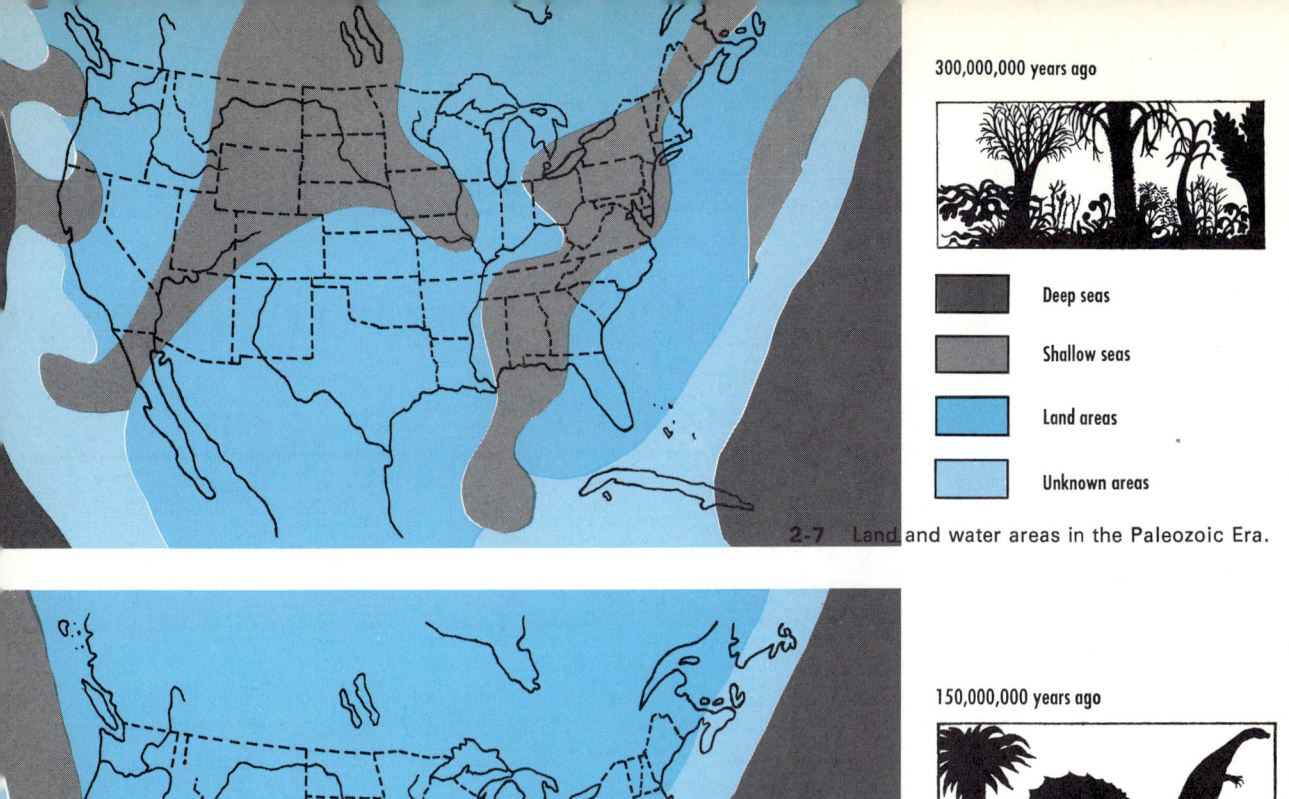

300,000,000 years ago

Deep seas
Shallow seas
Land areas
Unknown areas

2-7   Land and water areas in the Paleozoic Era.

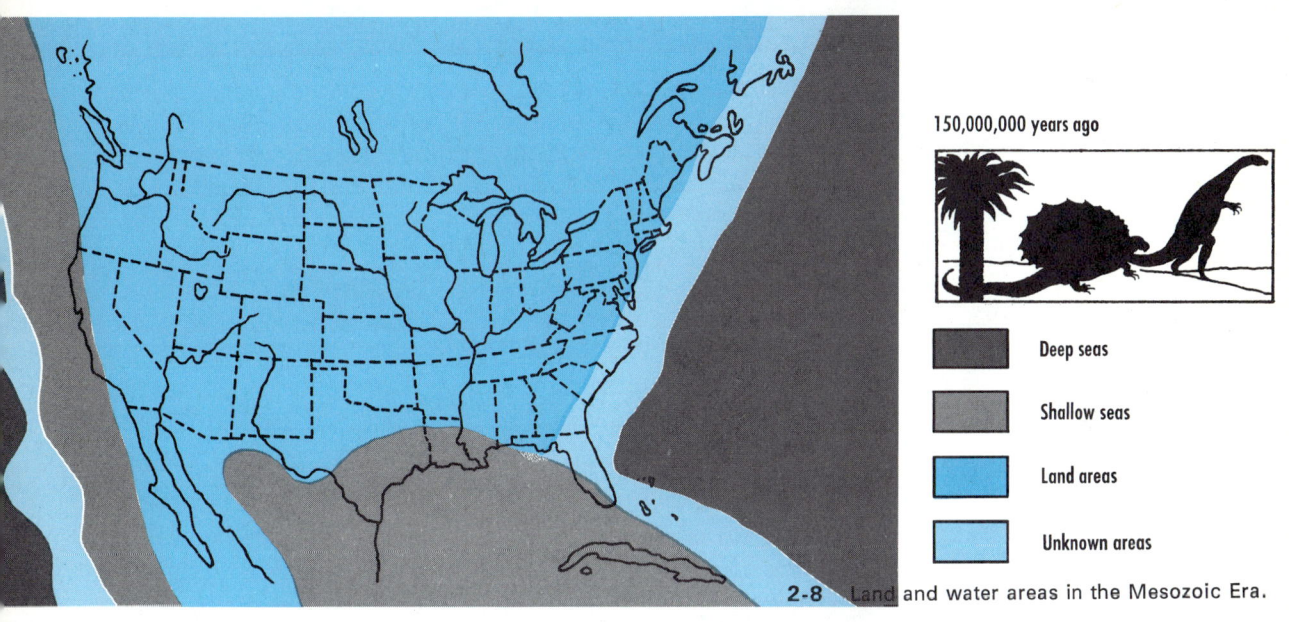

150,000,000 years ago

Deep seas
Shallow seas
Land areas
Unknown areas

2-8   Land and water areas in the Mesozoic Era.

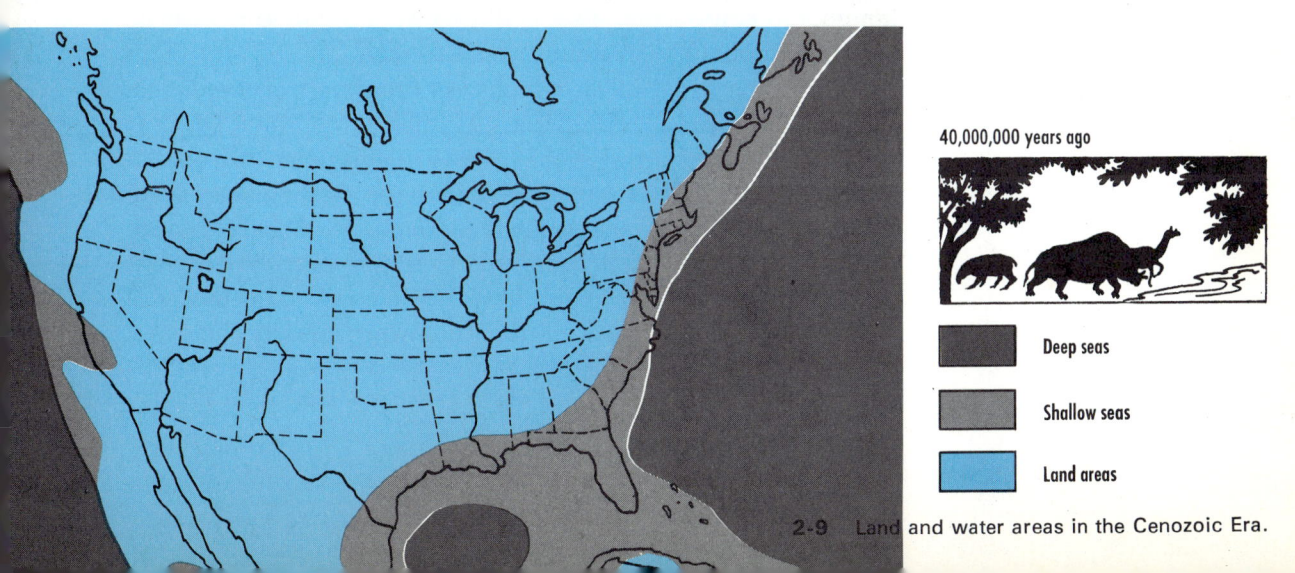

40,000,000 years ago

Deep seas
Shallow seas
Land areas

2-9   Land and water areas in the Cenozoic Era.

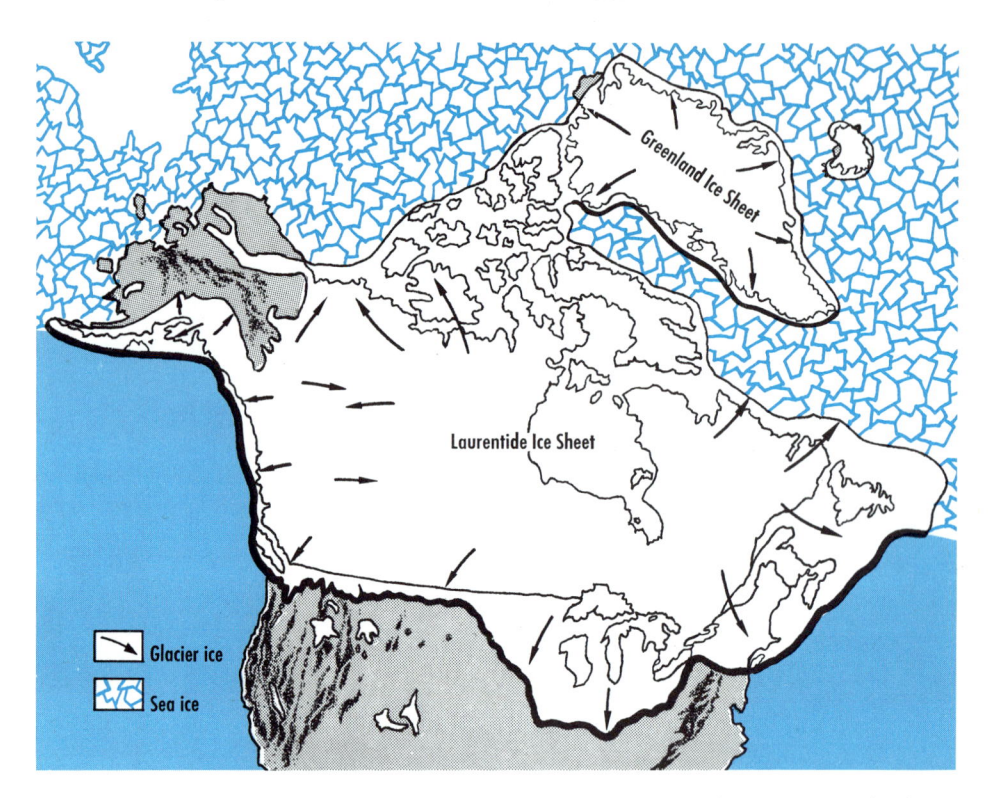

2-10 What were the causes of the ice sheets which once covered a large area of North America?

clearly understood. It is believed that some changes in ocean currents, cloud cover, and other conditions may have been the reasons for the long periods of very cold weather. Perhaps, too, a slight tilt of the earth's axis and the shape of its orbit around the sun occurs. This could cause a slight decrease in average temperature on the earth's surface.

The last great ice sheet, which was thousands of feet thick, covered the United States from Minnesota to New York. It extended as far south as the Missouri and Ohio Rivers, as shown in Fig. 2-10. It is believed that the Great Lakes were formed as this ice sheet melted. Using carbon-14 dating methods, scientists have been able to determine that this last ice sheet existed until about 11,000 years ago. The use of a "time line" as described in the following activity helps in visualizing the earth's history.

MEASURE

On a large sheet of cardboard, draw a line 30 inches long. This line represents three billion (3,000,000,000) years. How many years does one inch represent? Divide the line according to the number of years in each era shown in the geologic timetable on pages 28-29. Label each era. Identify the main plants and animals which appeared in each era.

REVIEW

1. Name two radioactive elements used in dating the age of rocks.
2. What is the age of some of the oldest rocks now known to scientists?
3. What is meant by a fossil?
4. Name the five main eras in geologic time.
5. About how long ago did the Ice Ages begin in North America? When did they end?

B/How Have Internal Forces Changed the Earth?

Pressures and strains have changed the earth's surface. Sedimentary materials are worn away from the earth's surface in the form of soil, sand, and gravel. These are constantly being carried by streams and rivers into the oceans where they build up layer upon layer. Over millions of years, the pressure, heat, and chemical actions in these materials form thick rock layers. As the weight of these layers increases, part of the earth begins to sink. Great pressures and strains are produced, as a result, in the mantle under the crust.

Geologists believe that some types of mountains were formed when these pressures caused parts of the earth's surface to rise and wrinkle. Rock layers were slowly uplifted and folded, as shown in Fig. 2-11. Mountains formed in this way are called *folded mountains.* Many of the great mountain ranges in our country, including the Appalachian Mountains, Rocky Moun-

2-11 A folded mountain in Montana. How was it formed?

36 / UNIT 1 MAN EXPLORES THE EARTH

tains, and Coast Ranges, were partially formed by this type of folding. The next activity helps you understand the folding process of mountain formation.

EXPLAIN

Spread some soft modeling clay in the bottom of a small glass dish. Press your hand on the clay at one end of the dish. What happens to the clay in the other parts of the dish? Let the clay represent soft rock. How would this explain the rise of mountains on some parts of the earth's surface?

The pressures and strains on the mantle may cause another kind of movement of the crust. Huge blocks of the crust, some of them hundreds of square miles in area, may move up or down. Crustal block movement usually occurs along breaks in a rock surface falled *faults*. If a large crustal area is raised a considerable distance, *block mountains* may be formed. The Wasatch Range in Utah, over a mile high, and the Sierra Nevada Mountains in California, over two miles high, are examples of block mountains.

Where the earth's crust develops a weak spot, the pressure of the material making up the mantle may force its way up toward the surface. This causes an area of the crust to rise (Fig. 2-12). *Dome-shaped* mountains are formed in this way. The mountains in the Black Hills of South Dakota and in parts of the Adirondack Mountains in New York are examples of this type of formation. You can observe how this kind of lifting takes place in the next activity.

2-12 What kind of mountains are formed in this way?

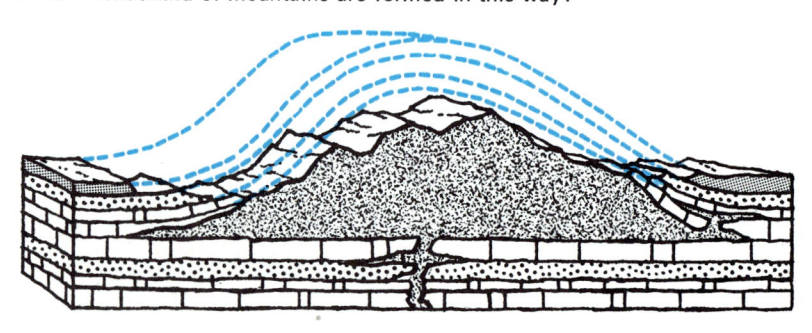

OBSERVE

Attach the mouth of a rubber balloon to a long piece of glass tubing. Put a balloon in a large pan as shown. Arrange the glass tubing so that you can blow up the balloon without moving it. Pour three or four inches of soil over the balloon and carefully blow through the glass tube. What happens to the soil that is in the pan? If the air remained in the balloon, what would be the effect on the soil? How does this activity illustrate dome mountain formation?

Balloon covered with soil

Volcanic action has changed the earth's surface. As we have already learned, the mantle produces a great deal of pressure on the crust. If the crust is not strong enough to contain this force, molten rock, steam, and gases may erupt to the earth's surface to form a *volcano*. The molten material that flows out from a volcano is called *lava*. Have you ever used a toothpaste tube that had a small hole in its side? As you squeezed the tube, the pressure forced the toothpaste out. Lava flow in volcanic action is similar to such a process. One of the best examples of lava formations is found in the Columbia River Plateau.

If volcanic eruptions take place repeatedly, the lava layers may build up a mountain. This process has been studied in a Mexican volcano named Paricutin. Lava and rocks exploded out of the middle of a level cornfield in 1943. Layer after layer of lava flowed out and cooled, until today, Paricutin is a mountain about 900 feet high. Mount Shasta in California, Mount Hood in Oregon (Fig. 2-13), and Mount Ranier in Washington are examples of ancient volcanic mountains in the United States. The only active volcanoes in this country are Lassen Peak in California and volcanoes in the states of Alaska and Hawaii.

Scientists are trying to learn more about volcanoes in order to develop warning systems. These systems will give advance notice when a volcano is about to erupt. Hawaii, which has several active volcanoes, is one of the best areas for this study. Records of the study are kept at the Volcano Observatory located in Hawaii National Park.

2-13 How was Mt. Hood formed?

38 / UNIT 1 MAN EXPLORES THE EARTH

Observatory scientists have recorded earth *tremors,* weak movements in the earth's crust, beneath the crater of Mount Kilauea on a special instrument called a *tiltmeter.* These tremors were in the plastic rock of the mantle at a depth of about 36 miles. They are believed to be movements of hot lava building up pressure. The fact that the volcano erupted shortly after the tremors were recorded partly proves these observations.

Earthquakes have changed the earth's surface. The result of earth vibrations produced by a sudden slipping of great crustal blocks along a fault is called an *earthquake.* There are many every day, but most are so slight that they register only on sensitive instruments. Usually these minor earthquakes take place deep in the earth and produce only very weak tremors. If an earthquake is severe enough to move the earth's surface, much damage may result. Earthquakes seem to occur most often along definite *fault lines* in the earth. The *San Andreas Fault* along the coast of California (Fig. 2-14) is such an area.

2-14 The San Andreas Fault which is located in California is shown. Describe the fault which is shown in the center of the photograph. How was the San Andreas Fault believed to have formed?

CHAPTER 2 THE CHANGES IN THE EARTH / 39

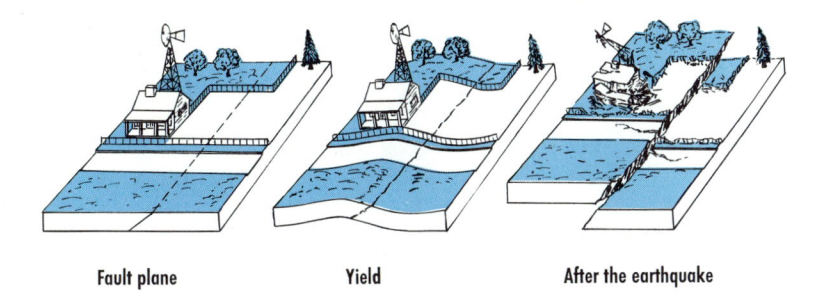

Fault plane   Yield   After the earthquake

2-15 Earthquakes may be caused by the straining and bending of huge blocks in the earth's crust. What produces the strains in these blocks?

Not all earthquakes are caused in the same way. Most of them are the result of physical forces that are constantly changing the earth's surface. The earth's crust may be compared to many huge blocks of rock piled up on each other. Over millions of years, each block settles a little as the crust cools and shrinks. In this slow movement, some blocks are strained and bent out of shape. Slowly, century after century, the strain increases, much like a metal spring that is bent farther and farther. Then, the breaking point is reached, as shown in Fig. 2-15. The huge block of rock snaps back to its normal shape and tremendous amounts of energy are released. The earth's surface begins to quake and roll, and great cracks, sometimes hundreds of miles long, open in the earth (Fig. 2-16). During severe earthquakes, mountains may split open and great tidal waves may occur.

Earthquakes produce shock waves in the earth. Three types of earthquake waves are produced. Each type has certain characteristics that help scientists to identify it from the others. One kind of wave compresses or squeezes the earth's material as it passes; it is called a *primary wave.* Primary waves (also called compressed waves) exist in *solids, liquids* and *gases.* A second kind of wave that is generated by an earthquake causes rock particles to slide over one another in a shearing or side-to-side motion; it is called a *shear wave* and is very destructive. A shear wave can exist *only in a solid.* A third type of wave, called the *surface wave,* moves along the earth's crust or surface; it is produced when either a primary wave or shear wave reaches the earth's surface. Surface waves are the weakest waves. These three waves are shown in Fig. 2-17 on page 40.

2-16 How was this crack in the earth's crust caused?

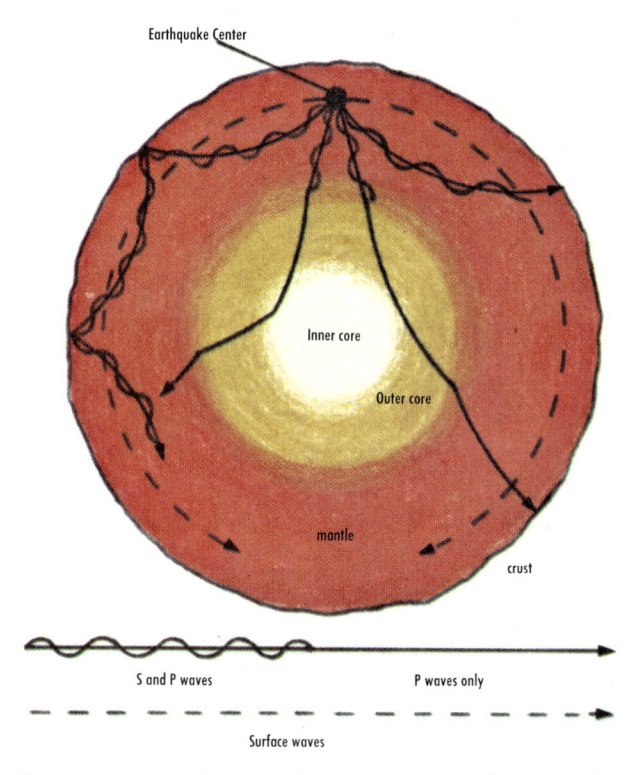

2-17. The three kinds of waves that are produced by an earthquake are shown in the diagram. Why is a shear or secondary wave more destructive than a surface wave?

2-18 What does a seismograph do?

Another important characteristic that helps to identify types of earthquake waves is the *speed* at which they move. The primary wave travels at a higher speed than the shear and surface waves. Because it arrives sooner at a distant point, it is called the primary wave and travels 3.7 miles to 6.45 miles per second. The shear wave arrives second, and so it is called a *secondary wave*. The surface wave is the slowest. It is important to note that even though primary and secondary waves travel at different speeds through rocks of varying densities, *the difference in speed always remains the same*. This is true of all materials. Thus, primary waves always travel about 1.7 times faster than secondary waves. It follows, then, that primary waves arrive at a distant point before secondary waves arrive. The speeds of primary and secondary waves are the best evidence concerning the internal composition of the earth.

These earthquake waves can be detected with an instrument called a *seismograph* (SYEZ-mo-graf) which records the shock waves (Fig. 2-18). The distance from the earthquake's center can be determined by noting how long it takes the secondary wave to reach the observer after the primary wave has passed. Make a simple seismograph in the next activity.

DESCRIBE

Hang a one-pound weight from a ringstand as shown. Tie a pencil to the weight so that the point just touches a sheet of paper placed on the base of the ringstand. Let the weight come to rest. Pound hard on the table; pull the sheet of paper as you pound. What happens to the weight? Put a clean sheet of paper under the pencil and let the weight come to rest again. Have someone walk across the room near the device, or jump up and down, as you pull the paper. What happens?

By studying earthquake waves, scientists have been able to learn a great deal about the composition of the mantle and the core deep inside the earth. Both primary and secondary waves are known to travel through the earth down to a depth of about 1800 miles. Scientists find that these waves travel faster through the mantle than through the crust (Fig. 2-17). Generally, wave speed increases with depth until the core is reached. This information indicates that the mantle is made up of denser material than the crust. At a depth of 1800 miles, primary waves are bent, and secondary waves disappear entirely.

Scientists have never found evidence of secondary waves in the core. It is believed, then, that secondary waves cannot travel through the core. What kind of material transmits only these waves? Since secondary waves cannot pass through a fluid, it believed that the core is fluid. But a puzzle remains. Other studies indicate that the *inner core* (See Chapter 1, page 5) is more like a true solid so the problem is not completely solved at the present time.

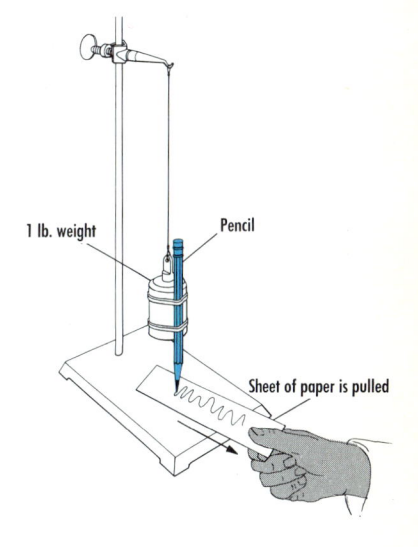

REVIEW

1. Name three kinds of mountains formed by the movement of the earth's crust.
2. How do volcanoes change the earth's surface?
3. Name three inactive volcanoes in this country.
4. Along what type of geological formation do most earthquakes occur?
5. Name three types of shock waves produced in the earth by an earthquake.

C/How Have External Forces Changed the Earth?

Weathering has changed the surface of the earth. *Weathering* is the breaking up of rocks by means of physical and chemical changes in nature. Water is one of the most active substances in producing weathering. Unlike most substances, water expands when it freezes, producing great pressure. You may have seen a water pipe that has burst because water inside it has frozen.

If water freezes in a crack in a rock, the resulting force is great enough to break up even large rocks. Figure 2-19 shows a rock slide that was probably caused by a breakup in the rocks due to freezing water.

Temperature changes also cause weathering of rocks. When heated by the sun, rocks expand and may break off at the edges. A rapid cooling, such as that caused by cool rain, may also cause a warmed rock to break.

The continual expansion and contraction of rocks over the years break up large rocks into smaller pieces. These smaller pieces are gradually broken down into still smaller ones until fine sand particles are formed. This material, mixed with important organic materials and minerals, forms the part of the crust known as *soil*.

Another way in which weathering occurs is by means of chemical reactions. Water, seeping through the ground, dissolves minerals from the soil as well as decayed plant and

2-19 How may rock slides occur?

animal materials. Ground water also contains dissolved carbon dioxide from the air and from the exchange of gases in plant roots. This solution is a weak acid. As the acid-containing water seeps into cracks in the rocks, various chemical reactions take place. These help weaken and break up the rocks. Explore the effect of water on soil in this laboratory experience.

EXPLORE

Fill a jar about half full with warm water. Pour some soil into the water and stir thoroughly. Filter the soil particles out of the water by first pouring it through a cloth. Then pour the clear water through a piece of filter paper. Put a few drops of the clear, filtered water in an evaporating dish and the same amount of tap water in another evaporating dish. Place the dishes on a tripod. Heat gently until the water in both dishes evaporates. Observe what is left in each dish.

Did you find any solid materials in the dish of tap water? In the dish of soil water? Where did the materials in the soil water come from? Does water affect soil minerals?

Another way in which weathering takes place is by the biological action of growing plants. As plant roots grow into the ground, weak acids are produced which help to dissolve and break down soil minerals. In addition, the growth in the size of the plant root system will sometimes split rocks. Observe the effect of plants on rocks in the next activity.

EXPERIMENT

To show how plants act in weathering rocks, fill two test tubes half full of water. Cork one test tube with a one-hole stopper and the other with a regular cork. Push the root end of a seedling plant through the one-hole stopper so that the roots are in the water. Seal the seedling with melted wax so that it is airtight in the tube. Put the test tubes in a rack and leave them on a sunny window sill for several days. Remove the stoppers and test the water in each tube with blue litmus paper. What does a change in the color of the paper to red or pink indicate?

44 / UNIT 1 MAN EXPLORES THE EARTH

Weathering may produce underground caves. In layers of limestone rock, ground water containing weak acids may dissolve and carry away the limestone. After thousands of years, this slow chemical action may form great caves and tunnels hundreds of feet in length. Some of the well known caves formed in this way are the Carlsbad Caverns in New Mexico, Mammoth Cave in Kentucky, and Wind Cave in South Dakota.

In caves where the ground water drips from the roof, icicle-shaped deposits, called *stalactites* (stuh-LAK-tites), may be formed. At the same time, as the drops of mineralized water fall on the floor under the stalactites, deposits grow upward in the form of rounded masses called *stalagmites* (stuh-LAG-mites). If this process continues long enough, these two formations may meet to form columns extending from the ceiling to the floor of the cave (see Fig. 2-20).

Mineral layers may also be deposited on the ground by water flowing out of hot springs. Among the most famous of these deposits are the very colorful layers formed by the Mammoth Hot Springs in Yellowstone National Park. Some of the color is due to tiny plants growing in the water.

Wind has changed the surface of the earth. You have probably seen small dust clouds being blown by the wind. Wind can move large amounts of soil and sand if the ground is bare, and the sand particles carried by the wind can cut away solid rock. This action is a form of erosion. Wind erosion usually occurs over long time periods. For example, in Bryce Canyon National Park in Utah, tall spirelike rocks with unusual shapes were formed by the eroding action of wind and sand (Fig. 2-21 and Fig. 1-4 on page 8.)

Wind blowing over bare soil may also pile the sand in hills called *dunes* (Fig. 2-22). Where there are large amounts of sand with the wind blowing steadily from one direction, dunes up to 300 feet high are formed. If the wind direction changes, the dunes may move from place to place, covering everything in their path. Trees and fertile land can be ruined by shifting sand dunes. Dune formation is controlled by planting grasses and other cover crops to keep the soil and sand from blowing.

Flowing water has changed the surface of the earth. We have learned how wind carrying fine sand can wear away rocks.

2-20 How were these stalacites and stalagmites formed?

2-21 What shaped these rocks?

2-22 Describe the process that formed these dunes.

CHAPTER 2   THE CHANGES IN THE EARTH / 45

Flowing water also causes erosion in a similar manner. Over thousands of years, even the hardest rocks will be eroded.

Moving water will wash away anything it can carry. Soil particles are more easily washed away, but even larger pieces of rock will be carried away by swiftly running water. The soil and the rocks in the water help to wear away the land by scouring the bottom and sides of the river bed (Fig. 2-23). Observe this type of erosion in the next activity.

EXPLAIN

Make a tray out of wood or metal and fill it with soil and pebbles. Direct a stream of water from a hose over the tray. Take the nozzle off the hose for best results. Try varying the amount of water flowing from the hose. After a stream bed has formed, what happens if you put a small rock or other object in the bed? Try using different soil mixtures in the tray. What were some of the factors which influenced the eroding action of the water?

As running water slows down sediment, particles settle to the bottom. The heavier pieces of rock and pebbles settle out first, and then the finer sand and soil particles. In some cases, the particles may be so small that they do not settle out at all, remaining suspended in still water. Study this process in the following experiment.

INVESTIGATE

A. Drill a hole in one end of the box and put a soil layer on the bottom. Place the two wooden blocks under each end of the box so that the part of the box with the hole is lower than the other. Arrange a funnel and filter paper at one end of the box and a beaker to catch the water that flows out at the other end which has the opening. Pour water through the funnel so that it runs through the soil in the box and flows out into the beaker. Remove the filter paper from the funnel and put it aside. Place a clear piece of filter paper in the funnel and filter the water collected at the end of the box into another beaker. Examine the two pieces of filter paper and compare the sediment on them.

2-23   How were the banks of this river formed?

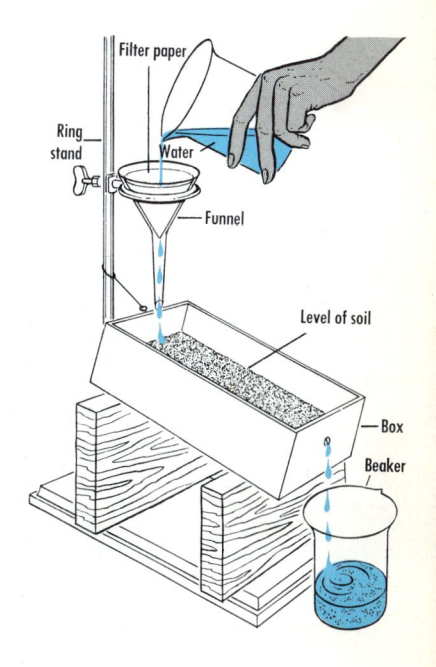

46 / UNIT 1 MAN EXPLORES THE EARTH

Was there any sediment on the first piece of filter paper? On the second piece? Where did this sediment come from? How does this experiment show that running water will pick up sediment.

B. Put a handful of each of these materials in a quart jar: gravel, coarse sand, fine sand, and garden soil. Fill the jar with water and stir the contents rapidly for about a minute. Allow the jar to stand for 10 minutes. Describe the appearance of the materials in the jar after they settle out. How does this activity demonstrate that sediments settle out of water when it stops moving?

When a river flows into a large lake or the ocean, its motion practically stops. Over many years, the sediment which settles out may build up a *delta* at the mouth of the river. A famous delta is found at the mouth of the Mississippi River at New Orleans where it flows into the Gulf of Mexico (Fig. 2-24). It has been estimated that the Mississippi River carries over 500 million tons of sediment into the Gulf each year. Deltas show how the earth's surface changes.

Ice has changed the surface of the earth. It does not seem possible that ice could change the earth's surface much, but along the shores of lakes and rivers ice has been known to move rocks, boats, and other heavy objects. Often, the ice carves out the banks along a lake. Heavy spring rains loosen the soil and wash it into the water. However, the changes along the shores of lakes and rivers are small compared to changes that large ice masses produce.

In the polar regions and in high mountains, many feet of snow fall each year. Because the temperature stays below freezing for long periods, the snow piles up year after year and packs down into huge, solid ice sheets called *glaciers* (GLAY-shuhrz). To show that ice under pressure tends to melt and then freeze again, do the next activity.

OBSERVE

Tie some weights to a metal wire and place the wire across a large cube of ice. Observe the position of the wire after about an hour. What has happened to the ice where the wire has cut

2-24 Why is there a delta at the mouth of the Mississippi River?

through? If the cube of ice lasts long enough, will the piece of wire pass clear through?

There are two main types of glaciers. *Mountain* glaciers are found in all parts of the world where mountain ranges are high. Here, the snow packs down and does not melt. These may vary in length from a few miles to as much as 75 miles and in thickness up to 1000 feet. Many small glaciers of this type are found in the Sierra Nevada Mountains, Rocky Mountains, and Cascade Mountains.

Continental glaciers are found in polar regions. Here, the temperature remains low enough during the entire year so that most of the ice does not melt. The two main continental glaciers are in Greenland and the Antarctic. These huge ice sheets cover thousands of square miles.

COMPARISON OF GLACIERS	
Mountain glaciers	Continental glaciers
Cover small areas	Cover large areas
Hundreds of feet thick	Thousands of feet thick
Move short distances	Move great distances
Tend to gouge out the ground	Tend to smooth off the ground
Direction of flow controlled by the underlying ground	Direction of flow controlled by force within the glacier

When enough snow and ice have piled up, pressure causes some of the ice at the bottom of the glacier to melt. The whole mass now slowly begins to move. As the glacier moves, rocks, trees, and soil in its path are picked up and carried along. These materials grind against the ground and loosen more rocks and soil to be carried along by the glacier. Over the years this action may dig out huge areas of the earth. The following activity will help you understand glacial movement.

PREDICT

Make a trough from two pieces of wood. Put some snow or shaved ice in one end to a depth of several inches. Put some blocks under one end of the trough to make it slant. Does the snow flow downward? Now, press your hand on the snow to pack it down. Does the snow begin to move in the trough when you press on it? How does this activity show glacial movement?

A glacier moves along until its front edge reaches a part of the land where the temperatures are warm enough to melt the ice (Fig. 2-25). As the ice melts, the material carried by the glacier is deposited in formations called *moraines* (maw-RAYNZ). A glacier may melt and drop its material in the same place over a long period of time, building up a *terminal moraine*. Some moraines, left by the great continental glaciers that once covered North America, are miles long and hundreds of feet high. (See Fig. 2-26.) The hilly land of northern Long Island in New York, and some hills along the eastern side of the mountains in Rocky Mountain National Park in Colorado are moraine deposits.

Glaciers have changed the earth's surface in many ways. The Great Lakes and many smaller lakes in Wisconsin, Minnesota, and northern New York were formed by glacial action. There, great ice sheets also carried huge boulders hundreds of miles and deposited moraines to form hilly land. In Ohio, Indiana, and Illinois, glaciers have built up the land to over 60 feet in height.

2-25 Explain how these ridges of soil were formed.

2-26 How were these moraine lakes and hills formed?

REVIEW

1. What factors cause weathering of rocks?
2. How do plants aid in the process of weathering?
3. What two factors increase the eroding action of running water?
4. What process forms a delta?
5. Name several ways in which glaciers have changed the earth's surface.

THINKING WITH SCIENCE

A. *On a separate sheet of paper write the numbers 1 to 15. After the number of each question, write the term that correctly completes the statement.* **Do not write in your book.**

1. The geologic era in which mammals first appeared on the earth is the _____.

2. A deep crack appearing in the earth's crust is called a (an) _____.

3. The breaking up of rocks by physical and chemical action is called _____.

4. The geologic era in which the first life on dry land developed is the _____.

5. The wearing away of the land by wind, water or ice is known as _____.

6. An opening in the earth from which gases and molten rocks erupt is called a (an) _____.

7. The geologic era in whose rocks the most ancient fossils are found is the _____.

8. A huge, slow-moving mass of ice and snow is known as a (an) _____.

9. Materials carried by wind and water from the surface of the earth are called _____.

10. The remains of ancient plants and animals found in rocks are known as _____.

11. The newest method to measure the age of rocks uses the half-time in the decay of radioactive _____.

12. Scientists who study the remains of ancient forms of life are called _____.

13. Carbon dating gives us a clue to the age of _____.

14. Layers of rock formed by deposits of sediments are called _____ rocks.

15. The Sierra Nevada Mountains are examples of _____ mountains.

B. *Write the answers to the following in your notebook. Be sure to use complete sentences and correct grammar.*

1. Explain the theory that accounts for the formation of water and air on the earth.

2. Describe how scientists use radioactive materials to estimate the age of the earth.

3. Why are most fossils found in sedimentary rocks.

4. Compare the ways the three main types of mountains were formed.

5. How do scientists predict volcanic eruptions?

6. What is one explanation of the cause of earthquakes?

7. Describe the physical factors that cause weathering in rocks.

8. How has wind changed the earth's surface?

9. How has water changed the earth's surface?

10. Give some examples of how glaciers have changed the earth's surface.

RESEARCH IN SCIENCE

1. Use clay to make models of different kinds of mountains.

2. Use blocks of wood to illustrate how the crust shifts along a fault.

3. Make a model to demonstrate a volcanic eruption.

4. On an outline map of the world, locate the areas where major earthquakes and volcanic eruptions take place.

5. Make a model to show the extent of the glacier covering North America during the last Ice Age.

6. Report on types of weathering found on buildings and land surfaces in your area.

chapter 3
The Resources of the Earth

A/How Are Soil Resources Conserved?

The earth has abundant resources. *Natural resources* are the valuable materials we find on and in the earth, like water, soil, minerals, and fuels. When many of the earth's resources are used up, they are gone forever. Since it has taken millions of years for some of these resources to be formed, the careful use of these materials is very important. In fact, our lives depend upon their wise use.

The wise use of the earth's natural resources is called *conservation*. This means that ways must be found to prevent wasting of natural resources. Ways of either reusing them or substituting other materials to make the resources last longer must constantly be discovered. Successful conservation requires everyone's cooperation. Our cooperation today aids future generations of Americans.

Soil is one of our most valuable resources. Soil is a mixture of small particles of rock, bits of decayed plant and animal products called *humus* (HYOO-muhs), and various forms of microscopic life. Different soils contain varying amounts of

CHAPTER 3 THE RESOURCES OF THE EARTH /53

these materials, affecting the kinds of plants that can grow. The most fertile type of soil is called *loam.*

Plants can grow well only in fertile soil; they are an important source of food to animals as well as man. Most of our land is covered with a soil layer varying in thickness from a few inches to several feet. Soil, like many of our other natural resources, takes a long time to form. Scientists estimate that it takes 500 years to form one inch of fertile soil. You can learn how to test soil samples in the next activity.

TEST

Obtain samples of soil from your yard or a nearby field. Fill a jar with each sample. Keep the jar tightly closed until the soil is examined. What is the color of each kind of soil? Rub a pinch between your thumb and forefinger. Is it coarse or fine? Squeeze a handful of moist soil. If a crumbly ball is formed, it is mostly clay. Heat a thin layer of the sample over a hot flame. If the soil smokes or blackens, it contains a good supply of humus. Make a chart showing the kind of soil and its composition for each of your samples.

3-1 Identify the topsoil in this photograph. Why is it important to conserve this layer?

The richest part of the soil, containing the greatest amount of humus, is usually the *topsoil* (Fig. 3-1). Scientists tell us that when our country was first settled, the average depth of this rich topsoil was about 10 inches. Now, the average depth is about 5 inches. You might say that this country is within 5 inches of becoming a desert. Of the two million square miles of land in the United States, only one-fifth or about 300,000 square miles can be used to grow crops. And of this amount, over half is poor land. What has happened to our fertile soil?

Soil resources are being used up. If crops are planted year after year with no attempt made to replace the minerals absorbed by the plants, the soil gradually wears out. This using up of the necessary minerals from the soil is known as *depletion* (dee-PLEE-shuhn). In addition, water seeping through the ground dissolves minerals, carrying them deep into the area below the topsoil. The loss of soil fertility is called *leaching* (LEE-ching).

Erosion is probably the most important cause of soil loss. As we learned before, erosion is the wearing away of the land by the action of water, wind, and ice. Each year millions of tons of valuable soil are washed away by floods or blown away by winds. Erosion has ruined over 50 million acres of farmland. However, erosion is a problem facing not only the farmer. Heavy rains may cause erosion when gullies are cut in empty lots and streets are washed out. You can study erosion in the next activity by visiting areas that are near your home and school.

OBSERVE

A. Make a survey of your school grounds to locate evidence of erosion. Are there any ways by which this erosion may be prevented? How can you help?

B. Make a thorough survey of your lot at home. Sketch the lot to show where erosion is occurring. Can you discover ways of preventing it?

C. Visit a city park or an area in your community where there is a river or stream. Make a sketch of a short part of the stream to show where erosion is occurring.

Sheet erosion takes place when a field is flooded and the water runs off slowly. The water carries with it a thin layer of topsoil as it flows away. Each time the field is flooded, some of the topsoil is removed. If flooding occurs often, the land may become worthless in a very short time. A barren surface of rock or subsoil replaces a once fertile field.

On hilly farms where rain falls on bare soil, tiny channels or *rills* are formed as the water flows downhill across a field. Unless this is prevented, *rill erosion* can occur. Each time water flows downhill, it tends to follow the same rills. These become deeper and wider as more soil is carried away. Finally, *gully erosion*, in which the land is cut into deep gullies, may ruin the field for future use (Fig. 3-2). The following experiment, with simple materials, will help you understand the factors contributing to erosion.

3-2 How can erosion, shown in this photograph, be prevented?

EXPLORE

Fill a box with fine sand, packing it evenly to about a depth of half an inch. Wet the sand and pack it tightly. Now make a hole in the side of a gallon can, approximately an inch from the bottom. (You will also need a bucket to catch the water.)

At one end of the sand box, make a small hole to allow the water to flow out. Keep your finger over the hole in the bottom of the gallon can and fill it half full of water.

Tilt your sand box so that it has a slope of about 30° to your desk. Now allow the end of the box to extend over the end of your desk and put an empty bucket on the floor to catch the excess water. Allow the water to flow out of the gallon can into the sand box. Does the water flow in a straight line or does it wander over the surface of the sand? What do you notice after the water has been completely emptied from the can?

Repeat the activity, but this time make a small groove in the center of your wet sand. What happens when you allow the liquid from the gallon can to flow into the container of sand? Does it now flow in a straight line?

Change the angle of the box to 45°, and repeat the activity. Does it make a difference?

One way in which the damage done in gully erosion can be repaired is by using check dams. These small dams slow the water flow so that the valuable soil being carried away has a chance to settle out (Fig. 3-3). Gradually, the gully may fill up and become productive land once again. The next activity will illustrate the difference in rate of erosion for different soils.

MEASURE

Find out whether runoff water from different kinds of ground contain different amounts of sediment. Collect the runoff water after a good rainstorm from a grassy lawn, a wooded area, a plowed field, and a steep slope in the same kind of glass jars. Fill each jar to the top. Label it so you will know where the runoff water came from. Let the jars stand for several hours. Measure the thickness of the sediment layer that settles out in each container. What kind of ground produced the greatest amount of sediment in the runoff water?

Various methods of planting can be used to control erosion. If hilly land is plowed up and down a slope, the furrows act as man-made rills and may lead to gully erosion. To prevent this from happening, the ground can be plowed and planted in curving lines across the slope of the hill. This keeps the water from flowing downhill rapidly. It is known as *contour planting* (Fig. 3-4).

Since barren soil is more easily eroded than soil in which plants are growing, many farmers plant strips of *cover crops*, like oats, clover, or alfalfa, between crop strips grown in rows, like corn, cotton, or potatoes. This is called *strip cropping* (Fig. 3-5). As water runs off from the strip of row crops, it is checked by the plants in the cover crop strip. In addition, plants like clover and alfalfa put nitrogen into the soil. Therefore, strip cropping is not only useful in preventing erosion, but it can actually be helpful in restoring soil fertility. Ways of preventing erosion are shown in the following activity which is found on top of page 58.

3-3 What is the purpose of this check dam?

3-4 How does contour planting aid in conservation of soil?

3-5 Contour planting is being used on this hillside. Why?

58 / UNIT 1   MAN EXPLORES THE EARTH

EXAMINE

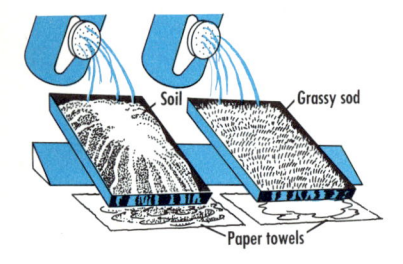

Obtain two large pans and fill them with soil. Cover the soil in one pan with a piece of grassy sod cut out to fit the pan. Tilt both pans as shown in the drawing and put a piece of paper towel under the lower end of each. Pour equal amounts of water from a sprinkling can over the soil in the pans. After the water has drained off, examine the paper towels. Which pan kept the most soil from running off? What does this show you about the value of growing cover crops as a means of preventing erosion?

Minerals can be restored to the soil. Fertile soil contains much humus rich in minerals needed by plants. Since plants use minerals from the soil in making food and growing, they will grow poorly after the soil minerals are used up, as seen in Fig. 3-6. Decayed plants and animal manure are plowed into the soil as two ways of restoring humus. Commercial *fertilizers* are chemical substances containing minerals needed by plants, like nitrates and phosphates. These commercial fertilizers may also be added to the soil to replace the minerals used up.

Crop rotation is another way of conserving soil fertility. By planting different crops each year, the soil loses different amounts of various minerals yearly. The soil may also recover some minerals once used up. For example, a common method of crop rotation used by many farmers is to grow corn, wheat, oats, clover, and then to plant corn again over a five-year period. Which of these helps the soil to recover some minerals? The choice of crops depends on the climate as well as other growing conditions.

Have you ever heard of the term "hidden hunger?" We assume that all foods we eat contain the necessary *nutrients* (NYOO-tree-uhnts). These are substances needed for proper growth and health. However, scientists have discovered that soil fertility affects the food value of the crops grown. This is because there are different minerals contained in the soil. For example, vegetables grown in rich soil furnish better food nutrients for the body than those grown in poor soil.

Winds cause soil erosion. As we have already learned, winds also cause erosion. The harder the wind blows, the more soil

3-6 Can you tell which corn was given a balanced fertilizer?

CHAPTER 3 THE RESOURCES OF THE EARTH /59

it carries away. If the wind blows over loose soil, dust storms may occur. When the wind blows over large land areas where the soil is dry and not held together by plant roots, much damage may result (Fig. 3-7). The great dust storms of 1934-1935 removed valuable topsoil from many parts of our southwestern states. The soil, carried by the wind, also covered up growing crops in other fields. Do the next activity outdoors to study wind erosion.

OBSERVE

Put some fine soil in a box and spread the soil to a depth of about three or four inches. Direct the stream of air from an electric fan over the box and observe the results. Now, place small cardboard "trees" or twigs in the soil and observe how the airstream affects the soil. Make drawings to show how the "trees" prevent the wind from blowing the soil.

As a result of such observations, scientists have learned that if the force of the wind is broken, soil will not be eroded and carried away easily. Rows of trees, known as *shelterbelts* or *windbreaks,* are planted along one side of open fields in many parts of the country (Fig. 3-8). Shelterbelts or windbreaks are important methods in preventing destruction by wind erosion.

3-7 How could wind erosion have been prevented?

3-8 Trees make good shelterbelts. How can they prevent erosion?

REVIEW

1. What is the typical composition of soil?
2. What was the average depth of topsoil when this country was settled? What is the average depth of topsoil at the present time?
3. Name at least three ways in which soil can be eroded by running water.
4. What are two methods of planting that are used to prevent water erosion?
5. How can wind erosion be prevented?

B/How Are Water Resources Conserved?

Water soaks into the ground. If the ground is bare and eroded, water is more likely to run off and be wasted. Ground that is covered with plants and topsoil, however, acts as a sponge in holding water in the soil. The water that soaks into the ground becomes ground water. As air spaces in the soil fill up with water, the water moves to lower levels until it reaches the *water table*. This is the level at which water is found in the ground.

The water table depth depends on the amount of rainfall, condition of the soil, and kinds of rock layers under the soil. If a well is drilled into the ground far enough to reach the water table, then water becomes available. Ground water may flow along an underground rock layer until it comes out on the side of a hill as a spring.

However, during a long drought or where there are many wells pumping water from the ground, the water table may be lowered far enough for the wells to become dry. This drop in the water table takes place because the ground water is brought to the surface and runs off in streams. In addition, it evaporates into the atmosphere as water vapor. Therefore, one method of water conservation is to reduce the amount of water lost through runoff and evaporation. This increases the water supply stored below the soil.

The water cycle restores water in the ground. Have you ever wondered why we seem to have a continuous water supply even though plants and animals are constantly using it? Normally, water movement takes place from the atmosphere to the ground in the form of rain, snow, and hail, and from the ground to the atmosphere by evaporation. This series of processes is known as the *water cycle* (Fig. 3-9).

We learned that water from rivers, lakes, and oceans evaporates from these bodies in great amounts. In addition, green plants and animals give off water into the atmosphere. All this water goes into the air around us as water vapor; it is the first step in the water cycle.

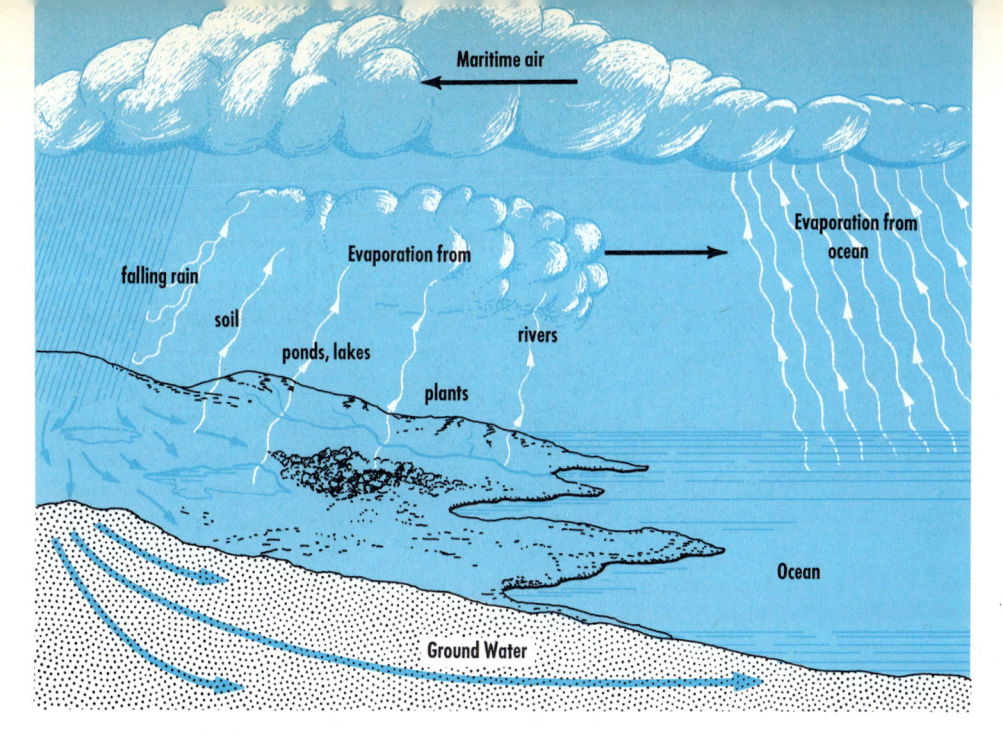

3-9 Trace the main steps of the water cycle in the diagram.

The warm, moist air rises and the water vapor condenses as tiny water droplets or ice flakes. As you learned, these droplets form the clouds high above the earth's surface to become the second step in the water cycle.

As the clouds become cooler, various forms of precipitation occur, like rain or snow, depending on weather conditions. This water soaks into the ground or runs off in streams and rivers into lakes and oceans; it is the third step in the water cycle. This continual movement of water in the cycle has been taking place for millions of years ever since water appeared on the earth's surface.

Soil water is constantly rising to the surface of the ground. *Capillary* (KAP-ih-*ler*-ee) *action,* the way water rises in a blotter, a paper towel, or in a tiny glass tube, is one way water works it way up to the surface of the ground. The amount of water brought to the surface by capillary action during dry weather and lost by evaporation may be great enough to prevent proper plant growth. To help prevent evaporation, a *mulch* can be made by covering the ground surface with leaves, grass, or straw. In fact, sawdust and even paper that is spread on the ground will help to protect the roots of plants from loss of soil water. A dust mulch, made by breaking up the top

3-10 How are aqueducts, like this one, used to conserve water?

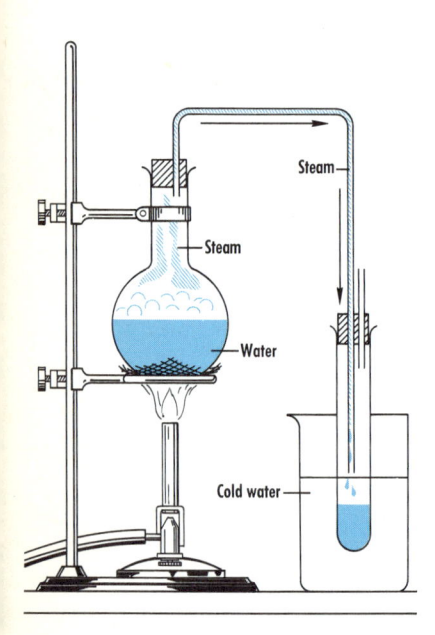

62 / UNIT 1 MAN EXPLORES THE EARTH

two or three inches of soil, will also help prevent this kind of water loss. The mulch breaks up the tiny tubes which form in closely packed topsoil. Where would a mulch be most useful? The next activity should help you answer this question.

EXPLAIN

Place two cubes of sugar in a shallow dish. Sprinkle a layer of powdered sugar on top of one cube. Slowly add a few drops of colored water to the dish, but be careful not to drop any on the sugar cubes. Does the water rise in the two cubes at the same rate? How does the action of powdered sugar illustrate mulch action on the ground?

Water for drinking and household use must also be conserved. In areas of the country where rainfall is not plentiful, huge storage dams have been built. The water is carried to cities and farmlands by using closed pipes or open ditches called *aqueducts* (AK-wah-duhkts). See Fig. 3-10.

Water is purified to make it safe to drink. There is a difference between pure water and water that is safe to drink. Pure water contains no dissolved mineral matter and no bacteria. We can make pure water by the process of *distillation* (dihs-tih-LAY-shuhn). Distillation is the evaporation of water and the condensation of vapor back to water. The following activity illustrates how water can be purified by distillation.

DEMONSTRATE

Assemble your apparatus as shown. Fill your flask half full of water. After assembling the apparatus, slowly boil the water. Observe the water changing into steam. Watch how the water vapor condenses in the test tube. Taste a little of this water. Does it taste different from the water you usually drink? Now pour the distilled water from the test tube into another several times and taste it. What is added to the water in pouring it from test tube to test tube?

Sediment is removed from water by *filtration* (fihl-TRAY-shuhn). In filtration, a liquid passes through a filter to remove

CHAPTER 3 THE RESOURCES OF THE EARTH /63

undissolved particles. As water moves through the ground, or as it remains in our reservoirs, a certain amount of sediment and harmful bacteria may enter the water. The next activity will help you understand filtration.

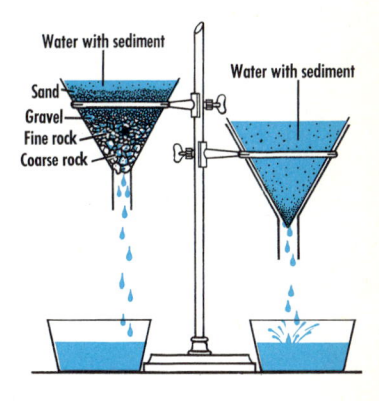

OBSERVE

Obtain a large funnel. In layers from the bottom up, place some washed coarse rocks, fine rocks, gravel, and fine sand. Your funnel should resemble the one shown in the drawing. Now, in another funnel, place a sheet of filter paper. After you have assembled your two funnels, mix some muddy water. Pour some of this water into both funnels. Is the water clear as it drips out of both funnels? Repeat the procedure a second time. If muddy water comes through the sand gravel filter, try plugging the neck of the funnel with asbestos or absorbent cotton.

3-11 What is the purpose of aerating drinking water?

Certain chemical compounds can be added to water to hasten the separation of particles from the water. One such compound, *alum*, reacts with water to form a jellylike mass. This traps the foreign particles and sinks to the bottom of the reservoir. This process is known as *coagulation* (ko-*ag*-yoo-LAY-shuhn).

Filtration and coagulation alone do not make water safe for drinking. In both these processes, the sediment is filtered out, but bacteria and other harmful substances may still remain.

Chlorine, a chemical element, is added to the water in very small amounts to be certain that the remaining bacteria are killed. In addition, the water is usually treated by *aeration* (ayr-AY-shuhn), the process of allowing the water to mix with the air (Fig. 3-11).

Fluoride salts are now being added to the water in some communities. Many dentists believe that *fluoridation* helps keep the teeth strong and free from decay. In 1945, an experimental study was carried out in two eastern communities. In one community, tiny amounts of fluoride salts were added to the water. The other community served as a control with no fluoride salts added to its water. The testing was continued for ten years, during which time children in both communities received annual check-ups. The results of the study showed that the children from the test community had fewer cavities than those from the control community.

Fluoride compounds are added to many commercial toothpastes. This helps to prevent cavities in the teeth of people living in areas where fluorides are not present in the water.

REVIEW

1. What is the main source of ground water?
2. Why do wells and springs usually dry up during a long dry spell?
3. Name the two main ways water vapor enters the atmosphere.
4. In what three main forms does water return from the atmosphere to the ground?
5. Name three ways to purify water for drinking.

C/How Are Mineral Resources Conserved?

Minerals are important in modern life. Minerals are the natural deposits of rocks or ores that can be mined commercially. The earth contains hundreds of useful minerals, and we have already mentioned some of the more important ones. Scientists are constantly experimenting to find newer and better ways of using our mineral resources. The main groups of useful minerals and useful products from them are shown in the table.

USEFUL MINERALS	
Type of mineral	Useful products
Metallic minerals	Iron, copper, aluminum
Nonmetallic minerals	Sulfur, asbestos, nitrates
Building minerals	Limestone, marble, sand
Precious minerals	Gold, silver, diamonds

There is no way of determining our total supply of mineral resources. We produce more metallic and nonmetallic minerals than any other country. However, we are using up our mineral resources rapidly. Although new mineral deposits are discovered yearly, it has been estimated that we have already used up half of our known deposits.

Unwise use of our mineral resources may reduce the living standard for future generations. Minerals were formed in the earth by slow physical and chemical processes over millions of years (Fig. 3-12). Once mined and used, they cannot be replaced in the earth.

Metallic minerals are obtained from ores. As you have learned, most metals are found in the form of *ores*. Ores are mixtures of minerals which contain at least one valuable metal. Mineral deposits may occur naturally formed in the earth, or they may present a changed appearance caused by weathering or changes in the rock structures in which they are found. Thus, metal ores may be the result of a long series of chemical and physical changes that concentrate the metal. In addition, most ores contain impurities which are of little value. These impurities must be separated from the ore before the metal can be obtained in a fairly pure form. Some common metallic ores are listed in this table.

3-12 Limestone is quarried in blocks. Why is this a conservation problem?

MINERAL RESOURCES IN ORES		
Metal	Mineral	Composition
Aluminum	Bauxite	Aluminum oxide
Copper	Cuprite Azurite Chalcocite	Copper oxide Copper carbonate Copper sulfide
Iron	Hematite Siderite Pyrite	Iron oxide Iron carbonate Iron sulfide
Lead	Galena	Lead sulfide
Mercury	Cinnabar	Mercury sulfide
Sodium	Halite (salt)	Sodium chloride
Tin	Cassiterite	Tin oxide
Uranium	Uraninite	Uranium oxide

66 / UNIT 1 MAN EXPLORES THE EARTH

Various methods have been developed to separate minerals from the impurities in the ores. One such process is *flotation* (flo-TAY-shuhn). The ore is first ground into a fine powder and then mixed with oil and water. After thorough stirring, the ore rises to the top of the mixture with the oil droplets, and the impurities settle out on the bottom of the water. You can observe the flotation process in the next activity.

DESCRIBE

Grind some clean sand into a powder with a mortar and pestle. Mix 15 grams of powdered sand with one gram of powdered galena (lead sulfide). Pour this into a tall glass jar containing 40 milliliters of mineral oil and 30 milliliters of water. Put your hand over the top of the jar and shake the contents to mix them thoroughly. Let the jar stand for a while so that the materials separate. What happens?

After most of the sand, rock, and other such impurities have been separated from an ore, the metal must be separated from the chemical compounds which make up the ore. These are usually in the form of carbonates, oxides, and sulfides. These are, respectively, compounds of the metal combined with carbon, oxygen, or sulfur. This process is called *smelting*. Some ores are first *roasted*, or heated in air. When carbonates are heated, carbon dioxide is given off and an oxide of the metal is left. This oxide must then be separated by heating with charcoal or coke. Heating removes the oxygen, leaving behind the metal. The carbon (charcoal or coke) combines with the oxygen to form carbon dioxide gas. We can show how copper oxide can be separated in an activity.

EXPLORE

Mix a small amount of copper oxide powder with a little powdered charcoal. Pour the mixture in a test tube. Heat it in a hot flame for several minutes and observe what happens. Allow the tube to cool and pour the contents into a beaker of water. Stir the mixture for a minute and carefully pour off the water.

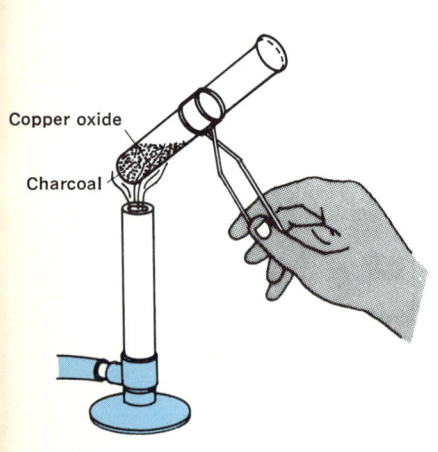

CHAPTER 3   THE RESOURCES OF THE EARTH /67

What is left in the bottom of the beaker when the water is poured off? Where did this substance come from? What was removed from the copper oxide? How does this activity show ore separation?

Iron is obtained by heating ore in a blast furnace. Since steel is one of the basic materials used in industry, let us see how this material is obtained. Iron, from which steel is made, can be separated from its ores by separation. This process requires a great deal of heat. Therefore, a blast furnace of some type must be used (Fig. 3-13). The iron ore, usually an oxide, is mixed in the furnace with coke (which is obtained from coal) and with limestone. A hot air blast is then blown through the heated mixture. The coke, which is mostly carbon, combines with the oxygen from the hot air to form carbon dioxide gas. This is reduced (separated) to carbon monoxide by contact with the hot coke. The carbon monoxide then reduces the iron oxide of the ore and sets free the molten iron.

This molten iron settles to the bottom of the furnace where it is drawn off as a fiery mass of liquid metal (Fig. 3-14). The iron is not pure. It consists of 92 to 95 percent of the metal, with the rest as impurities. Called *pig iron* at this stage, it can be melted and cast into parts for heavy machines that are not exposed to great strains.

Some of the ore impurities will combine with the limestone to form *slag*, which floats on top of the molten iron. Slag is drawn off from the side of the furnace and usually discarded as waste material.

To produce steel, the remaining impurities must be removed from pig iron. This is done by pouring the molten pig iron into an *open-hearth furnace* and blowing hot gases over the liquid iron. The open-hearth process is controlled so that the quality of the steel can be checked at each production stage.

One disadvantage of the open-hearth process is that it takes between 6 to 12 hours to produce a batch of steel. The *basic-oxygen process*, which produces steel in 50 minutes, is used in many steel mills. This process was first developed in Austria as a method of producing small amounts of high-quality steel. In this process pure oxygen is mixed with the mol-

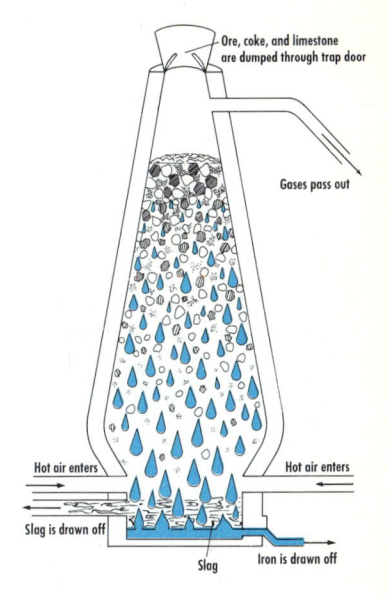

3-13 Trace the process by which iron is separated from its ores in this blast furnace.

3-14 Hot metal from a blast furnace is poured into an open-hearth furnace. What is the product called?

68 / UNIT 1 MAN EXPLORES THE EARTH

ten metal at great pressure. Compared with the open-hearth process, steel is produced eight times as fast by the basic-oxygen process (Fig. 3-15). Since this process produces high-grade steel faster and cheaper than the open-hearth process, it is rapidly replacing the open-hearth process in our country.

Steel is actually an *alloy,* a mixture of iron with carbon and various other elements. Different types of steel are produced by adding carbon, tungsten, nickel, manganese, and other elements in measured amounts to the molten steel in the open-hearth furnace or basic-oxygen process. The highest grades of alloy steel are produced in *electric furnaces* in which temperatures can be carefully controlled (Fig. 3-16). Steel produced in electric furnaces is expensive, however.

3-15 How does the basic oxygen process differ from the open hearth method of producing steel?

3-16 What kind of steel is produced in this electric furnace?

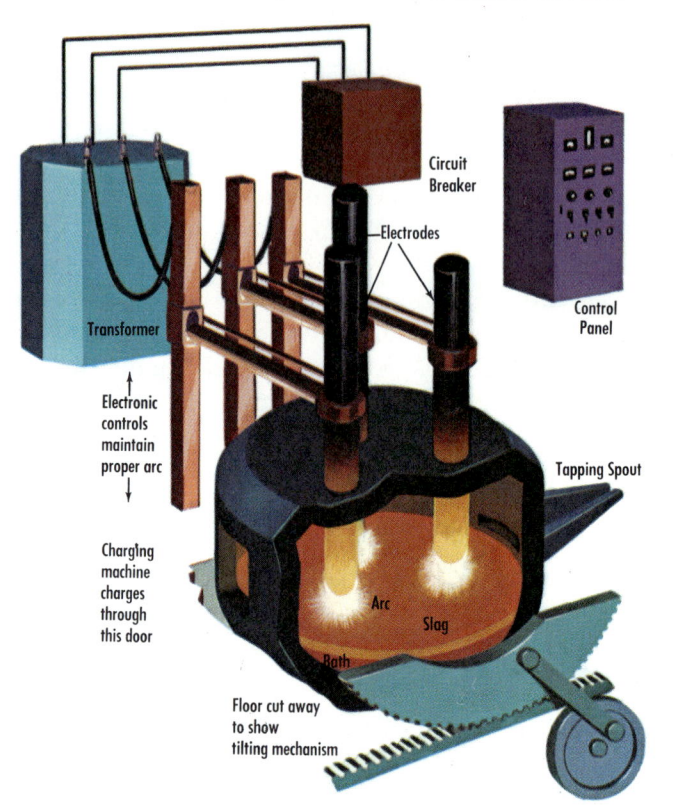

CHAPTER 3 THE RESOURCES OF THE EARTH /69

Coal and petroleum are sources of many useful materials. As steel-making became an important industry, there was a great demand for coke. *Coke* is produced by heating bituminous coal in a closed oven. A gummy substance, called *coal tar,* is produced as a side product. Scientists, who experimented with coal tar, found that many useful compounds, for home and industry, could be made from it. Today, coal-tar products are probably much more valuable than the coke which is produced at the same time.

Petroleum is in the form of a thick, crude oil when it is pumped from the ground. To obtain useful products from this substance, the oil must be refined (Fig. 3-17). Kerosene, fuel oil, and gasoline are important products produced in this process. Crude oil is heated to a high temperature and the gases are treated with various chemicals to separate the different parts.

In addition to fuels, other important substances are also produced from petroleum. For example, most detergent powders are not soaps, but grease-dissolving chemicals made from petroleum. Fertilizers, paints, medicines, cosmetics, and even the ink used in printing this page all come from petroleum. Can you think of other ways in which substances like these are important in your daily life?

3-17 What products come from a modern oil refinery?

Mineral resources must be conserved. There has been waste in mining and converting minerals into useful products. Since it is cheaper to use better grade ores, these have been removed, leaving poorer grades behind. Helium is found, together with natural gas deposits, in some parts of the country. Helium is wasted in large amounts because it is not separated from the the natural gas as the natural gas is burned. Yet, helium is one of the most valuable materials in the study of low-temperature science and space exploration, aside from its commercial and industrial uses.

To help in conserving mineral resources, scientists have developed new methods for using low-grade ores (Fig. 3-18). More complete use of by-products, such as the separation of silver from lead and copper ores, has added a rich supply of useful minerals. Iron, copper, zinc, and lead are just a few of the metals that can be melted down and reused.

3-18 Low-grade iron ore being crushed, screened and washed.

Another way in which our mineral resources are being conserved is through the development and use of man-made materials. For example, plastics have been used to replace many articles once made of metal (Fig. 3-19). *Lexan,* one type of plastic, can be produced in a number of attractive colors. *Polyceram* (*pahl*-ee-sE-ram), which is made from glass, is harder than steel and lighter than aluminum. It is used in space vehicle nose cones, and it could be useful in piston heads, brake shoes, and other automobile parts requiring high-temperature resistant materials. The following activity should be done next to learn more about conserving our valuable minerals.

3-19 How does the use of plastics conserve mineral resources?

EXAMINE

Bring in objects that use materials to replace metals. These may be made of plastic, fiberglass, or other similar substances. Explain how each conserves our natural mineral resources.

Materials have been developed that are better in some ways than natural minerals. *Borazon,* a man-made substance that is very hard, has been made from nitrogen gas and boron, a soft element. These methods of producing materials to replace minerals are very costly at present. But some day, man will be able to make special materials to replace the scarce minerals that are being used up so rapidly from our mineral resources.

REVIEW

1. Name the four main groups of useful minerals.
2. What are the two main steps in heating an ore to obtain the metal in it?
3. What is removed from iron ore when it is heated with: (*a*) limestone, (*b*) coke?
4. What two substances are produced when soft coal is heated in a closed oven?
5. Name three fuels that are obtained from petroleum.

THINKING WITH SCIENCE

A. *On a separate sheet of paper write the numbers 1 to 15. Some of the statements are true and some are false. Rewrite the false statements, changing the terms in italics if necessary, to make them all true. Write the word true if the statement remains unchanged.* **Do not write in your book.**

1. The movement of water from the ground to the atmosphere and back again is called the *water cycle*.

2. The process of separating the various products in petroleum by heating and other methods is known as *refining*.

3. *Depletion* is the loss of soil minerals by water seeping through the ground.

4. The planting of alternate strips of row crops and cover crops to reduce erosion is called *crop rotation*.

5. The loss of topsoil when fertile soil is flooded time and again is called *rill erosion*.

6. *Erosion* is the using up of valuable minerals by plants growing in the soil.

7. The planting of different crops in successive years to conserve soil fertility is called *contour plowing*.

8. A *blast furnace* is usually used to separate iron from the impurities in its ores.

9. The gummy substance produced when coal is heated to produce coke is called *coal tar*.

10. *Conservation* is the wise and considerate use of the earth's natural resources.

11. The most fertile type of soil is *humus*.

72 / UNIT 1 MAN EXPLORES THE EARTH

12. Decayed bits of plant and animal products in soil are known as *loam*.

13. Loss of soil fertility is called *strip cropping*.

14. The damage that is caused by gully erosion can be repaired by using *check dams*.

15. Oats, clover, and alfalfa are called *cover crops*.

B. *Write the answers to the following in your notebook. Be sure to use complete sentences and correct spelling and grammar.*

1. Why must everyone make an effort to conserve our natural resources?

2. What happens to soil and the minerals in it during a heavy rainfall?

3. Explain how check dams are used to prevent erosion that will result in useful soil.

4. Describe the main steps in the water cycle.

5. Why is water one of our valuable natural resources?

6. Describe how the flotation process is used to separate impurities from some ores.

7. Describe the process of making steel from pig iron.

8. How is coal tar obtained from soft coal?

9. Describe the process in which various fuels are obtained from petroleum.

10. What are some ways in which our mineral resources are being conserved?

CHAPTER 3   THE RESOURCES OF THE EARTH /73

RESEARCH IN SCIENCE

1. Make a soil profile of the soil in your community by gluing soil samples on a sheet of cardboard.

2. On a map of the United States, locate the major deposits of coal, iron, copper, and aluminum.

3. Make an exhibit of some important products obtained from coal tar or petroleum.

4. Make a model of a farm showing the proper methods of soil conservation.

5. Report on the characteristics of cover crops and row crops, explaining why cover crops prevent erosion.

6. Make an exhibit of some of the more important chemical fertilizers, indicating the particular elements each one replaces in the soil.

7. Write a report on the conservation measures used in your community to keep air and water clean.

READINGS IN SCIENCE

Bascom, Willard, *Waves and Beaches*. Anchor Books, 1964. The story of "the struggle for supremacy between sea and land" is told with enthusiasm and authority.

Bullard, Fred M., *Volcanoes in History, in Theory, in Eruption*. University of Texas, 1962. A clear and well-illustrated study of the development, principles, and applications of volcanoes. It is very complete, yet needs no previous knowledge of the subject.

Clayton, Keith, *The Crust of the Earth*. Natural History Press, Doubleday and Co., 1967. A fascinating account of the history and thinking of geology from Greek times to the present. Included in the list of topics are: continental drift, polar wandering, sea-floor spreading, mountain building and earthquakes.

Cromie, William J., *The Living World of the Sea.* Prentice Hall, 1966. The author traces the fascinating world of the sea from its beginnings to the present day. The book, too, presents a vivid introduction to life beneath the surface of the sea.

Gallant, Roy A., *Discovering Rocks and Minerals.* Doubleday and Co., 1967. Very useful as a guide for the collection and identification of rocks, minerals and fossils. The writing is informal, and many large illustrations of excellent quality are present.

Gallant, Roy, *Exploring Under the Earth.* Doubleday, 1961. This book explains the past work and the methods of the present-day study of geology. It has many color pictures and diagrams plus a geologic time chart at the end of the book.

Matthews, William, *Fossils — An Introduction to Prehistoric Life.* Barnes and Noble, 1962. The book describes and illustrates many vertebrate and invertebrate fossils. It includes a brief history of paleontology and evolution.

Matthews, William H., *The Story of the Earth.* Harvey House Publisher, 1968. The materials, processes, and history of the planet earth are presented with numerous illustrations. Definitions are clear and concise.

Poole, Lynn, and Poole, Gray, *Volcanoes in Action.* Whittlesey House, 1962. The book tells the story of volcanic eruptions. The causes of eruptions and their effects on the earth are described. Good photographs are included.

Rhodes, Frank H. T., Herbert S. Zim, and Paul R. Shaffer, *Fossils: A Guide to Prehistoric Life.* Golden Press, 1962. A handbook full of useful information for the fossil collector. Introductory material on fossil hunting is followed by a survey of life of the past, then invertebrate and vertebrate animal fossils are described and a brief account of fossil plants given.

Richards, Leverett, *Ice Age Coming?* John Day, 1960. The book presents a general description of prehistoric glaciers and tells about important glaciers in the world today.

Roberts, Elliott, *Deep Sea, High Mountain.* Little, Brown, 1961. The book describes a series of actual incidents in the work of the United States Coast and Geodetic Survey. Featured are many accounts resulting from the study of the earth and oceans.

Roberts, Elliott. *Our Quaking Earth.* Little, Brown, 1963. Surveys famous earthquake disasters and various myths and superstitions concerning them. The author explains their actual causes, distribution, and the various scientific and nonscientific attempts to explain and predict their occurrence.

Stone, A. Harris and Dale Ingmanson, *Rocks and Rills: A Look at Geology.* Prentice Hall, 1967. Simple procedures are given to demonstrate 28 chemical and physical processes affecting the earth's surface. This is an excellent reference for geology related studies.

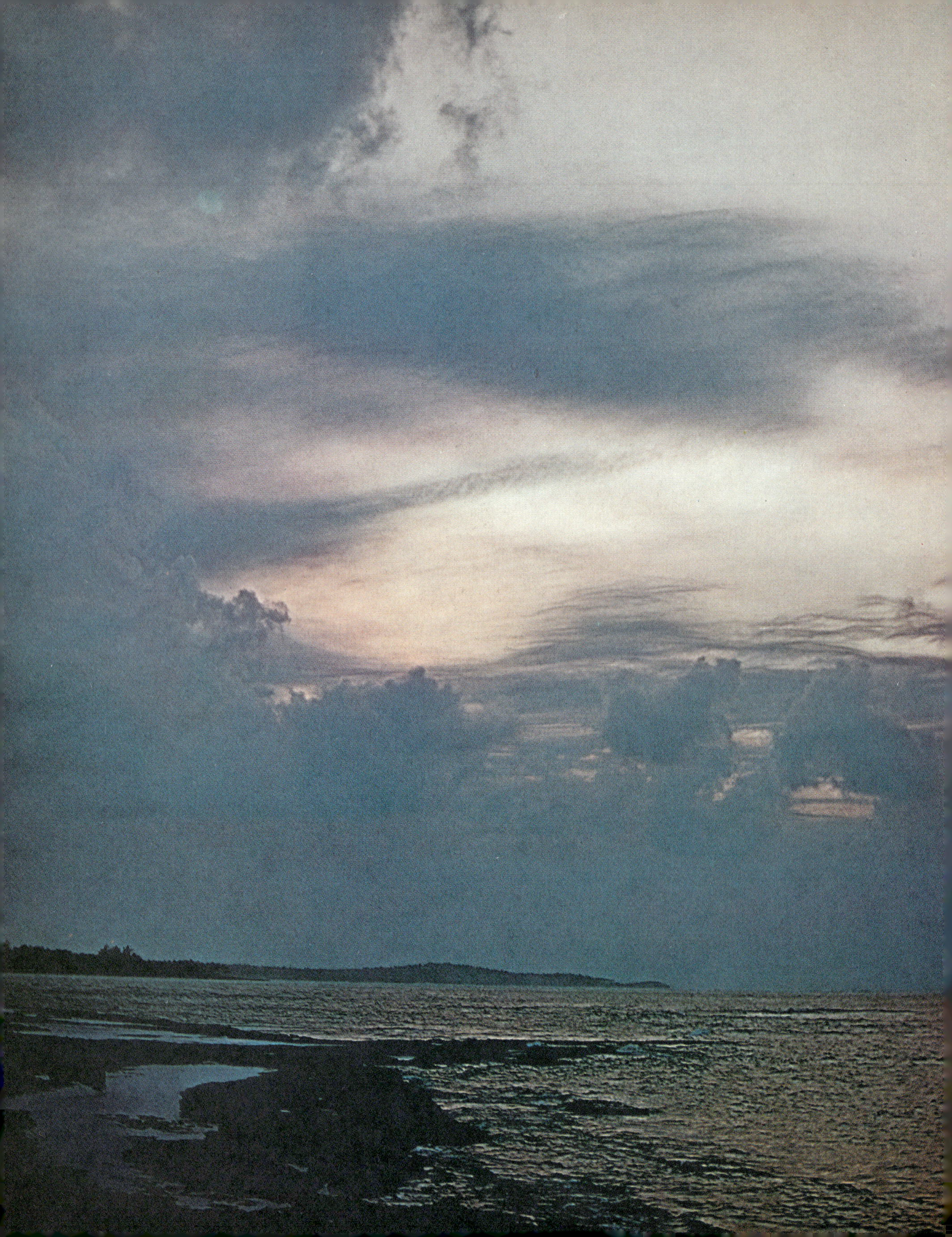

unit 2

Man Explores His Environment

The first weather satellite, Tiros I, was launched into orbit on April 1, 1960. For the first time, weather scientists actually saw the patterns of weather as they formed. More than 14,000 photographs were transmitted of the earth's weather by the television cameras of Tiros I. Valuable photographs of cloud cover, all over the world, were taken. The information was used by weather scientists to make more accurate weather predictions.

Man has come a long way from the beliefs of primitive man who thought that changes in the atmosphere were caused by their gods. Supposedly, angry gods produced bad weather and the kind gods produced good weather. In early times, priests who attempted to influence the gods to produce needed rain or to stop floods were considered very important people. Stories

78 / UNIT 2   MAN EXPLORES HIS ENVIRONMENT

and myths from ancient Greece and Rome tell of the different gods of weather.

In the astrology of the ancient Babylonians and Egyptians, atmospheric conditions were believed to be influenced by the position of the "wandering stars" or planets in the sky. They left weather records in the tables and charts carved in stone in many of their buildings. Some people, today, believe in these forecasts, in spite of the lack of scientific evidence to support them.

The accuracy of the observations of the early scientists suffered, however, because they had no scientific instruments with which to measure the changes in the atmosphere influencing weather. It was not until *Galileo Galilei (1564–1642)* invented the first practical air thermometer, *Gabriel Fahrenheit (1686–1736)* the mercury thermometer in 1714, and *Anders Celsius (1701–1744)* the Celsius scale in 1742, that accurate observations of temperature changes of air became possible. Other instruments useful in measuring conditions in the atmosphere, like air pressure, were the mercury barometer invented by *Evangelista Torricelli (1608–1647)* in 1643, and the anemometer (measures wind speed) invented by *Robert Hooke (1635-1703)* in 1667.

Scientists now have these and additional instruments to make accurate observations of atmospheric conditions. Other factors which influence weather changes are studied, making accurate weather predictions possible. Weather satellites, like the Tiros and Nimbus series, and others, are probing new heights and recording valuable data.

Important discoveries of the nature of air were made by *Joseph Priestley (1733–1804)*, an Englishman, and *Antoine Lavoisier (1743–1794)*, a Frenchman. Priestley was the first to observe that when a certain red powder (actually the compound mercuric oxide) was heated, it yielded a gas. Lavoisier performed a similar experiment shortly after, also using the same powder. He concluded that the newly discovered gas was only one of the substances found in air, and that air was not an element but a mixture of gases.

In this unit, we shall study some of the factors which are responsible for our weather and climate. In order to understand these, it is also necessary to understand the properties of air and water, and heat as an energy source. All of these factors are interrelated, and man is dependent upon each of them for survival.

Galileo

Torricelli

Lavoisier

chapter 4
The Changing Atmosphere

A/How Do Atmospheric Conditions Change?

Man is adapted to live in a sea of air. Lift man out of his protective atmosphere and he will die just as a fish dies when taken out of water. What we call the atmosphere (AT-muh-*sfeer*) is the layer of gases surrounding the earth. The only way man can survive in space is to build machines that will enable him to take his earthly living conditions along with him. What conditions does man need if he is to survive in outer space?

Mountain climbers at an altitude that is above three miles begin to suffer from a condition in which too little oxygen reaches the body tissues. Unless these people use oxygen masks to prevent this condition, they gasp for breath, become dizzy and sluggish, and begin to lose muscular control. Unconsciousness may result.

A somewhat similar result happens to the body's internal organs when the air pressure that is around the body is lowered greatly. At very high altitudes, the air that is trapped in your intestines and lungs, for example, begins to expand. The following activity on air pressure may help you to understand what happens.

EXPLORE

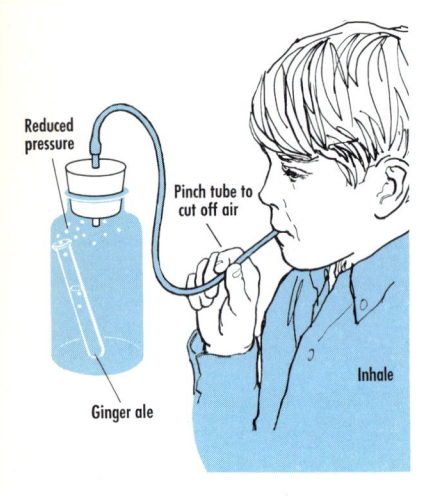

Push a glass tube through the stopper. Place it securely in the wide-mouth bottle so that the stopper is airtight. Attach a length of rubber tubing to the glass tube so that by pinching it with your fingers you can control the air flow from the bottle. Fill a test tube about half full with ginger ale or other carbonated beverage from a freshly opened bottle. Stand the test tube in the wide-mouth bottle, as shown. Draw your breath through the rubber tube to remove air from the bottle. After each breath, pinch the tube to prevent air from flowing back into the wide-mouth bottle.

What happens to the air pressure in the bottle as you draw air out? What is the effect of this change on the gas dissolved in the ginger ale? What happens to a gas dissolved in a liquid if the surrounding air pressure is lowered?

4-1 Atmospheric conditions found at 60,000 feet exist in this pressure chamber. Why is the blood in the beaker boiling?

If you were to ride rapidly to a high altitude in the open cabin of a balloon without special protection, you would soon begin to feel the effects of the lowered air pressure. At about six miles in altitude, nitrogen gas, which under sea level pressures is dissolved in the bloodstream, begins to bubble out of the blood. These dangerous gas bubbles collect in the body's joints, nerve tissue, brain, muscles, and other body tissues. They can cause a condition called the *bends*. Extreme pain, difficulty in breathing, blurred vision, dizziness, and unconsciousness are symptoms of this condition. Death sometimes results from the bends.

If you were to go still higher, the air pressure would drop even more. At about ten miles in altitude, the air pressure is so low (1.69 pounds per square inch) that normal breathing adapted to a sea-level air pressure of *14.7 pounds per square inch*, becomes impossible. A condition where insufficient oxygen is present to support life would affect you. You would lose consciousness quickly. At about 12 miles in altitude, the fluids in the body actually begin to "boil" due to the greatly lowered air pressure, as shown in Fig. 4-1, and death results quickly. The next activity illustrates the effects of lowered air pressure on the lungs.

CHAPTER 4   THE CHANGING ATMOSPHERE / 81

EXPLORE

Insert two pieces of glass tubing in a two-hole stopper. Fasten a small balloon, which has been blown up two or three times to let it stretch easily, to the lower end of one glass tube. Fasten a length of rubber tubing, attached to a vacuum pump, to the upper end of the other glass tube as shown. Place the stopper in a glass bottle so that the seal is tight. Blow into the tube to which the balloon is attached to inflate partly the balloon. Seal the tube with a piece of soft clay. Start pumping the air out of the bottle. What happens to the balloon?

What happens to the air pressure in the bottle as the air is removed? What happens to the size of the balloon? Can you give the reason for this?

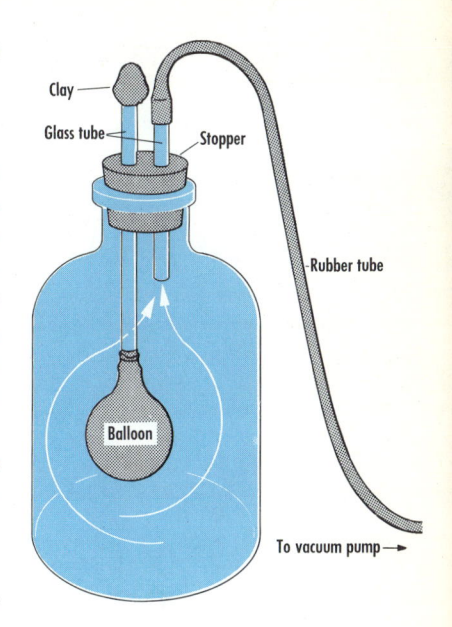

4-2 What is the purpose of sending this balloon aloft?

Scientists use different methods and instruments to study the atmosphere. Man is constantly trying to understand the relationships among the different factors such as temperature, air pressure, wind, and humidity. Clearer understanding of these relationships will allow more accurate weather predictions. It is possible that man might eventually be in a position to control weather conditions by varying the factors that determine it.

From your previous study of science, you are already familiar with some of the instruments used in atmospheric study. The *thermometer*, for example, measures air temperature, and the *barometer* measures air pressure. Can you list some other such instruments along with their uses?

Today, scientists use balloons, airplanes, satellites, and rockets to aid in atmospheric study. In Fig. 4-2, a balloon carrying a *radiosonde* (RAY-dee-o-*sahnd*) which contains recording instruments, is released. The instrument sends back information by means of radio signals to a receiving station located on the ground. It floats up to approximately 15 miles and bursts. The instruments then fall back to earth by means of a small parachute.

You have learned that satellites are used to carry a variety of specialized instruments for the collection of information about the atmosphere. As a satellite circles the earth, it might record the number of meteors entering our atmosphere, the

4-3 The parts of a Tiros satellite. What does this satellite do?

4-4 How are the layers of the atmosphere classified?

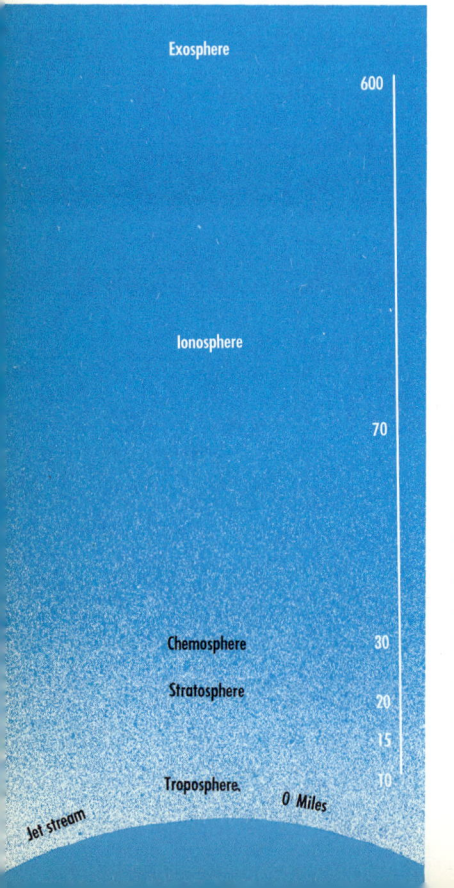

amount of radiation, the concentration of various gases, and other important data (see Fig. 4-3).

The atmosphere extends to the edge of space. Anyone can see and measure the upper layer of our oceans, but no one can locate exactly the upper layer of the atmosphere. As we go higher and higher, the air gets thinner and thinner until it finally becomes nearly empty space. We can fill a glass half full of water, but can we fill a glass half full of air?

At 100 miles altitude, there is so little atmosphere that meteors entering the atmosphere have not yet begun to heat up and glow. At about 120 miles altitude there is not enough air left to carry sound. At 125 miles altitude, heat can no longer be transferred. A person would be cooked by the sun on one side and frozen solid on the shaded side. At 150 miles, the atmosphere is so thin, that for all practical purposes, there is no drag on an object in flight. As a result, a satellite will stay in orbit above this altiude. Scientists believe that the last traces of air molecules disappear above 650 miles.

The layers of the atmosphere have different properties. The earth's atmosphere has regions or layers, each with particular features. Scientists classify the layers of the atmosphere according to temperature changes with altitudes. Of these, the *troposphere* (TRO-puh-*sfeer*) is the layer of the atmosphere closest to the earth. The troposphere is about five miles high above the poles of the earth and about 11 miles high at the equator (see Fig. 4-4). Over most of the United States, it is approximately seven and one-half miles in height. What accounts for this difference?

The troposphere is where most of our clouds and masses of air are constantly mixing. Our weather is the result of changing conditions in this atmospheric layer. Practically all the water in the atmosphere may also be found here.

At an altitude of approximately 6 to 9 miles, the *jet-streams* may occur (Fig. 4-5). These are enormous currents of air in the upper atmosphere. They are believed to be the "heat exchangers" of the atmosphere. Wind speeds in a jet stream sometimes reach a speed of 200 miles per hour and usually blow from west to east. Why would an airplane traveling from west to east usually make better time than another airplane traveling east to west?

4-5 Why do jet streams blow from west to east in our hemisphere?

The *stratosphere* is a layer of air which extends upward from the troposphere to an altitude of about 19 miles. The temperature of the air in this layer is usually fairly constant. It is about −40°C at the upper boundary of the layer. The air is rather thin in the stratosphere and storms do not develop in this air region.

From 19 to 50 miles above the earth is the layer called the *mesosphere* (MEZ-us-*sfeer*). Temperature in the mesosphere ranges from 10°C to lower than −90°C, depending on the season and the location. In this layer, chemical changes brought about by the radiation of the sun begin to be noticeable.

Above the mesosphere we find the *thermosphere* (THUHR-muh-*sfeer*), the outermost of the temperature layers. Here temperature increases and varies widely with location, time of day, and the activity of the sun. When a solar flare takes place, temperature in this layer may rise to over 1700°C.

The *chemosphere* (KEM-uh-*sfeer*) is a different type of layer from the others. It extends from 10 to 120 miles above the earth, overlapping the stratosphere, mesosphere, and the thermosphere. In the chemosphere, there occur chemical processes often started by the absorption of certain kinds of solar radiation. Most meteors approaching earth burn out

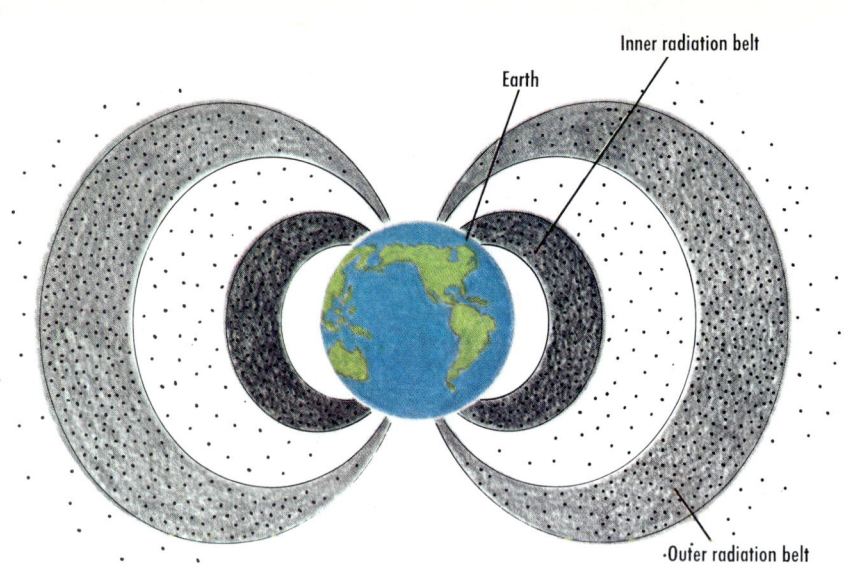

4-6 What is the composition of the Van Allen radiation belt?

when they reach this layer. The chemosphere also filters out harmful rays from the sun, thus protecting living things.

The *ionosphere* (eye-AHN-uh-*sfeer*), still another type of layer, which can be identified in Fig. 4-4, extends from about 35 to 600 miles above the earth and overlaps the chemosphere. Different levels of this layer reflect radio waves from the earth and shield us from the sun's harmful rays. The lower parts of the *Van Allen radiation belt,* a layer of charged particles in space around the earth, extend into the upper portions of the ionosphere (Fig. 4-6).

Finally, beyond the ionosphere lies the *exosphere* (EKS-o-*sfeer*). Here the composition of the atmosphere changes to helium and then to hydrogen. Beyond this region, the atmosphere, as we know it, fades into space.

REVIEW

1. Name the different layers of the atmosphere.
2. Give the approximate boundaries for each layer.
3. What are the jet-streams?
4. How does the chemosphere protect us?
5. In what layer of the atmosphere is the Van Allen radiation belt?

CHAPTER 4 THE CHANGING ATMOSPHERE / 85

B/How Is Our Earth Heated?

Our atmosphere is heated by the sun. The sun, either directly or indirectly, is the source of all of our heat, except heat produced by atomic sources. If one goes up in the atmosphere to a height of two miles, there is a drop of about 20°C in temperature from that at the earth's surface. For every rise of 1000 feet, therefore, there is normally a drop of about 1.8°C. The earth's surface also varies in temperature from season to season and from day to night. Try the next activity to see the effect of sunlight upon the temperature of water.

COMPARE

Take two tin cans and paint one black. Fill each can with the same amount of cold water. Put a thermometer in each can and place both cans in the sunlight. Record the temperature at five-minute intervals. Compare your results with those of your classmates. What conclusions can you draw from the results?

The heat from the sun reaches the earth by means of *radiation* (*ray*-dee-AY-*shuhn*). Radiation is the giving forth and the transfer of energy in waves through space. Energy waves, coming from the sun, travel at the speed of light or about 186,000 miles per second. When the energy waves strike the earth's outermost atmosphere, some of the heat energy is absorbed by the air molecules. The more molecules in the air, the more energy that is absorbed. However, since the densest part of the atmosphere is closest to the earth's surface, the largest amount of heat energy is absorbed directly by rocks, soil, buildings and oceans.

The best heat absorbers are surfaces that are rough and dark. The earth, being both dark and rough, absorbs much of the sun's radiant energy and becomes warmer. Thus, close to the earth's surface, the air is heated not only by the sun but also by the radiation of heat back from the earth. The earth warms up much more quickly than the surrounding air.

Heat energy also can be transferred by conduction and convection. Heat energy moves from a warmer substance to a

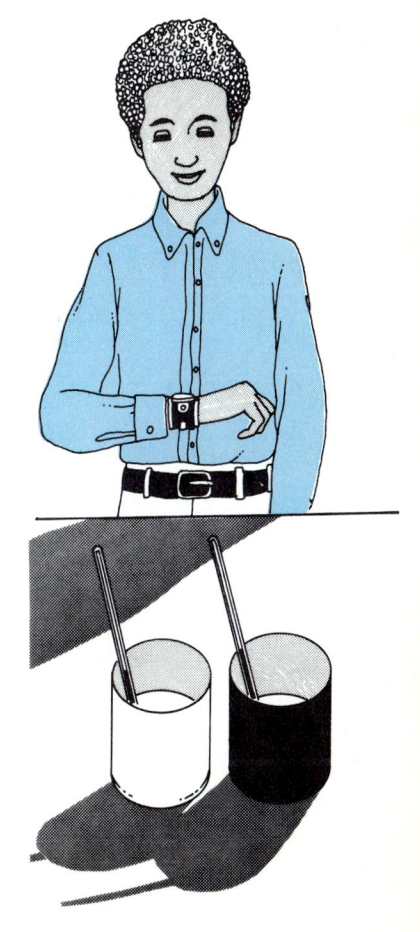

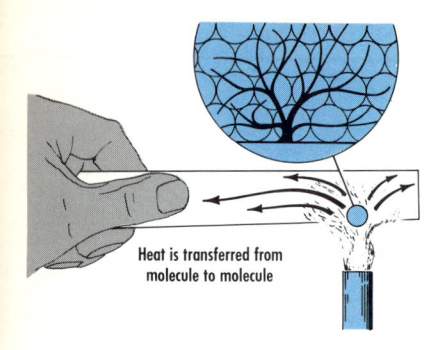

4-7 What term is given to the type of heat transfer shown?

cooler one until they are both at approximately the same temperature. Heat can be considered as a measure of the motion of the molecules within a substance. The warmer the object becomes, the faster the molecules move. When a solid is heated at one end, the molecules at that end begin to move at a faster rate and bump into those near them. As the molecules collide with one another, heat energy is transferred from one molecule to the next by the process of *conduction* (kuhn-DUHK-shuhn). Since heat is associated with the movement of molecules, those in the neighboring area begin to increase their speed. By this chain reaction, heat is carried along a rod from one end to the other as seen in Fig. 4-7.

In general, metals are good heat conductors while liquids and gases are poor conductors. For example, iron conducts heat about 100 times as well as water and about 2500 times as well as air.

When the molecules move at a faster rate, the space usually increases between one molecule and the next. Therefore, as an object gets hotter, what happens to its size? Have you ever used this principle to open a jar cover that was too tight to open? Perhaps you held the cover under hot water for a few seconds and then found it was easier to turn. You can study the conduction of heat in different materials very easily in the next activity.

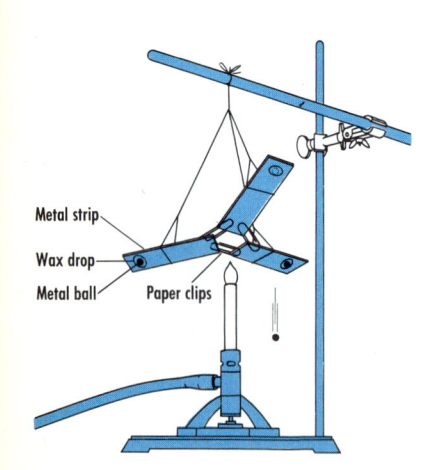

OBSERVE

Obtain three similar strips of copper, steel, and aluminum, about ½″ by 6″ in size. (All the strips must be the same length and thickness. Why?) Using a candle, drop a bit of wax on the end of each strip and press a small steel bearing or lead shot into the wax before it hardens. Fasten the other ends of the strips together with paper clips. Suspend the strips from a rod as shown. Light a Bunsen burner under the ends of the metal strips and note which wax drop first melts enough to release the metal ball. List the metal in the order in which the bearing or shot dropped. Can you explain the reason for this order? Compare your results with those of your classmates. How would you explain different results?

Heat cannot be transferred very rapidly by conduction because of the slow rate of transfer of energy from one molecule to another. In addition, very large numbers of molecules are involved in the process. For heat energy to pass from one end of a short piece of wire to the other end of the wire, it would have to be transferred across billions and billions of molecules of metal. However, heat is also transferred in liquids and gases by *convection* (kuhn-vek-shuhn). When you go to bed on a warm night, you may open the windows both from the top and bottom. This allows the cool air to come in through the bottom. We know that air and water expand when they are heated and become less dense. In the window, the more dense, cooler air moves in below, forcing the warmer air in the room to rise. This sets up *convection currents* in the air or water which move the heat from one place to another. You can see this in the following activity.

EXPLAIN

Pour cold water into a beaker and add some drops of ink. Heat one side of the beaker gently as shown. What do you observe? Can you explain the reason for the circulation of water in the beaker?

Convection in air takes place in the same way as it does in liquids. Many homes are heated by convection currents. A radiator gets hot and heats the air in contact with it by conduction. However, since warm air is lighter than cold air, it rises, and colder air comes in to take its place. As this air is heated, it also rises, and this way, a convection current is begun. Because scientists have learned that heat is transferred by *radiation, convection,* and *conduction,* many new types of heating systems have been developed. How do some of these function?

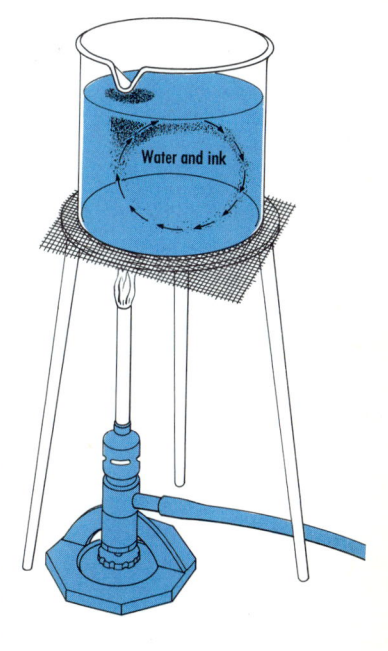

Temperature is a measure of the energy of the molecules of a substance. The temperature of a substance is not the same as the heat the substance contains. Heat is the amount of energy that a substance can give out or absorb; it is difficult to measure directly. One unit that is used to measure the quantity of heat is the *calorie* (KAL-uh-ree.). One calorie is the amount of heat needed to raise the temperature of one gram of water one degree Celsius.

The heat given off by hot objects equals the heat received by cold objects. Thus, calories lost equal calories gained. For example, an iron nail which is heated might have a temperature of 155°C. Water in a bathtub might have an approximate temperature of 37°C. There is no doubt that the iron nail has a higher temperature than the water, but which has more heat? The nail, placed in the water, would not change the temperature of the water in the tub very much. However, the temperature of the nail would be cooled to about 37°C in a few seconds. Since heat is the amount of energy a substance can give off or absorb, it is obvious that the tub full of water contains more heat than the iron nail. The following activity allows you to make an accurate measurement of heat exchange.

EXPLORE

Put some rock wool or styrofoam in the bottom of a large can. Then put two smaller cans inside the large outer can, and pack rock wool or styrofoam between the two inner cans as shown. Wrap the large outermost can with cardboard. Stand the can on another piece of cardboard. Put the stirrer through one hole in the cardboard cover and the thermometer through the other hole as shown. This apparatus is now a crude *calorimeter* (kal-aw-RIH-mih-tuhr), an instrument used to measure heat exchange.

Fill the innermost can with 60 grams of water heated to 60°C. Into the large outer can, pour 120 grams of water heated to 20°C. Cover the cans with the cardboard cover. Stir the water frequently and record the temperature of the water in the innermost can every five minutes.

Make a record of the temperatures. How do your results compare with those of your classmates? At what temperature did your thermometer reading stop changing?

Scientists sometimes predict their probable results mathematically before actually performing an experiment. You could have done this for the investigation in the following way:

CHAPTER 4   THE CHANGING ATMOSPHERE / 89

---

60 g. at 60° = 3600 cal.
+120 g. at 20° = 2400 cal.

---

180 g. at ?° = 6000 cal.
Temperature = 33.3°C

How close did you come to this in your observation? Can you check mathematically to determine if this answer is correct? Can you explain why your observation may be different from the calculated temperature?

REVIEW

1. How many degrees Celsius does the temperature drop with the rise of 1000 feet of altitude in the troposphere?
2. Why does the earth heat up more rapidly than the surrounding air?
3. What is the difference between the processes of conduction and convection?
4. What is heat?
5. What is meant by a calorie?

C/What Are the Properties of Air?

Air exerts pressure. From your previous experiences in elementary science, and from your observations of the environment around you, you know that air has weight and exerts *pressure*. Pressure is the force applied to a unit area.

Early experiments to determine the atmospheric pressure were performed by *Evangelista Torricelli (1608–1647)* an Italian scientist. Torricelli was given this problem by his famous teacher, Galileo. In order to understand pressure better, let us do the following activity which is effective and can be performed with simple materials.

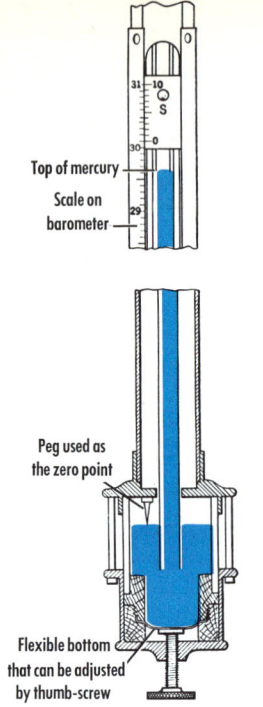

90 / UNIT 2 MAN EXPLORES HIS ENVIRONMENT

OBSERVE

Obtain a flat-sided can with a capacity of one gallon that has a screw cap. Place the gallon can on a hot plate and remove the screw cap. Place two tablespoonsful of water in the can and boil for several minutes with the **screw cap off the top of the can.** Remove from the heat and stopper the can opening immediately. Observe what happens to the can as the steam cools. Can you explain your observations?

Torricelli, the brilliant pupil of Galileo, is given credit for inventing the barometer which is used to measure the pressure of the atmosphere.

Changes in air pressure can be measured. A mercury barometer is mounted in a frame as seen in Fig. 4-8. If air pressure decreases, mercury will flow into the well. If it increases, mercury will flow back into the tube. However, because it is difficult to use a liquid barometer while traveling or climbing, scientists developed a barometer which could be easily moved from place to place.

An *aneroid* (AN-uhr-oid) *barometer,* as shown in Fig. 4-9, does not contain any liquid, and therefore can be moved easily. Its essential part is a shallow box with a thin metal cover. Air is removed from the box to produce a partial vacuum. This container is very sensitive even to slight changes in atmospheric pressure.

A *barograph* (BAR-o-graf) is a self-recording aneroid barometer. This instrument is built upon the same principles as the aneroid barometer, but instead of a pointer, it has a pen which records the air pressure on a rotating drum. This gives a permanent record. It can be studied by weather scientists looking for patterns or relationships between pressure and other factors.

Blaise Pascal (1623–1662), a famous French scientist who read of Torricelli's work, reasoned that if air pressure supported the mercury column in Torricelli's barometer, then the height of the mercury should change if he climbed a mountain with the barometer. What do you think happened to the height of the mercury column in the barometer atop a mountain 3000 feet above sea level?

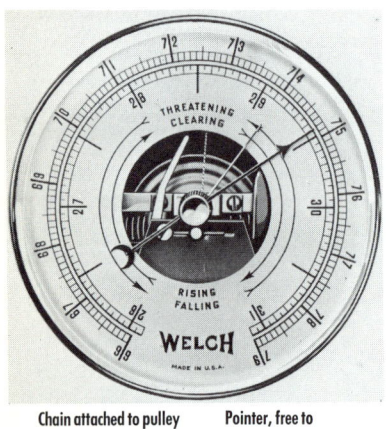

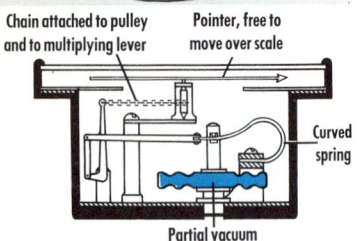

4-8 This mercury barometer mounted in a frame has a sliding scale which permits a more accurate reading. How does it operate?

4-9 How does the aneroid barometer differ from the mercury barometer?

4-10 The altimeter (insert) is an aneroid barometer marked to read in feet of altitude instead of inches in mercury. Why is the aneroid barometer used and not a mercury barometer?

An *altimeter* (al-TIH-mih-tuhr) is used to determine height above sea level. The number of molecules in a given volume of air is greatest at sea level. The higher the altitude, the fewer the molecules of air that are found per given volume. Since pressure is associated with the number of molecules in a given volume, the higher the altitude, the lower the pressure. We know that the barometric pressure at sea level is approximately 30 inches of mercury, and that for each 900 feet above sea level, it drops one inch. Thus, a pilot can determine his altitude by reading his altimeter. If a plane were flying at 18,000 feet, the instrument would read about 10 inches. To be certain that pilots do not make errors, the altimeters are marked to read in feet of altitude instead of inches of mercury (see Fig. 4-10). You can determine the effect of unequal air pressure in the activity that follows.

EXPLORE

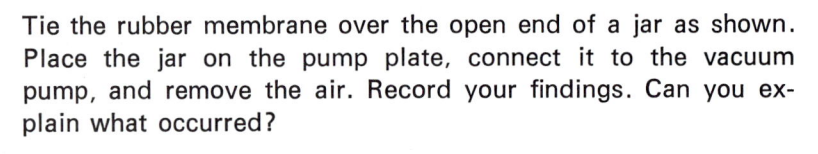

Tie the rubber membrane over the open end of a jar as shown. Place the jar on the pump plate, connect it to the vacuum pump, and remove the air. Record your findings. Can you explain what occurred?

The weight of a given volume of air affects its density. We have learned that warm air rises because it is lighter than cold air. The reason for this is that air is made up of molecules of several gases, which are in constant motion. When the molecules are heated they move faster and spread out. Thus, a gas expands when it is heated because the molecules move farther apart. The fewer molecules there are in a given volume of gas, the lighter the gas is. Warm gas has a lesser density than a cold gas and weighs less for a given volume. We can illustrate how temperature affects the density of a gas by doing the activity below.

EXPLORE

Hang two bags open-end down on opposite ends of a yardstick and balance them as shown. Light the candle and put it under the open end of one bag. **Be careful not to burn the paper.** Observe what happens to the bags. Remove the candle and note what happens. Put the candle under the other bag, repeat the procedure and note again what happens.

What happens when the candle is put under one bag? What causes this result? What happens when the air in the bag is allowed to cool? What is your conclusion about the weight of warm air when you compare it to that of cold air?

The air around you behaves just as the air in the bag did. As the air is heated by radiation from the ground, the gas molecules begin to move faster and faster, and the air expands. This lowers the density of the air, and it becomes lighter than the surrounding air. The lighter air now rises, and the cooler, heavier air flows in to take its place. When this occurs over a large area of the earth, you have air in motion, or *wind*.

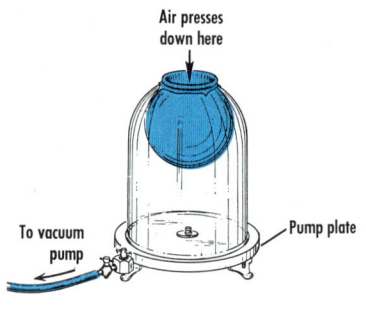

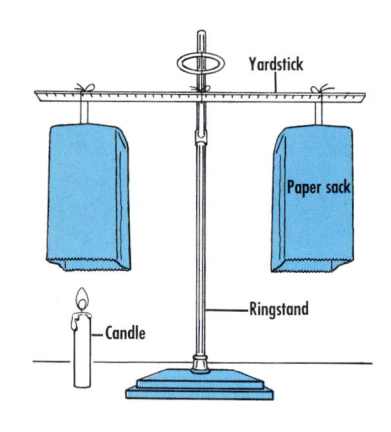

A gas can readily be compressed. Liquids and solids can only be compressed very slightly, but a gas, under pressure, can be compressed a great deal. The space that is occupied by the substance is known as its *volume*. In mathematics, we determine the volume of a solid by multiplying its length by its width by its height. We use volume to measure gases also. But this measurement does have a disadvantage since the volume of a gas varies with the conditions of pressure and temperature. Do the next activity to see how temperature affects a gas.

EXPLAIN

Fill a balloon with air and seal its mouth. Measure the balloon's circumference. Float the balloon in a pan of hot water for a few minutes. Now measure its circumference again. How can you explain your results?

You will notice from these activities that there is a relationship between temperature, pressure, and volume. *Robert Boyle (1627–1691)*, a British physicist and chemist, was the first person to perform experiments related to the compressibility of air. He discovered that the volume of a gas varied *inversely* (or in the opposite direction) with the pressure exerted upon it, provided the temperature remains constant. Thus, if the pressure is doubled, the volume is halved. If the pressure is reduced to one-fifth, what happens to the volume?

A French scientist, *Jacques Charles (1746–1823)*, demonstrated that all gases expand the same amount when heated one degree, if the pressure remains the same. Thus, increase in volume is directly proportional to the temperature change (if pressure is constant).

Air pressure is used in a number of ways. Compressing air into a football makes it harder but does not increase its size greatly. Thus, as more and more air molecules are forced into it, the football becomes harder because of the increase in air pressure. The same explanation also holds true for automobile tires.

4-11 Why does compressed air make the tire hard?

When taking a long car trip, it is a good idea to fill the tires to the proper pressure before starting out, not after you have traveled 100 miles or more. What is the reason?

Ordinary air pressure is approximately 14.7 pounds per square inch at sea level. If five cubic feet of air are compressed into one cubic foot of space, the pressure becomes about 75 pounds per square inch. This means that there are 75 pounds of pressure being exerted on every square inch of surface. If a tire is inflated to a pressure of 30 pounds as measured by a service station pressure gauge, what is the actual pressure of the air inside the tire? (See Fig. 4-11).

Pneumatic (noo-MAT-ihk) appliances are those which make use of compressed air. Some examples of them are the air brakes on trucks and trains, riveting machines, compression drills, and air hammers used to break up concrete or asphalt pavement. Can you think of others?

Caissons (KAY-sahnz) and similar devices make use of compressed air to keep out water. Caissons allow men to work safely far below the surface of the water. Deep-sea divers use pressurized suits for diving. The air pressure is increased as they descend; it is controlled from the ship's deck. What causes the pressure to increase on the outside of the suit as the divers go farther down into the water? Why is it important to try and equalize the outside and inside pressures?

Unbalanced air pressure is used to operate certain appliances. Since air has weight, it exerts a force when it is in motion. The medicine dropper, a very familiar item, works because of differences in air pressure. Can you explain how it operates? Study the effect of unbalanced air pressure in the next activity.

EXPLAIN

Obtain a U-shaped glass tube that has one tube-end longer than the other. Now place a beaker full of water on your desk. Fill the tube with water and keep it in by holding your finger over the opening of the long end. Place the short end in the beaker. Hold a larger vessel underneath the longer end. Now remove your finger. Why does the water flow from one container to the other? This device is called a *siphon* (SYE-fuhn).

CHAPTER 4 THE CHANGING ATMOSPHERE / 95

The *lift pump* also makes use of unbalanced air pressure. Perhaps your teacher will demonstrate the use of the lift pump. Look at Fig. 4-12 and give an explanation for the operation of the lift pump.

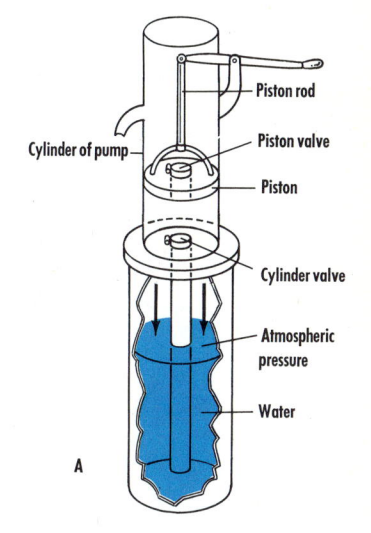

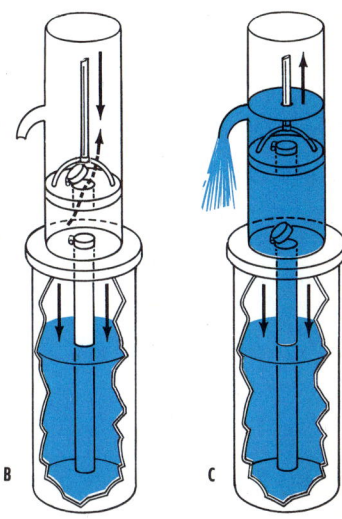

REVIEW

1. Give a definition for air pressure.
2. What is the reason for the use of mercury in most barometers?
3. What method did Torricelli use to measure air pressure?
4. Give an explanation for the operation of an altimeter.
5. How does a siphon remove liquid from a tank?

4-12 What is taking place in each of the diagrams of a lift pump?

D/What Is the Composition of Air?

Air is a mixture of gases. When we think of the composition of air, we are thinking really of the gases in the troposphere. We know most about the composition of the troposphere, but the same mixture of gases is found in the other layers of the atmosphere.

AVERAGE COMPOSITION OF AIR

Gases present	Approximate percentage
Nitrogen	78%
Oxygen	21%
Argon, helium, neon, krypton, radon and xenon	0.97%
Carbon dioxide	0.03%

In addition to the above gases, water and dust are also present in varying quantities, as well as gases of industrial origin. It will be useful to you later to learn the properties of three of these important atmospheric gases: oxygen, nitrogen, and carbon dioxide.

96 / UNIT 2 MAN EXPLORES HIS ENVIRONMENT

Oxygen is the most abundant element on earth. About 50 percent of the earth's crust of rocks and sand is made up of oxygen. Your body contains 65 percent oxygen. Water contains about 89 percent, and common mineral substances, like sandstone and limestone, nearly 50 percent oxygen. You can prepare oxygen in the following activity.

OBSERVE

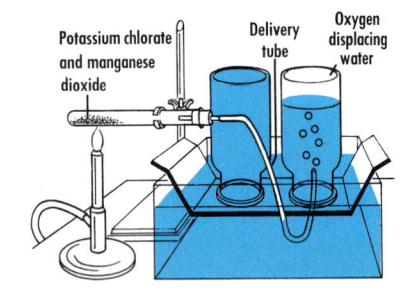

Mix eight grams of fine potassium chlorate and four grams of powdered manganese dioxide in a clean evaporating dish or a beaker. Transfer the mixture to a Pyrex test tube. Gently push a bent glass tube through a one-hole stopper and put the stopper in the mouth of the test tube, as shown. Fill four collecting bottles with water. Invert them in an aquarium or trough half-filled with water. Gently heat the mixture in the test tube. Note what happens in the bottle. Why do we collect oxygen in this way? Is oxygen soluble in water? (Remove the delivery tube from the water before you stop heating the mixture.) Light a wood splint and blow out the flame. Put the glowing splint in a bottle of oxygen. What happens? Make a list of the properties of oxygen which you can determine from the above activity.

Oxygen is one of the most active elements. It combines readily with other elements forming *oxides* (AHK-syeds). The union of oxygen with another substance is called *oxidation* (ahk-sih-DAY-shuhn). It may be *slow*, as in the rusting of iron, or it may be *rapid*, as in the burning of a fuel. Perform the next activity, to illustrate oxidation.

OBSERVE

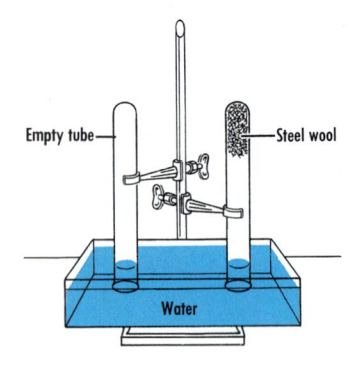

Arrange the equipment as shown, and leave for 24 hours. Would moistening the steel wool before placing it in the test tube speed up the reaction? What happens to the water level in each test tube? How can you test to show what gas is used up from the air in the test tube containing the steel wool? Describe what must have happened to the oxygen in the air in the test tube with the steel wool. How does this show the approximate amount of oxygen in air?

Oxygen combines with most materials, but this process is usually so slow that no burning results. However, heat is always given off in the process of oxidation. The following activity will show you how this takes place.

RECORD

Put some moistened steel wool in the bottom of a test tube. Push a thermometer through a one-hole stopper. Insert this in the test tube so that the thermometer bulb rests in the moist steel wool. As a control, put dry steel wool in another test tube with another thermometer. The next day, record the temperature. Take readings for a period of time. Check your findings with others in your class.

Painting reduces oxidation or rusting. Coating the surface of a metal with paint or varnish prevents moisture and oxygen from reaching the surface. This slows down the oxidation process. Automobiles and ships are painted in order to avoid the process of oxidation (rusting) as well as to make them look more attractive (Fig. 4-13).

Oxygen is necessary to support life. Most of the food we eat combines with oxygen to yield *energy* as well as heat. The heat resulting from this process maintains the body temperature. The importance of oxygen to man is illustrated if you consider that man must take it with him when he goes up into space or down into the sea, as seen in Fig. 4-14.

Nitrogen dilutes the oxygen in the air and helps to decrease the activity of the oxygen. Nitrogen, like oxygen, is colorless, odorless, and tasteless. It is essential to plant and animal life, yet few living things can can use it *directly* from the air. The *nitrogen cycle,* involving living organisms and nitrogen from the air, is essential to living things.

Nitrogen from the air is taken by soil bacteria and combined with other substances into useful minerals in the soil. Plants use these compounds to build proteins which in turn are used by animals to build animal proteins. As man and animals eat plants, they excrete wastes containing nitrogen compounds. These wastes and organic matter are acted upon by other

4-13 What chemical process is prevented when a metal is coated with paint?

4-14 Describe the life support system of this diver.

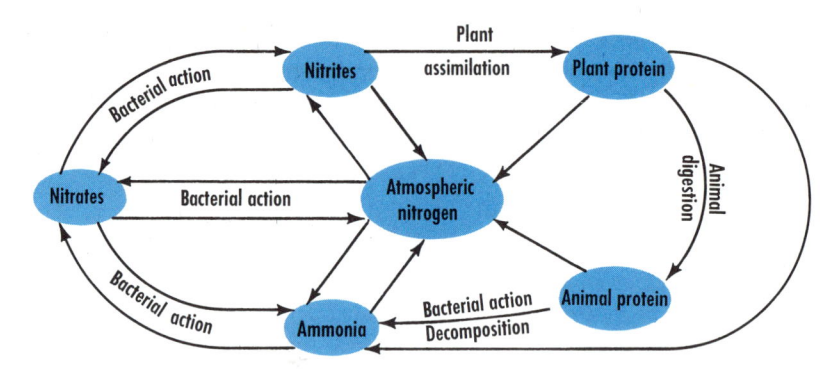

4-15 The Nitrogen Cycle. Why is it essential to all life?

bacteria. These bacteria convert them into ammonia, which contains nitrogen and hydrogen. Bacterial action and combustion return nitrogen to the air and convert ammonia to more useful nitrogen compounds; this completes the cycle (see Fig. 4-15).

Nitrogen is a necessary element for all living things. It is combined with oxygen, hydrogen, and carbon to form proteins. Proteins enable animals and plants to build and repair their bodies.

Carbon dioxide is a compound of carbon and oxygen. A steady stream of carbon dioxide enters the atmosphere from the burning of wood, coal, and petroleum products, decay of vegetable and animal matter, and animal respiration. Carbon dioxide does not increase in the air because the action of growing plants uses this compound and the oceans absorb carbon dioxide. Even so, scientists are concerned that the amount of carbon dioxide in our atmosphere is on the increase. It has been found that the relative amount of carbon dioxide in the atmosphere is increasing at the rate of about one per cent a year. This may increase the temperature on earth due to a "greenhouse effect." This is the trapping of the sun's rays beneath an atmosphere rich in carbon dioxide. A final result could be the slow melting of the Polar ice over the years and the raising of the ocean level.

Carbon dioxide is a colorless and odorless gas. It is heavier than air and can easily be prepared in the laboratory. Although carbon dioxide is not poisonous, it is dangerous since it may produce suffocation if present in the air in large amounts. The amount of carbon dioxide in your blood is one factor that helps to determine your rate of breathing. The next activity will allow you to make carbon dioxide and to observe its properties.

EXPLORE

Put several pieces of marble in some water in the bottom of a flask. Insert the funnel and the bent glass tube in a two-hole stopper, as shown. Pour dilute hydrochloric acid slowly into the flask through the funnel. Collect the gas in a bottle. What happens when a burning wood splint is thrust into the collecting bottle? What happens when limewater is poured into the bottle?

The noble gases make up only about one percent of the total atmosphere. These rare gases are *argon, helium, neon, krypton, xenon* and *radon*. They do not normally react with other elements.

Neon is used in special lamps which light up store windows and advertising signs. An electric discharge passing through this gas under reduced pressure produces an orange-red light. Since helium is extremely light and yet does not explode or burn, it has replaced hydrogen in filling balloons and airships (Fig. 4-16). Helium is also used to make an artificial atmosphere. Helium replaces nitrogen in the air of deep sea laboratories because nitrogen, under pressure, dissolves in the blood affecting brain function. Argon, sometimes mixed with nitrogen, is used in electric light bulbs to increase the life of the filament. It also may be used in fluorescent light tubes.

Burning substances cause air pollution. Particles resulting from burning rise into the air and are known as *pollutants* (puh-LYOO-tuhnts). With the increase in population, as well as in industrial plants, the amount of pollutants entering the air is increasing at an enormous rate. Any substance that burns in the air produces solid particles, seen as smoke, and invisible gaseous compounds.

If man lived by himself on a remote mountain top, air pollution would hardly be a problem. Here, the natural movements of the air would quickly carry away the small amounts

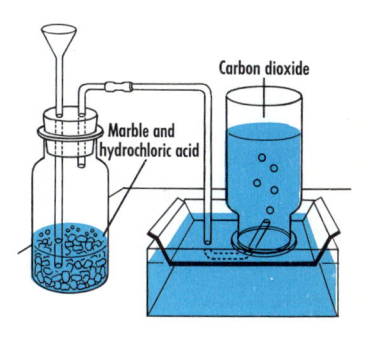

4-16 Why is helium now used in all airships?

of pollutants. Most people, however, live in towns and cities where the pollutants from homes, automobiles, and industries are concentrated in a relatively small volume of air. Therefore, air pollution and its control are serious problems. What can be done to reduce air pollution? The following activity may give you some ideas.

EXPLAIN

Pour a little water in a clean beaker and dry the underside. Hold the bottom of the beaker over a burning candle. What forms on the bottom of the beaker? What is the source of this material? What happens to this material when it is given off into the air?

To understand why air pollution has become a major problem, let us consider the amount of pollutants given off daily by a community of 10,000 homes. If each home in the community burned an average grade of soft coal for heating and cooking, about 30,000 pounds of pollutants would be given off into the air every day. The table below shows the types of materials making up the smoke and gases produced by burning the coal. Just the backyard incinerators from these same 10,000 homes would produce about 4000 pounds more of air pollutants from the trash burned daily.

| POLLUTANTS FROM BURNING COAL ||
Pollutants	Pounds
Unburned coal (smoke)	20,000
Oxides of sulfur (gases)	4,200
Organic acids (gases)	3,000
Other materials (solids and gases)	2,800

Automobile and truck exhausts produce a special group of pollutants. If there were 10,000 automobiles and trucks in your community, they would probably burn about 20,000 gallons of gasoline per day. Burning this much fuel would

CHAPTER 4 THE CHANGING ATMOSPHERE / 101

create about 6400 pounds of carbon monoxide gas, 6000 pounds of unburned hydrocarbons (compounds of hydrogen and carbon), 2000 pounds of nitrogen oxides and smaller quantities of other pollutants in the air every day.

Man's industries are a source of air pollution. However, they may not be the most serious offenders because many industries spend millions of dollars to control air pollution. Industrial pollutants take three forms: (1) visible particles like smoke, (2) nonpoisonous gases, and (3) poisonous and irritating gases, like carbon monoxide, oxides of sulfur and oxides of nitrogen.

Air pollution can be controlled. Air pollution from homes can be decreased by switching from soft coal to hard coal, oil, or natural gas for heating. This has been done in many communities. Control of industrial air pollution is a more difficult problem, but several methods for removing solid particles from smoke have been developed (Fig. 4-17). Another method is to burn coal with low sulfur content.

4-17 A city scene before and after smoke control. How are American cities trying to control air pollution?

Most difficult of all to control is air pollution caused by automobile and truck exhausts. Gasoline engine construction is such that it is impossible to obtain complete burning of the fuel in the engine. The unburned products are given off through the exhaust and become serious air pollutants.

Automobile manufacturers have spent much time and money searching for a way to reduce this type of pollution. Special devices, called *after-burners,* have been designed to burn the exhaust gases after they leave the engine. Another device that is inexpensive and simple to install has recently been developed. It is called a *blowby system,* and experiments have shown that such a device will reduce air pollution from gasoline engines by as much as 25 per cent. Most new cars are equipped with a crankcase blowby system by the manufacturers. Some states are requiring after-burners in all new cars licensed in those states. By 1970, all new cars in the U.S. must have a device to control this pollution.

REVIEW

1. Name the gases found in the air.
2. Tell the difference between oxidation and burning.
3. Explain the nitrogen cycle.
4. Why does a runner breathe more rapidly during a race?
5. Name three factors which contribute to air pollution.

THINKING WITH SCIENCE

A. *On a separate sheet of paper, write the numbers 1 to 15. On the basis of your experience, state whether the following are True, Probably true, Insufficient evidence to decide, or False.* **Do not write in your book.**

1. Oxygen, one of the important gases composing the air, is an inert gas.

2. Carbon dioxide is lighter than air.

3. Aluminum is the best conductor of heat.

4. Materials which will burn in air will burn more rapidly in oxygen.

5. Water would rise twice as high as mercury if used in a barometer (under the same atmospheric pressure).

6. The percentage of nitrogen remains constant in the atmosphere because of bacterial action in the soil.

7. Carbon dioxide, another important gas that is found in the air, supports combustion.

8. Carbon dioxide is a compound of carbon and oxygen which turns limewater milky white.

9. The aneroid barometer uses air instead of a liquid, like mercury.

10. Liquids rise in a straw because air pressure in the straw is reduced.

11. The chemosphere is a layer that filters out harmful rays of the sun.

12. The best absorbers of heat are surfaces that are light and smooth.

13. A calorie is the amount of heat needed to raise the temperature of one gram of water one degree Celsius.

14. Air expands when heated.

15. Painting a surface will reduce the formation of rust.

B. *Write the answers to the following in your notebook. Be sure to use complete sentences and correct spelling and grammar.*

1. Why does a gas completely fill any container into which it is introduced?

2. Why do you *not* feel air pressure as you sit and answer these questions?

3. Why does air pressure decrease as one goes up farther from the earth.

4. Describe how an altimeter is used to determine a plane's altitude.

5. How does nitrogen from the atmosphere enter your body?

6. How does compressed air make it possible for men to work under water?

7. How can we increase or decrease the rate of oxidation?

8. Explain why it is sometimes difficult to open a door between the classroom and the hallway.

9. Based on the laboratory preparations of oxygen and carbon dioxide, what properties of each of these gases were you able to identify?

10. Weather balloons usually burst when they reach 25,000 feet above sea level. Why does this happen?

RESEARCH IN SCIENCE

1. Set up an experiment to determine what factors affect the rate of flow of a siphon.

2. Examine the following devices and explain how they work by using atmospheric pressure: vacuum cleaner, atomizer, suction cup, aerosol bomb, medicine dropper.

3. Determine whether the amount of moisture present in the air has any effect upon the oxidation rate.

4. Carbon dioxide is sometimes called "fixed air." Write a historical explanation as to how this came about.

5. Collect and arrange pictures of machines which are operated by compressed air. Below each picture, write a brief description of how each works.

chapter 5
The Changing Oceans

A/How Do Scientists Study the Ocean?

Oceanography deals with the study of the oceans. An *oceanographer* (*o*-shuhn-AH-gruh-fuhr) is a scientist who studies the ocean (Fig. 5-1). Oceanographers may be geologists, physicists, chemists, or biologists who are conducting investigations in geophysical oceanography or marine biology or both. Study current developments in oceanography in the following activity.

CHECK

Watch the newspapers and magazines that you read for articles about the oceans, oceanography, and oceanographic research. Keep these articles in your science notebook. You may want to place some of them on your classroom bulletin board. What are some oceanographic problems currently under investigation? What are some of the organizations involved in oceanographic research? Where is the research taking place?

To find the answers to their questions, oceanographers have developed many special methods, techniques, and instruments

5-1 With what areas of science is an oceanographer familiar?

CHAPTER 5 THE CHANGING OCEANS / 107

for their work. Much oceanographic research and collection of data takes place aboard specially-equipped ships.

Ships can be used to gather information about the ocean, but some data can only be collected from beneath the surface. In 1930 *William Beebe (1877–1962)* descended more than 3000 feet below the surface of the ocean in a *bathysphere* (BATH-is-sfeer) (Fig. 5-2). This heavy metal ball took man deeper into the ocean than he had ever been before. Photographs and observations were made. Since the bathysphere was suspended from a surface vessel on a long metal cable and had no means of propulsion, its exploration area was very limited.

The situation was improved in 1948 with the invention of the bathyscaphe, which was designed by *Auguste Piccard (1884–1962)*, a Swiss scientist. Under its own power, the bathyscaphe is capable of both vertical and horizontal movement. The United States Navy bathyscaphe, *Trieste* (Fig. 5-3), has carried a two-man crew to a depth of greater than 35,000 feet (nearly seven miles below the surface of the ocean). When it reached this record depth, the *Trieste* was subjected to a water pressure that was equal to the weight of two huge battleships.

5-2 Why is the bathysphere shaped much like a ball?

5-3 The bathyscaphe *Trieste*. How is this underwater vessel an improvement upon the bathysphere?

The pressure in the ocean increases with depth. If scientists are so interested in what is found on the bottom of the seas, why don't they just put on some underwater diving gear and go down to take a look? Self-Contained Underwater Breathing Apparatus (SCUBA for short) makes it possible for divers to remain under water for considerable lengths of time. Deep-sea diving, however, is not as simple as it sounds. We must take into account the operation of a few physical laws.

Under normal conditions of life, man is adapted to live at an atmospheric pressure of 14.7 pounds per square inch at sea level. We are not aware of this force because the pressure inside the body equals the pressure outside it. A person becomes aware of this pressure when it changes rapidly, as going up in the air or down in the ocean. The occasional "popping" of the eardrums reminds us that we live in a sea of air at a certain pressure. You can observe the relationship between depth and pressure in the following activity.

OBSERVE

Cover the wide, open end of a thistle tube with a piece of rubber from a balloon. Partially fill the thistle tube with water and lower the covered end into a container of water. What do you observe happening to the liquid in the narrow portion of the tube as it is lowered deeper into the water? What does this show about the relationship between water pressure and depth?

This activity helps you understand that *the pressure exerted by a liquid is directly proportional to its depth.* The greater the depth, the greater the number of molecules pressing on an object submerged in the water or against the side of a container.

SAMPLE PROBLEM

The pressure under water depends on the depth to which an object is submerged. What would be the pressure on a diver 200 feet below the surface of the ocean?

Solution:
1. Pressure of 1 cubic foot (ft³) of *ocean water* is 64 lb/ft².
2. 200 × 1 = 200 ft³
3. 200 × 64 = 12,800 lb/ft²

Pressure is a force that is spread uniformly over an entire area. It is usually recorded as pounds per square inch (lb/in²) or pounds per square foot (lb/ft²).

If you have a container which is 1 foot square, and you pour water into it 1 foot deep, the container now holds 1 cubic foot (ft³) of water. The water weighs 62.4 pounds. Thus the water is pressing down on the bottom of the container with a force of 62.4 pounds per square foot.

Now suppose we have a container which is twice as high, but with the other dimensions unchanged, as in Fig. 5-4. If this is filled with water, what is the pressure on the bottom?

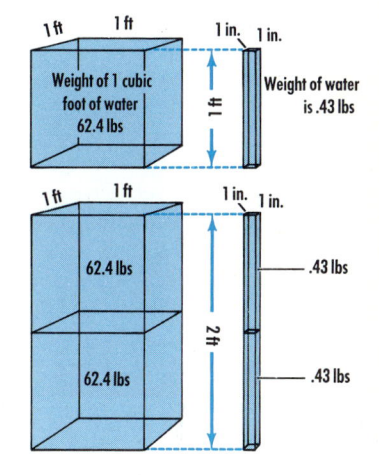

5-4 How does the height of a liquid affect the pressure on the bottom of a container?

SAMPLE PROBLEMS

1. Determine the pressure of water on the bottom of a one-foot square container that is two feet deep.

Solution:
1. Pressure of 1 cubic foot (ft³) of water is 62.4 lb/ft²
2. 2 × 1 = 2 ft³
3. 2 × 62.4 = 124.8 lb/ft²

2. Determine the pressure of water on one square inch of surface at a depth of one foot.

Solution:
1. 1 square foot = 144 square inches
2. Divide 62.4 pounds by 144
3. Pressure = 0.43 lb/in²

5-5 Why is the pressure of any liquid not affected by the shape of its container?

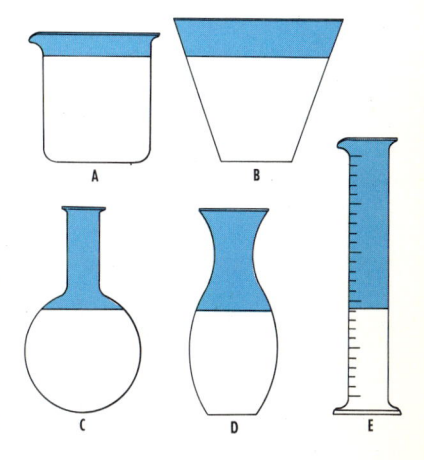

Would the amount of pressure on the bottom change if we use a differently shaped container? Since the pressure depends on the depth or height of the liquid, the shape of the vessel can be disregarded (Fig. 5-5).

Liquids push upon objects with an upward force. This force that a liquid exerts upon a body placed in it is called the *buoyant* (BAW-yuhnt) *force*. Archimedes (287?–212 B.C.), a

Greek philosopher and scientist, discovered that the buoyant force which a liquid exerts upon an object under water is equal to the weight of the amount of liquid the object displaces. The next activity will demonstrate this principle.

MEASURE

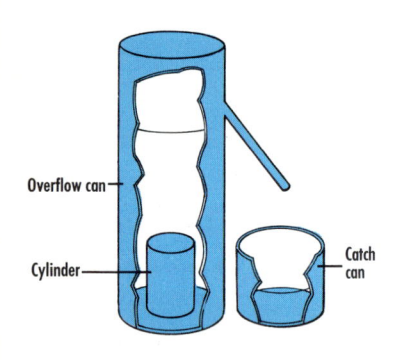

Fill an overflow can to the spout, as shown. Now weigh a metal cylinder. Weigh the catch can and place it under a spout. Put the cylinder in the water and allow it to sink to the bottom. Weigh the catch can after the water stops flowing. What are your results?

Why does an object under water weigh less than it does out of water? An object sinks in water if the weight of the water it displaces is less than the weight of the object. It will float, however, if the weight of the water it displaces is equal to or greater than its own weight. The activity below illustrates this.

COMPARE

Tie a stone to a string and record the weight of the stone. Place a small tank of water on a balance which registers its total weight. Slowly lower the stone into the tank without touching the sides or the bottom of the tank.
 What is the reading on the balance? Can you explain the results you obtained? Try other objects like wood, metal, etc., and note the results.

Deep diving affects the body's fluids. Now, let us put on the SCUBA equipment and go for a short dive. As we go down in the water, the pressure increases. (At 33 feet below the surface, the pressure is equal to two atmospheres.) The air in our lungs is compressed, and at the pressure of two atmospheres the air volume will have decreased to one-half. Fortunately, our SCUBA has an arrangement that supplies air at the same pressure as that of our dive, and we can breathe naturally. If this did not happen, we would be in trouble. Our lungs would be either over- or under-inflated (Fig. 5-6).

5-6 What are the benefits and limitations experienced by the diver using SCUBA equipment?

So far so good; now, it is time to surface. But coming up from a depth of 33 feet presents a very serious problem. The outside pressure compared to the pressure in our lungs and body cavities will be one-half. Unless we breathe deeply several times as we rise to the surface, our lungs may swell up like a pair of balloons.

At sea level, there are always small amounts of various gases dissolved in the blood from the air we breathe. In a dive to 150 feet, the pressure now is almost six atmospheres (six times the atmospheric pressure at sea level). Nitrogen, the gas which makes up about 79 percent of the air, is forced into solution in the blood and tissues. What would happen now if we surfaced rapidly? Visualize this by opening a bottle of highly-charged soda. The sudden release of pressure causes a violent fizzing as the confined gas in solution bubbles very rapidly out of the water.

This same kind of action can occur in our bodies with very serious results. The nitrogen bubbles, resulting from the rapid

5-7 Why are heavy suits used for underwater work?

5-8A How are soundings taken?

decompression, form in the blood and tissue too rapidly to be eliminated gradually. They may lodge in the joints and muscles to cause the severe pain usually called the "bends." If bubbles lodge in the brain and spinal cord, paralysis and death may result.

The way to avoid the "bends" is to surface slowly so that the dissolved gas has a chance to come out of the blood and be exhaled. The deeper the dive, the longer the decompression period. This system applies not only to skin divers using SCUBA, but also to deep-sea divers and men working in underwater tunnels or diving bells (Fig. 5-7).

From this discussion, we can see that man is limited in his ability to explore the ocean bottom without special protection. Therefore, the study of the ocean is carried on from the surface by specially equipped oceanographic ships. In addition to the studies carried out on ships operated by the United States Coast and Geodetic Survey and the Hydrographic Office of the United States Navy, present-day scientific advances are being carried out by many industrial and educational institutions.

Man can measure the ocean's depth. When oceanography was first developing in the nineteenth century, measurements of the ocean's depth (soundings) were made by using a rope with a heavy weight tied to one end. This was lowered into the water until the weight struck the bottom. Later methods involved using a wire cable instead of rope. Today, soundings are made by using sound waves which are reflected off the surface of the ocean floor. Since the speed of sound waves in water is known, the distance from the ship to the ocean floor can be calculated (Figs. 5-8 A and B).

Sound waves sent out in this way may bounce off a sediment layer and not actually indicate where the real ocean floor is located. In order to determine the depth of the real ocean floor, *seismic* (SYEZ-mihk) *soundings* must be made by using the powerful sound waves produced by surface explosions. These seismic waves penetrate any surface sediment. They are reflected from the real ocean floor to *hydrophones* (HYE-dro-fones). The depth of the real ocean floor can then be calculated readily.

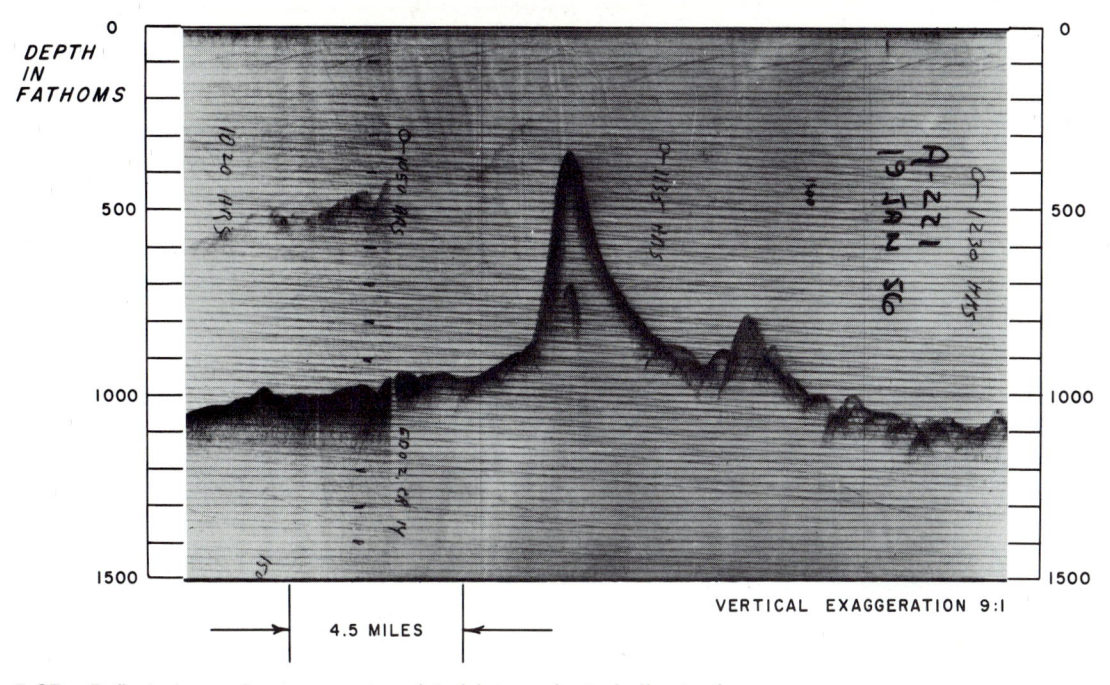

5-8B Reflected sound waves are translated into a chart similar to the one shown. Can you determine the deepest part of the ocean?

SAMPLE PROBLEM

Sound waves travel in ocean water at a speed of about 4800 feet per second. A wave sent out from a ship's transmitter is reflected off the ocean bottom and picked up on the ship's receiver. The time from transmission to reception is five seconds. How deep is the ocean at this point?
Solution:
1. Speed = 4800 feet per second
2. Time = 5 seconds for total time (2.5 seconds for ½ distance)
3. 4800 × 2.5 = 12,000 feet

Oceanographers can sample the ocean. A wide variety of tools are being used to obtain samples of the ocean bottom and of the living things in the ocean. A *dredge* attached to the ship can be scraped along the ocean bottom to obtain samples of the bottom as well as the plant and animal life living there. *Trawls* and *plankton townets* can be dragged behind the ship to catch fish, plants, and plankton, which are very small plants and animals (Fig. 5-9). Collect some water plants and animals in the following activity.

EXAMINE

Drag a fine mesh net along the top surface of a pond, lake or the ocean. This is the same kind of operation performed by oceanographers in sampling the oceans with trawls and plankton townets. What kind of things did you collect in your net? Did you observe any plants? Any animal life?

Grab buckets, much like steam shovels used on land, are able to scoop up samples of the ocean bottom. *Coring tubes* are driven into the ocean floor deep enough to remove samples of sediment, hard bottom, and remains of plants and animals deposited thousands of years ago. You can learn how to collect and observe core samples in this activity.

5-9 What use do scientists have for this townet?

OBSERVE

A simple *core sampler* can be made from a length of pipe or copper tubing. Using a hammer, pound your piece of pipe down into the soft bottom at the edge of the ocean, pond, lake, or some other body of water. Pull out the pipe and, using a round stick, carefully push your sample out of the tube for examination. What do you observe? Do you observe any living things in your sample? Do you observe any differences between the top and bottom of your sample? Can you think of any problems involved in generalizing from an insufficient number of samples?

Samples of ocean water can be obtained from a desired depth by using *Nansen bottles*. These devices consist of metal cylinders. They are open at both ends with a mechanism which automatically closes the ends at the depth which the oceanographer wants to sample. Water flows freely through the tube until the desired sample is sealed inside. Then, the Nansen bottle is brought up to the ship for examination. A special thermometer inside the tube records the water temperature at the depth where it was obtained. Study temperature changes in a body of water in the next activity.

MEASURE

Use a maximum-minimum thermometer to determine the water's temperature in the ocean or some other body of water. Lower a weighted thermometer, tied to a piece of rope, to different depths in the water. Make sure that you bring the thermometer readings back to the air temperature readings after each trial. Leave it at the depth you are measuring for about five minutes. What do you predict will happen to the temperature as the thermometer goes deeper in the water? What temperatures do you observe at different depths?

REVIEW

1. What is the pressure in lb/ft^2 on a diver at a depth of 300 feet?
2. How is the buoyant force related to the amount of liquid an object displaces in water?
3. What happens to dissolved nitrogen in the blood if a diver surfaces too rapidly?
4. What two agencies of the United States Government are carrying on present day studies of the oceans?
5. What devices are used to obtain bottom samples from the ocean floor?

B/What Do We Know About the Ocean?

The waters of the oceans have changed over the years. Some scientists believe that when the earth was first formed it was a mass of extremely hot gases. These gases gradually cooled until the outer crust of the earth was formed. The earth was still surrounded by gases which included water vapor. When the earth's surface was cool enough, it began to rain. According to this theory, the oceans were formed by the tremendous amounts of rain which fell over many, many centuries. The water collected in vast depressions that had been previously formed in the earth's surface.

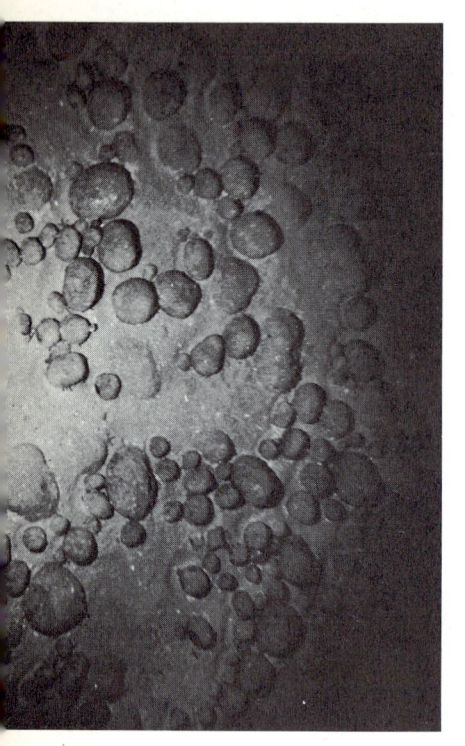

5-10 How were these deposits on the ocean floor formed?

When the oceans were first formed, the composition was probably quite different from what it is today. Investigations have shown that today about 3.5 percent by weight of sea water is made up of dissolved minerals. Do you think that the percentage was greater or less when the ocean was first formed? Remember that many of the materials washed away from the earth's surfaces by running water are finally deposited in the seas. As shown in Figure 5-10, these deposits contain useful minerals like manganese.

Most substances will dissolve to some degree in water. Substances which are able to dissolve to some extent are called *solutes*. The liquid in which these substances dissolve is called the *solvent* (SAHL-vent). A solute dissolves in a solvent to form a *solution*.

The concentration of a solution depends upon the amount of solute dissolved in the solvent. Thus, you have heard of *dilute* solutions and *concentrated* solutions. A solution which cannot hold any more solute at a particular temperature is said to be *saturated* (SA-choo-ray-tuhd). Prepare different concentrations of sugar and water solutions as directed in the next activity.

OBSERVE

Measure 50 milliliters of water into a large beaker. Add 5 grams of sugar to the water and stir. Did all of the sugar dissolve? Add 10 grams more of sugar to the same beaker. Stir carefully and record your results once again. Now, continue to add 5 grams of sugar at a time until you find that you can no longer dissolve any by just stirring or mixing. Take the temperature carefully of your sugar solution.

Heat the solution slowly to 10°C higher than the original temperature. Try to dissolve the sugar which had settled to the bottom of your solution by stirring the liquid again. Were you successful? Continue the process until the temperature of the solution is 80°C. Make a graph showing the effect of every 10°C rise in temperature on the number of grams of sugar that dissolve in the solution. What effect does an increase in temperature have on the amount of solute in a solution?

CHAPTER 5 THE CHANGING OCEANS / 117

When sea water is evaporated, about three-fourths of its mineral content is found to be common *table salt* (NaCl). If one cubic yard of sea water were evaporated, we would find, in addition to NaCl, the salts of magnesium, sulfur, calcium, potassium and bromine in lesser but significant amounts. Other minerals, like gold and iodine, would also be found in smaller amounts. Some minerals, like magnesium and bromine, are already being obtained in commercial quantities from sea water. Someday, in the future, man may have to return to the the sea in order to obtain many of the scarcer minerals.

Sea water also contains the gases oxygen, nitrogen, carbon dioxide, argon, helium, and neon in solution. The amounts of these gases found in the sea water vary considerably when compared with the amounts found in the atmosphere. Can you explain this?

Saltiness varies from place to place in the ocean. Using Nansen bottles, scientists can sample the ocean at different depths. They have discovered an interesting phenomenon, the saltiness of the ocean varies with its depth. Do the next activity to discover why.

PREDICT

Place a fresh egg in a container of fresh water. What do you observe? Does the egg float? Place the same egg into a container of saturated salt (NaCl) solution. What do you observe? Does it float? Which solution is denser — the fresh water or the salt solution? Weigh equal volumes of fresh and salt water. What do you observe? Would you expect the ocean to be saltier near the surface? Near the ocean floor? Why?

In addition to the variation of the salt content with depth, it also varies from place to place on the ocean's surface. In the Red Sea, it reaches a high of about 40 parts per thousand (weight of salt to weight of water ratio). Other places in the ocean vary between 33 and 37 parts per thousand.

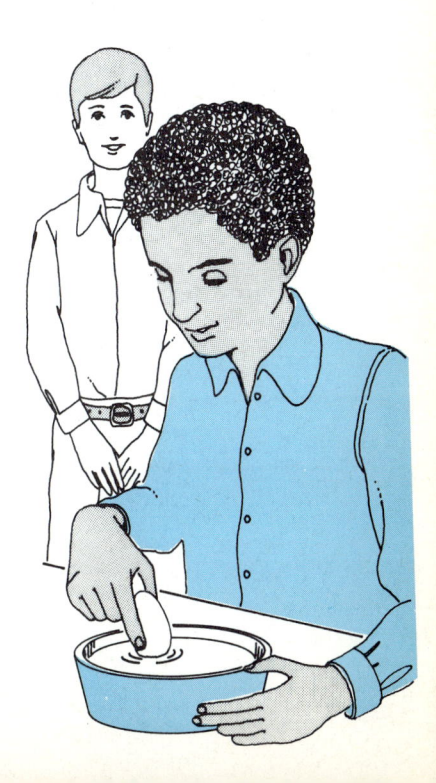

Anyone who has ever looked at the ocean has seen a wide variety of colors: dark green, almost black, white foaming sea spray, and many shades of blue (Fig. 5-11). If you were to look at a sample of ocean water in a glass jar, however, you would observe that it is colorless. The colors that we see, as we view the ocean from ships or the shore, are produced by the reflection of the sky and the light scattered from finely divided matter in the water.

Some ocean currents are caused by differences in density due to salinity or temperature. An important phase of oceanographic research involves studying the currents in the ocean. The simplest and most common method used to do this involves the use of *drift bottles*. Bottles with a note inside are thrown into the ocean. The finder of the bottle is asked to return the note to the oceanographer. In this way, the direction in which currents have moved the bottle, can be determined. Deep water current detectors can be attached to buoys to radio back information concerning the movements of currents at any depth in the ocean.

Ocean currents flow like rivers in the seas. There are two kinds of ocean currents in the seas. One type flows near the surface and can be readily observed. The other type flows deep in the ocean. In general, the surface currents flowing toward the poles are narrow and fast, but the currents flowing toward the equator are wide and slow. One of the best known of the ocean currents, the *Gulf Stream*, carries between 25 and 50 million tons of water per second northward from the equator.

Many interesting discoveries have been made about water circulation in the oceans. A great current far beneath the Gulf Stream was found to be moving in a direction opposite to that of the surface current. Another deep current, called the *Cromwell Current*, was found to flow in the opposite direction from the *Pacific Equatorial Current*. This deep current is about 250 miles wide and 1000 feet deep. It moves eastward along the equator for at least 3500 miles. It is so vast that it is believed to carry 1000 times as much water as the Mississippi River.

Ocean currents receive their energy from the sun and the wind. As you have already learned, when water is heated, its density decreases, and it rises to the top of the body of water.

5-11 What causes these color variations in ocean water?

CHAPTER 5   THE CHANGING OCEANS / 119

The cold water surrounding this area flows under the rising warm water. Since the average weather conditions at the equator are warmer than they are in other parts of the world, the ocean water is warmed more and becomes less dense. The cold water from the poles flows in toward the equator, and the warm water is forced away from this area of the earth. Study the principles of this circulation in the following activity.

EXPERIMENT

Fill a beaker almost to the top with water and place it on a ringstand. Adjust the burner so that it produces a small flame under one side of the beaker. Be sure the flame remains in the same spot. Let the water heat up for a few minutes. Drop four or five potassium permanganate crystals into the water on the side over the flame. Note the results.

What happens as the potassium permanganate dissolves? On which side does it go down?

As ocean currents flow north and south from the equator, the earth's rotation forces the currents toward the right in the Northern Hemisphere and toward the left in the Southern Hemisphere. Compare this with the circulation of the winds on the earth. That means that in the northern oceans, the currents move to the right in a clockwise circle. In the southern oceans, the currents move to the left in a counterclockwise circle (Fig. 5-12).

5-12   How do ocean currents in the Northern and Southern Hemispheres differ?

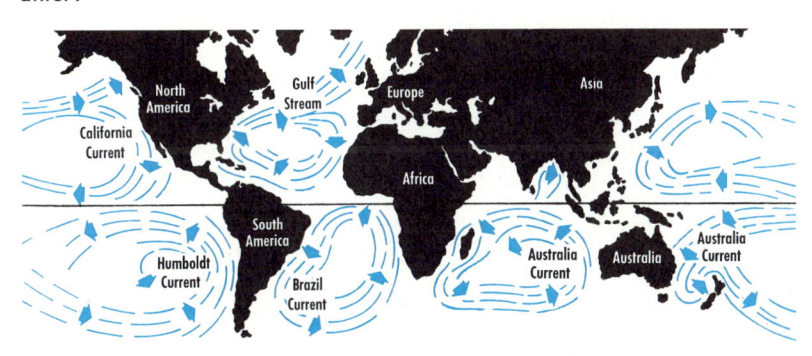

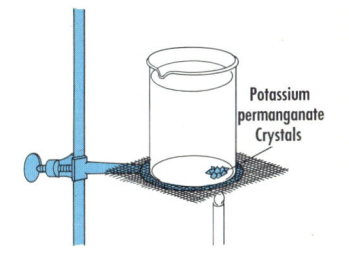

120 / UNIT 2 MAN EXPLORES HIS ENVIRONMENT

Land masses, tides, and waves affect the movements of ocean currents. Would the currents off the coasts of North and South America be flowing in the direction they are, if the continents were not there? In addition to land masses above the ocean's surface, underwater mountains and valleys also affect the movements of currents.

In addition to the currents flowing in the ocean there are other movements of the ocean water. There are the regular movements of the tides. There are the waves which sometimes violently, sometimes gently, wash ashore on the land masses rising from the ocean floor.

As you know, tides are caused mainly by the gravitational attraction between the earth and the moon. Ocean waves, however, are produced mainly by wind action on the surface of the ocean. The ocean surface, pushed by moving air, forms *crests* and *troughs* (TRAWFS). The crest is the high point on the wave, and the trough is the low point. Wave height is measured from crest to trough (Fig. 5-13).

An important thing to remember about waves is that they are movements, not objects. As you watch ocean waves, it may appear that water is moving from one place to another. It really isn't. A similar happening occurs when we make waves in a rope fastened to a stationary object. Waves travel from one end of the rope to the other, but the rope particles remain where they are. An activity will help you understand the movements of ocean waves.

5-13 The crests and troughs of ocean waves. Describe the motion of the water molecules in them.

OBSERVE

Fill a sink, aquarium tank, or large pan with water. Float a piece of wood or cork on the surface of the water. Produce waves in the container of water. Carefully observe the floating object. What do you observe? How do your observations compare with the discussion of ocean waves? How does a sailboat move from one place to another? Do the ocean waves contribute to its movement?

In the ocean, waves generally are no higher than about 50 feet. During storms, however, waves higher than 100 feet have been observed. The size of ocean waves depends on three fac-

CHAPTER 5 THE CHANGING OCEANS / 121

tors: (1) the force at which the wind is blowing, (2) the length of time the wind blows, and (3) the distance over which the wind has been exerting its force on the water. Larger waves are produced in the ocean than in any inland bodies of water. When ocean waves approach a shoreline, *breakers and surf* are produced because the waves are slowed down by friction with the shallower ocean bottom (Fig. 5-14).

When you throw a rock into a small pond, waves are produced which move through the water to both sides of the pond. The same kind of thing happens in the ocean. There, the waves, pounding ashore along the coastline, may have been caused by a storm located 5000 or 6000 miles away.

During 1957–1958, the International Geophysical Year (IGY), scientists from 66 countries worked together on many problems. Their studies showed that a great ridge, the *Mid-Atlantic Ridge* exists. It is almost the size of a continent and runs north and south in the Atlantic Ocean. Another great ridge in the Pacific Ocean extends from the Aleutian Islands in the north past the Hawaiian Islands. Strangely enough, the deepest valleys are close to the land masses, rather than in the middle of the oceans. A deep valley, thousands of miles in length, runs along the Atlantic Ocean floor, around the tip of Africa, and into the Indian Ocean. The greatest depths, up to seven miles, have been found near Guam and the Philippines.

Occasionally, the surface of the sea is broken by islands. In some places the tops of underwater mountains, known as *seamounts,* are just below the surface (Fig. 5-15). For example,

5-14 How does the motion of surf differ from waves offshore?

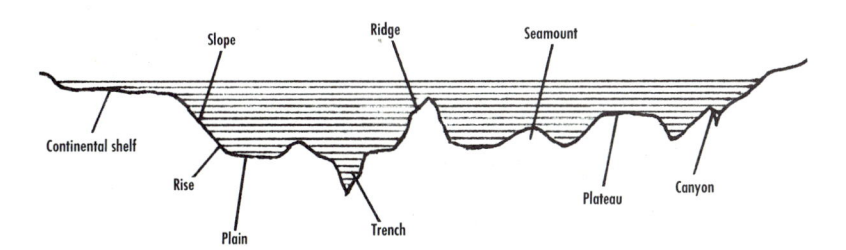

5-15 A cross section showing formations of the ocean floor.

a huge seamount, rising over three miles from the ocean floor to within 120 feet of the surface, has been discovered in the South Atlantic. It is believed that many seamounts were islands during the last *Ice Age,* 10,000 to 15,000 years ago. As the ice sheets melted, the ocean level rose about 200 feet and covered many islands.

Most of the earth's islands and seamounts are found in the Pacific Ocean. The Atlantic Ocean, by comparison, has few islands. Many islands in the Pacific Ocean are volcanic mountains. Mountain-building by means of volcanic action still seems to be taking place in and around the Pacific Ocean (Fig. 5-16). For example, Hawaii, Alaska, Japan, and islands in the East Indies have many active volcanoes.

The bottom of the ocean is usually divided into three different areas. The *continental shelf* is the relatively shallow area surrounding all of the continents. It usually slopes very gently away from the land. Off the coast of North Carolina, the continental shelf extends about 150 miles from the shoreline. Along the Pacific coast of the United States, the average width of the shelf is about 20 miles. In some places, such as parts of Florida, there is practically no shelf at all. But in general the continental shelf along the east coast of the United States is wider than that along the west coast.

Where the gradual slope of the continental shelf ends, the continental slope begins. Here, the gradual slope inclines sharply, continuing to the deep sea floor called the *abyssal* (UH-bihs-uhl) plain.

Almost all of the ocean floor is covered with a layer of sediment. This sediment consists of material which is carried into the ocean by rivers, rocks and other substances remaining from glaciers. Dust and other particles from volcanoes and land also are transported by the wind and deposited into the ocean. Here, they slowly sink to the bottom. The remains of the many living things which inhabit the ocean, finally, contribute to the building up of the sediment layer on the ocean floor. In a few places, the bare rock of the true ocean floor appears above the sediment layer, but these are unusual. In the Pacific and Indian Oceans, sediment layers, ranging from 100 to 1000 feet thick, have been discovered. Beneath the

5-16 How may volcanic activity change the bottom of the ocean?

layer of sediment lies the true ocean floor, the crust of our planet earth, which is generally thinner under the ocean than in any other place on earth.

REVIEW

1. What is meant by a: (*a*) solute, (*b*) solvent?
2. Name several minerals obtained from sea water.
3. Name three main ocean currents that flow in the seas.
4. What factors affect the size of ocean waves?
5. Into what three areas do scientists usually divide the ocean bottom?

C/How Does the Ocean Affect Us?

The oceans supply most of the water vapor that enters the atmosphere. You learned in elementary science that the process by which a liquid becomes a vapor or gas is called *evaporation* (ee-*vap*-uh-RAY-shuhn). Figure 5-17 illustrates how water evaporates from a number of sources. Can you name any other sources from which water enters the air by evaporation?

With as much as three-fourths of the earth's surface covered with water, you will not find it difficult to understand why most of the water in the atmosphere evaporates from the oceans.

Since molecules of a substance are in constant motion, each molecule possesses a certain amount of energy. As we increase the temperature of a substance, molecules move at a faster rate of speed. Molecules close to the surface of the water absorb energy from the sun, causing an increased rate of molecular motion. As they move faster, and continue to absorb energy from the sun, the molecular energy increases. When the energy of these molecules becomes great enough, some of them escape the forces which hold molecules together. These enter the atmosphere.

5-17 What is one process taking place in this scene?

Pressure affects the rate of evaporation. The pressure of any gas slows the evaporation of more molecules of the same gas. That is, the higher the pressure of a gas on the surface of the liquid, the lower will be the rate of evaporation of that gas from the liquid. An increase in pressure results in an increase in the number of molecules in a given unit of space. The greater the number of molecules in a given space, the more difficult it is for other molecules to enter.

As we have already learned, the faster air molecules move, the greater the amount of energy they possess. As they collide with the molecules on the surface of the liquid, some of the energy is transferred from the air molecules to the water molecules. Thus, the water molecules gain energy, enough in many cases to increase the rate of evaporation.

The air always contains some water vapor. The amount of water vapor per cubic foot of air is called *absolute humidity* (hyoo-MIH-dih-tee). However, we do not often speak in terms of the absolute amount of water vapor in the air. We usually refer to *relative humidity*. This is a ratio of the actual amount of water in the air to the total amount of water it could hold at a particular temperature. Relative humidity is usually given as a percentage. If the temperature rises and no more water vapor enters the atmosphere, the relative humidity is lowered. Cooling the air, without removing any water vapor, will increase the relative humidity. Find out in the next activity at what temperatures water comes out of the air.

MEASURE

Fill several types of containers (tin can, aluminum cup, porcelain cup, glass) with water. Record the room temperature. Start with the water at room temperature and slowly cool it by adding ice. Stir the mixture gently with a stirring rod until you notice a water film forming on the outside of the container. Record the temperature of the ice water. Where did the water on the outside of the containers come from? Do your temperature findings agree with those of your classmates? If your classmates record different temperature readings, what do you think might account for these differences?

CHAPTER 5 THE CHANGING OCEANS / 125

The temperature that you recorded in the activity is the *dew point*. This is the temperature to which the air must be cooled in order for water vapor to change back to a liquid. *Condensation* (kahn-den-SAY-shuhn) is the process in which a vapor is changed back into a liquid.

As you learned, the sun's energy heats the earth by radiation. On a clear night, the earth and the air close to its surface both lose a great deal of heat by this same process of radiation. This results in an increase in relative humidity because cooler air cannot hold as much water vapor as warmer air. If there is not very much wind, the relative humidity of the air that is close to the earth increases, until the dew point is reached. When this happens, moisture begins to condense, especially on the surface of cooler objects. For example, many mornings you find the grass wet even though there has been no rain during the night. This moisture is dew which condensed during the night (Fig. 5-18). On most days, relative humidity is highest in the cool hours of night and early morning. Relative humidity is lowest in the afternoon hours.

Condensation also may occur by contact cooling as you saw in the previous activity you performed. Because the water in the container was cooled, the air that is in contact with the outside of the cup now became colder. This increased the relative humidity until the dew point of the surrounding air was reached.

The relative humidity of the air can be measured. A *hygrometer* (hye-GRAH-muh-tuhr) is used to measure relative humidity. One type of hygrometer consists of two thermometers, one a dry-bulb and the other a wet-bulb. As seen in Fig. 5-19, the lower end of the wet bulb thermometer is wrapped in a piece of wet cloth. The wet-bulb is cooled by evaporating water. The evaporation rate depends on the humidity of the surrounding air. Thus, the drier the air, the greater the rate of evaporation and the lower the temperature shown on the wet-bulb thermometer.

The relative humidity may be determined by using the hygrometer readings along with the table which is shown on page 126. Why do you think a wet-and-dry-bulb hygrometer must be fanned before it is read?

5-18 Why is moisture or dew forming on the grass?

5-19 Why do both thermometers show different readings?

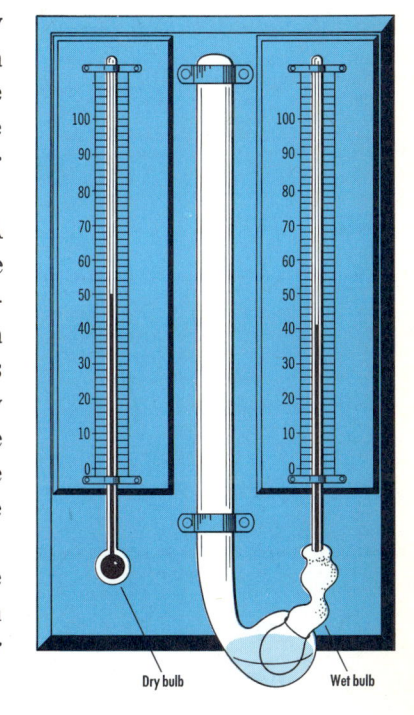

Difference Between Dry- and Wet-Bulb Thermometers

°F	1	2	3	4	5	6	7	8	9	10	11	12	13	14	15
Reading of dry-bulb thermometer, °F					Relative humidity (percent)										
63	95	89	84	79	74	69	64	60	55	51	46	42	38	33	29
64	95	89	84	79	74	70	65	60	56	51	47	43	38	34	30
65	95	90	85	80	75	70	65	61	56	52	48	44	39	35	31
66	95	90	85	80	75	71	66	61	57	53	49	45	40	36	32
67	95	90	85	80	76	71	66	62	58	53	49	45	41	37	33
68	95	90	85	81	76	71	67	63	58	54	50	46	42	38	34
69	95	90	86	81	76	72	67	63	59	55	51	47	43	39	35
70	95	90	86	81	77	72	68	64	60	55	52	48	44	40	36
71	95	91	86	81	77	72	68	64	60	56	52	48	45	41	37
72	95	91	86	82	77	73	69	65	61	57	53	49	45	42	38
73	95	91	86	82	78	73	69	65	61	57	53	50	46	42	39
74	95	91	86	82	78	74	70	66	62	58	54	50	47	43	40
75	95	91	87	82	78	74	70	66	62	58	55	51	47	44	40
76	96	91	87	83	78	74	70	67	63	59	55	52	48	45	42
77	96	91	87	83	79	75	71	67	63	60	56	52	49	46	42
78	96	91	87	83	79	75	71	67	64	60	57	53	50	46	43
79	96	91	87	83	79	75	71	68	64	60	57	54	50	47	44
80	96	91	87	83	79	76	72	68	64	61	57	54	51	47	44

Suppose the dry-bulb thermometer reads 75°F, while the wet-bulb thermometer reads 65°F. The difference in readings is 10°. Find the number 10 in the row of numbers that is across the top of the table. Now, run your finger down the column until you find the number opposite the dry-bulb reading of 75°F? What will be the relative humidity in this case? What is the relative humidity if the dry-bulb thermometer reads 80°F, while the wet-bulb thermometer reads 76°F? Which example shows air to be drier? You can make an hygrometer in the next activity.

CONSTRUCT

To make a simple hygrometer, soak a human hair in alcohol to remove any oil film. Arrange the hair as shown, using a drop of model airplane cement to fasten the hair. Be sure the hair is stretched tight. Cement a paper pointer to the hair and place a pivot pin through the pointer. Now construct a scale to indicate the position changes in the pointer.

What happens to the hair as it absorbs moisture from the air? Observe the different readings of the pointer each day for a period of a week. Make a chart of these readings.

Water moves through the soil. When rain falls to the earth, it may land in oceans, rivers, streams or on the earth's surface. In all cases, however, it gradually disappears. It may evaporate back into the air, run off into rivers and streams, be absorbed by plants, or sink into the ground to form the ground water reservoir (Fig. 5-20).

Fresh water can be obtained from sea water. *Desalinization* (dee-*sal*-ih-nih-ZAY-shuhn) is the process of removing salt from sea water to increase the fresh water supply. Ancient sailors learned to boil their drinking water from the sea, which left the salt behind. Scientists have continually been studying many desalinization methods. In principle, the methods involve removing water from the salt or taking the salt out of the water. Scientists now have been able to obtain fresh water from the sea by distilling, freezing, or by filtration. Filtration uses a material that allows the water but not the salt to pass through easily.

5-20 What is the source of underground water?

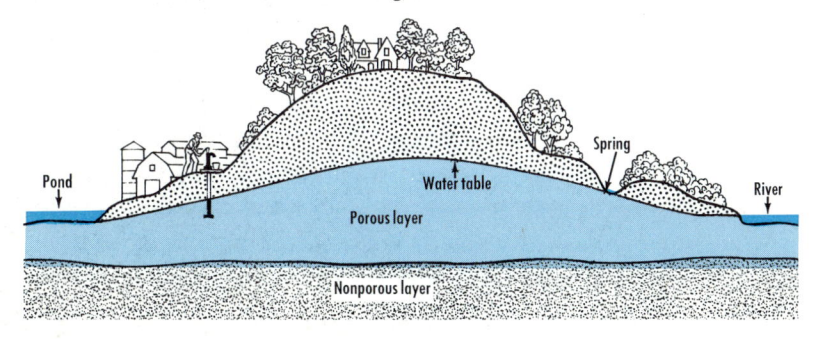

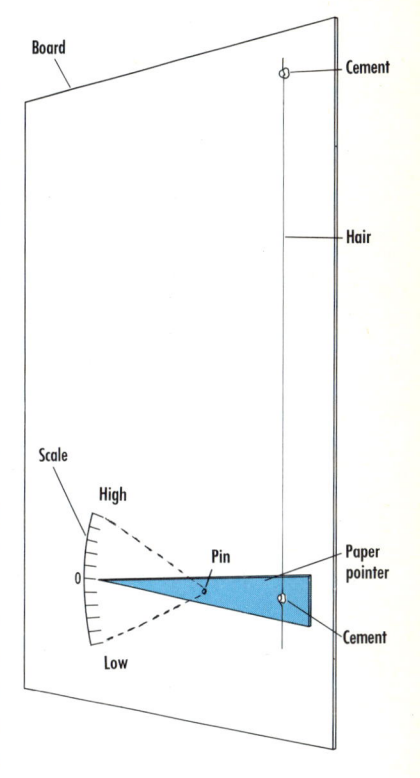

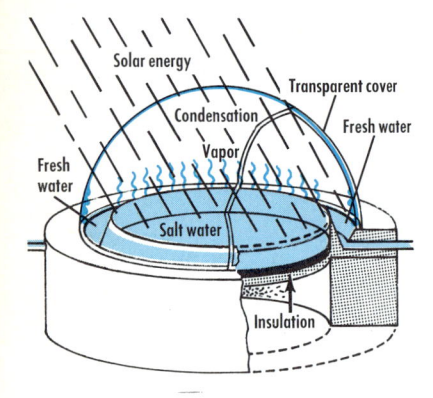

5-21 Describe how fresh water is produced from salt water in a solar still.

When freezing is used to take the salt out of water, ice crystals form separately from the brine. These can be melted down as fresh water.

A number of desalinization plants use the *distillation method* for desalting sea water. The water is changed to steam and then cooled to form fresh water. About three and one-half gallons of sea water produce one gallon of fresh water. Distillation systems can use electricity, natural gas, or nuclear energy for their power.

In Fig. 5-21, a solar still is shown. The sea water evaporates from a large tank and condenses on the transparent cover. The water runs down the sides of the cover into catch areas from which it is removed. In this process, the salt and other minerals remain in the tank. Construct a solar still and see how it works in the next activity.

CONSTRUCT

Place a heat lamp 8-12 inches above a glass bowl which is inverted over a saucer of salt water. The saucer of salt water lies on a pie plate. See the diagram.

Place the glass bowl on a table and observe what happens over a period of a day. Where is the water coming from that is condensing on the inside of the glass bowl? Taste the water that is collecting in the pie plate. Compare it with fresh water.

The major problem facing scientists is finding a process which produces fresh water cheaply. It now costs about one dollar for desalinization of 1000 gallons of sea water. Improved methods will reduce this cost in the future. In the past we have taken our cheap water supply for granted. There is no doubt that in the near future many communities will have large industrial plants for converting sea water into purified water for drinking and irrigation (Fig. 5-22).

Some other experimental methods for producing fresh water from ocean water are: the *electrical method* where electricity is used to remove dissolved salts; the *membrane method* where salt water interacts with a semiporous material which allows only fresh water to pass through; the *freezing method*

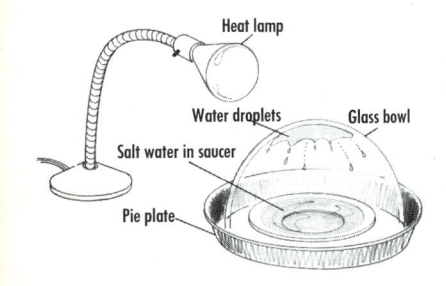

5-22 Why would a community need a modern sea water conversion plant similar to the one shown?

which works on the principle that as salt water freezes, the ice formed is composed of fresh water; and the *biological method* where living organisms are used to remove the salt from ocean water. Make pure water from salt water in the following activity.

EXPERIMENT

Fill one ice tray with fresh water and other one with salt (NaCl) solution. Place both containers in a freezer. In which container does the ice form first? Does ice form in both containers when they are left in the freezer overnight? Examine the contents of both trays carefully. Taste them. What do you observe? What problems can you think of that might be involved in using a method like this to produce fresh water?

At the present time, huge conversion plants are in operation in a few places to convert salt water into fresh water. They make use of the evaporation-condensation method. Most scientists agree, however, that some day many more of these plants will be in operation throughout the world, especially in areas of the world which do not have enough fresh water to meet the demand.

5-23 How is this device used to mine the ocean floor?

Over the years the ocean has been an important source of food. Man has caught and eaten different kinds of fish and shellfish from the earliest times. In the Orient, many different varieties of ocean plants are eaten. Many scientists and others are concerned about the rapid growth of the world's population. They feel that in the future, we will have to turn more and more to the vast food resources of the ocean. In order to do this, more efficient methods of "harvesting" the plant and animal life of the ocean must be devised. Although this engineering problem is not strictly the concern of oceanographic scientists, much that they discover about life in the ocean will contribute to a new field known as *aquaculture* (AK-wuh-*kuhl*-choor).

The sea is an untapped resource of wealth. It has been estimated that in a cubic mile of sea water there is more than 90 million dollars worth of gold and about 8 million dollars worth of silver. The costs, however, in removing these precious minerals at the present time is greater than the value of the recovered minerals.

Most scientists feel that before this happens we will be "mining" other metals from the ocean. Huge quantities of various manganese minerals in the form of nodules, containing varying amounts of cobalt, nickel, iron, and copper have been discovered on the ocean floor (Fig. 5-10). These nodules are being raised from the floor by a device similar to the one shown in Fig. 5-23.

REVIEW

1. How does the motion of molecules in a liquid affect the rate of evaporation?
2. How does the pressure of a gas on the surface of a liquid affect the rate of evaporation?
3. What is meant by: (*a*) absolute humidity, (*b*) relative humidity?
4. Name three methods used to produce fresh water from sea water.
5. What minerals make up most of the mineral nodules discovered on the ocean floor?

THINKING WITH SCIENCE

A. *On a separate sheet of paper write the numbers 1 to 15. Some of the following statements are true and some are false. If the statement is true, write true. If the statement is false, change the italicized term to make it true.* **Do not write in your book.**

1. As the pressure on the body in a deep dive is increased, the amount of gases that is dissolved in the blood is *increased*.

2. A solution which cannot hold any more solute at a particular temperature is *saturated*.

3. The temperature at which water vapor condenses out of the air is the *relative humidity*.

4. Today scientists measure the ocean depth by means of *seismic soundings*.

5. Samples of ocean water can be obtained by using *grab buckets*.

6. Drift bottles are a common method of studying the *ocean bottom*.

7. Ocean water is considered *colorless* regardless of the way it appears.

8. An instrument used to measure the relative humidity of water is the *Nansen bottle*.

9. The Cromwell Current flows in the opposite direction to the *Gulf Stream*.

10. A *solute* is a liquid in which a substance is dissolved.

131

132 / UNIT 2 MAN EXPLORES HIS ENVIRONMENT

11. A deep sea exploration submarine is a *bathyscaphe*.

12. The pressure exerted by a liquid is directly proportional to its *temperature*.

13. The upward push exerted by water on an object floating in it is called *buoyant force*.

14. The greatest concentration of mineral content dissolved in sea water is *manganese salt*.

15. In the Northern Hemisphere, ocean currents circulate in a *counterclockwise* direction.

B. *Write the answers to the following in your notebook. Be sure to use complete sentences and correct grammar and spelling.*

1. Name the various methods used by scientists to study the oceans.

2. In what ways do the oceans affect our lives?

3. What is the relationship between solute, solvent, and solution?

4. Compare the processes of evaporation and condensation.

5. How do ocean currents get their energy?

6. Describe the behavior of molecules during the process of evaporation.

7. Explain why the ocean currents flow clockwise in the northern hemisphere and counterclockwise in the southern hemisphere.

8. Describe different processes to change salt water into fresh water.

MATHEMATICS IN SCIENCE

Work the following problems in your notebook.

1. Find the pressure per square inch at the bottom of a container which is filled with water to a height of six feet.

2. If the reading of the wet-bulb thermometer is 64 and the reading of the dry-bulb thermometer is 72, what is the relative humidity of the air?

3. The density of ocean water is 64 pounds per cubic foot. A submarine is at a depth of 500 feet. (*a*) What is the water pressure per square inch on the submarine? (*b*) If the surface area of the closing hatch is 80 square feet, what is the total force being exerted on the hatch?

RESEARCH IN SCIENCE

1. Place self-addressed stamped postcards inside sealed bottles and let them be carried out by the tide in the ocean or the current of a river. Report on any cards returned.

2. Make a report on how your community receives its drinking water. Be sure to include the processes used in making it safe.

3. Make a model of an undersea laboratory in which man might be able to live and study the ocean depths. Explain its features to the class.

4. Determine which liquid has the greatest buoyant force: fresh water, salt water, rubbing alcohol, mineral oil.

5. Make a report on how minerals are now being extracted from sea water in commercial quantities.

chapter 6
The Changing Weather

A/What Causes Winds?

Many factors affect the motion of air. In the two previous chapters, many scientific concepts were presented about air and water, gases and liquids. You are now going to study relationships which exist between the earth, the atmosphere, the oceans, and the sun.

Whenever there is a difference in temperature between two nearby bodies of air, convection currents are set up. The warm air expands, becomes less dense, and rises, pressing down on the earth with less force. The cooler, denser neighboring air pushes down under the warmer air. Since the atmosphere is warmer over the equator, the air there expands and rises. As the air rises and spreads out, it increases the atmospheric pressure in the neighboring regions and the pressure over the equator decreases.

If winds were caused only by the uneven heating of air masses, it would be simple to understand their circulation. However, winds are deflected or change direction because of the earth's rotation on its axis. The direction from which the wind is blowing is usually given in describing air movements.

Thus, a wind that blows from the north is called a *northerly wind*. One that blows from the southeast is called a *southeasterly wind*. What would you call a wind blowing from west to east? Weather forecasters use a wind vane to determine the direction of the wind.

As you know, the earth rotates from west to east on its axis at a speed of approximately 1000 miles per hour at the equator. Since the circumference of the earth is 25,000 miles at the equator and zero at the poles, the speed of rotation on the surface of the earth is different in various parts of the world. For example, in most parts of the United States, it has been found that the surface speed on the earth's rotation is less than 1000 miles per hour.

Suppose that air movement is in a southerly direction moving at 800 miles per hour. The earth's surface speed, however, is about 1000 miles per hour. There is a difference, therefore, between the speed of the air and the surface speed of the earth. If the earth did not turn, we would have only north and south winds. However, since the earth does turn on its axis, the winds are deflected to the *right* in the Northern Hemisphere and to the *left* in the Southern Hemisphere, as illustrated in Fig. 6-1. Thus, we find that the prevailing winds in our part of the world are westerly winds. To summarize the discussion above, winds are caused by moving air masses. The direction of winds is determined also by the rotation of the earth.

Anemometers are used to measure the speed of the wind. You have probably seen an *anemometer* (*an*-uh-MAH-mih-tuhr) spinning in the wind at an airfield or weather station. The anemometer, shown in Fig. 6-2, has several small cups that are attached by spokes to a shaft that is free to rotate in all directions. Other types have a propeller that spins easily in the wind. The harder the wind blows, the faster the instrument turns. By measuring the speed with which the cups (or the propeller) on the anemometer turn, a weather forecaster is able to determine accurately the wind speed. An arrangement which produces a record of wind speed on a graph is very useful, too. You can construct an anemometer by following the directions in the next activity.

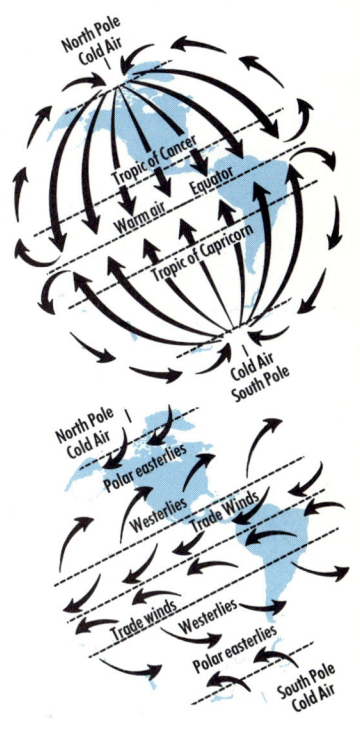

6-1 Why are winds on earth deflected in different directions?

6-2 How is an anemometer used to determine wind speed?

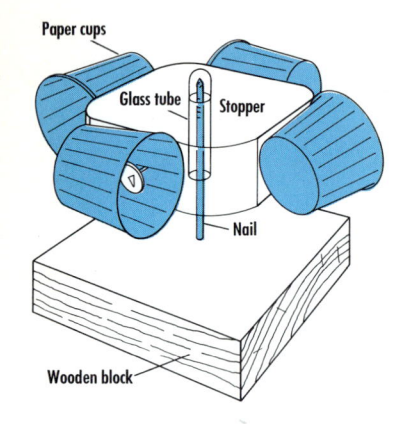

CONSTRUCT

Cut the sides from a large one-hole stopper with a sharp knife so that you have a square block. Close the end of a short piece of glass tubing in a Bunsen burner flame. Push the glass tube into the bottom of the stopper. Use thumbtacks to fasten four paper cups to the block as shown. The cups should face in the same direction. Hammer a large thin nail into a block of wood and slip the glass tube over the point of the nail.

Your anemometer should now turn freely. Place it outside to see how the wind causes the anemometer to turn. Explain how the anemometer is used.

You can learn to judge the speed of the wind by noting the rising of smoke, the movement of leaves, the swaying of branches and the whitecaps on waves. The different wind speeds and a description of their effect are found in the *Beaufort wind scale* shown on page 137. The speed of the wind is organized according to a scale of numbers from 0 to 17.

Weather maps often use little flags or symbols to show the speed of the wind according to the Beaufort scale. The slant of the symbol on the map indicates the wind direction. Practice using the wind scale in the next activity.

OBSERVE

The speed of wind can be determined roughly from the Beaufort wind scale shown. Observe the wind conditions daily and keep a record of the wind speed and direction. Is this method as accurate as that used by the Weather Bureau? Check your observations with the official weather reports found in your local newspaper.

Uneven heating of the earth's surface produces wind. You probably know that the earth is tilted on its axis of rotation. Therefore, during some parts of the year, certain areas of the earth are tilted *toward* the sun. At other times, these areas are tilted *away* from the sun. You can demonstrate this by doing the activity on page 138.

BEAUFORT WIND SCALE

Terms used in forecasts	Miles per hour	Beaufort number	Wind effects observed on land
Calm	0–1	0	Smoke rises vertically.
Light	1–3	1	Direction of wind shown by smoke drift; but not by wind vanes.
	4–7	2	Wind felt on face; leaves rustle; ordinary vane moved by wind.
Gentle	8–12	3	Leaves and small twigs in constant motion; wind extends light flag.
Moderate	13–18	4	Raises dust, loose paper; small branches are moved.
Fresh	19–24	5	Small trees in leaf begin to sway; whitecaps form on inland waters.
Strong	25–31	6	Large branches in motion; whistling heard in telegraph wires; umbrellas used with difficulty.
Gale	32–38	7	Whole trees in motion; inconvenience felt walking against wind.
	39–46	8	Breaks twigs off trees; impedes progress.
	47–54	9	Slight structural damage occurs.
	55–63	10	Seldom experienced inland; trees uprooted; considerable structural damage occurs.
Whole gale	64–74	11	Rarely experienced; wide damage.
Hurricane	75–136	12–17	Rarely experienced; wide damage.

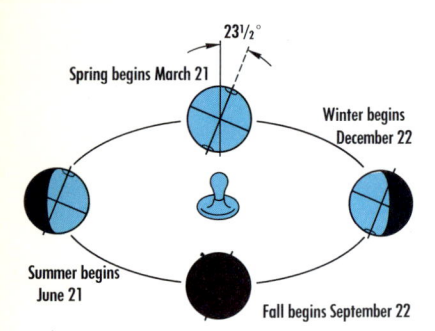

EXPLAIN

Put a light bulb in the center of a table to represent the sun. Push a knitting needle through the center of an orange to represent the axis of the earth. Tilt the needle about 23½ degrees from the vertical. Move the orange into different positions, as shown. Darken the room. Observe the amount of light that covers the surface of the orange in each position.

At which position does the Northern Hemisphere of the earth receive the greatest amount of light from the sun? The least? When is the amount of light received exactly equal in both hemispheres? Why are winter days shorter in length than summer days? Why is it colder in winter than in summer?

The tilt of the earth on its axis results in uneven heating of the earth's surface. We know that perpendicular or direct rays of sunlight produce more heat than rays which hit the earth at an angle or slant. The sun's rays travel to the earth in a straight line and strike the earth perpendicularly at or near the equator. Because the earth's surface is curved, the rays strike the earth at an angle in other areas. The following activity will help you understand how the angle of the sun's rays affects the earth's temperature.

COMPARE

Fill two pans with the same amount of soil. Put a thermometer in each pan, as shown. Be sure that the thermometers placed in both pans are barely covered with soil because soil is a good insulator. They should both be at the same depth and *not* touching the bottom of the pans. Put the pans on a sunny window ledge with one lying flat. The other is propped up so that the surface of the soil is perpendicular to the sun's rays. Leave both of the pans in the sunlight for 30 minutes. Read the thermometers without taking them from the soil. Which pan has the warmer soil? Which pan received the more direct rays?

The weather is cooler in our part of the world than at the equator. This occurs because the sun's rays strike our part of the world at a greater angle, spreading out the rays. Therefore, the earth absorbs less heat energy from the sun and the air is cooler.

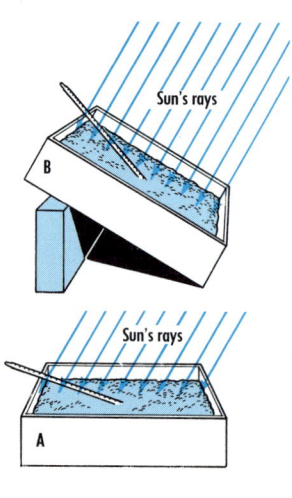

CHAPTER 6   THE CHANGING WEATHER / 139

Unequal heating of land and water produces wind. Land and water absorb different amounts of heat energy from the sun. In much the same way, there is a difference in the amount of heat that is lost by land and water areas. We can demonstrate this in the following activity.

EXPLAIN

Fill a one-gallon glass jug with soil and another with water to the same level. Put the thermometers in two one-hole stoppers. Place the stoppers in the jugs so that the bulbs of the thermometers are about one-half inch below the surface of the materials in each jug. Put the two jugs in the sunlight and record the temperatures every ten minutes. Make a graph comparing the changes in temperature of the water and the soil. At the end of 40 minutes, put the jugs in a cool closet. Continue reading the temperatures every ten minutes. On the same chart, show the changes in temperature by using a different colored pencil.

Which material absorbed heat quicker? Which lost heat quicker? How does this show the heating effects of land and water on the air over them? How would this affect the direction of winds?

This change in the amounts of heat absorbed and lost causes differences in air temperature during the daytime and at night. During the day, the air over the land is heated more and

6-3 Why does the wind blow from sea toward land during the day?

6-4 Why does the wind blow from land toward the sea at night?

becomes lighter. The warm air is forced up by the cool air moving in from the water, causing a *sea breeze*. The air is moving from the sea to the land (Fig. 6-3), at night, the land loses its heat more quickly than the water, and the air over the land becomes cooler than the air over the water. The warm air is forced up by the cool air from the land, causing a *land breeze* as the air moves from the land to the water, as in Fig. 6-4.

Surface features affect the wind. It is easy to understand how hills and mountains can affect moving bodies of air. For example, let us follow a large body, or mass, of air that has been out over the Pacific Ocean. While there, it has absorbed heat and water vapor from the ocean; it is now relatively warm and moist. As it moves inland, pushed by a westerly wind, it appears as part of an ocean breeze. When this air mass moves across the surface of the land, it begins to rise to higher altitudes, pushed up by the hills. As the air mass rises to higher altitudes, there is less pressure on it from above, and it expands. As we have already learned, when a gas expands, it becomes cooler. As the warm, moist air becomes cooler, part of the water comes out as tiny droplets, forming a rain cloud. At still higher altitudes in the high mountains, the air becomes thin

6-5 What happens when air currents rise and become cooler?

and cold. As a result, most of the remaining water vapor comes out as snow which covers the tops of the mountains. When the cold dry air comes over the mountains and down into the lower plains, the air mass becomes compressed and heated once more. However, the air mass has left behind nearly all its original water on the western sides of the mountains. The air mass can no longer produce any rain. This is the reason why the western slopes of the Rocky Mountains are moist and fertile, and their eastern slopes are hot, dry deserts with little rainfall. These events of the discussion above are shown in more detail in Fig. 6-5.

REVIEW

1. How are wind directions named?
2. Compare the density of warm air and cool air.
3. How does an anemometer measure wind speed?
4. What is the effect of the earth's tilt on the heating of its surface?
5. How does the warming of the air on land compare with the warming action of water: (*a*) by day, (*b*) by night?

142 / UNIT 2   MAN EXPLORES HIS ENVIRONMENT

B/What Causes Clouds?

Cloud formation depends largely upon the humidity and temperature of the air. Most clouds form when rising air cools. There are three distinct types of clouds: *cirrus* (SIHR-uhs), *stratus* (STRAY-tuhs), and *cumulus* (KYOO-myoo-luhs). Each of them has a different appearance and is easily recognized.

Cirrus clouds are white and feathery. They are seen high in the sky, usually seven to ten miles above the earth. They look like thin wispy curls, as shown in Fig. 6-6. Cirrus clouds are always made up of ice crystals. Do you know why? Mare's tails, the highest of these clouds, are usually ten to twelve miles high. They frequently indicate that a rainstorm will follow in a day or two. The following activity tells how you can photograph clouds.

6-6 At what altitude do cirrus clouds occur?

6-7 What kind of weather do cumulus clouds indicate?

OBSERVE

Use a camera to photograph various cloud formations. Try different filters on the camera to see which produces the best pictures of clouds in black-and-white photography. Make an exhibit of different cloud formations. Name the types you were able to photograph.

Cirro-stratus (SIH-ro STRAY-tuhs) are thickened cirrus clouds which have merged. They have the appearance of a white veil of uneven and often threadlike texture. *Cirro-cumulus* (SIH-ro KYOO-myoo-luhs) are small fleecy cumulus clouds, five to seven miles above the surface of the earth. They are often present in very large numbers, often giving a picture of a "curdled" sky.

Cumulus clouds are usually dome-shaped and billowy. They are often called "wool pack" and are found from two to four miles above the earth. They appear as a huge mass, flat on the bottom but piling up to a great height. Cumulus clouds (see Fig. 6-7) are big and fluffy and usually indicate fair weather. However, if the rising warm air mass contains enough moisture, the clouds may grow into a towering dark mass and become a "thunderhead." This type of cumulus cloud will produce thunderstorms with flashes of lightning and heavy rain. Cumulus clouds can also be formed by convection, which takes place over a large heated surface, like a forest fire. Can you explain the process that causes this?

Cumulonimbus clouds are any thick, extensive layers of formless clouds from which rain or snow is falling (Fig. 6-8). They are produced mainly by some type of forced convection, the upward deflection of winds which cool the upper air by their contact. Their undersides are usually about a mile above the ground.

Stratus clouds are low and flat and form in layers. Sometimes they may be close to the ground to be mistaken for fog. Stratus clouds are formed in layers and usually indicate calm weather conditions (Fig. 6-9). These clouds form when water-laden air rises slowly. They contain many heavy droplets of water. As these clouds become thicker and darker, a **slow steady rain** may begin to fall. This type of rain may last for one or two days. You can compare the weather that usually accompanies or follows various cloud conditions in the next activity.

6-8 Explain how a cumulonimbus cloud forms.

EXPLAIN

Make a chart in the form of a calendar. Observe the cloud conditions each day for a month and make a note of the daily clouds. Show the precipitation on days when it rains or snows. Compare the precipitation with the clouds you observed for the day before and during that day. How can rain or snow be forecast by studying the clouds?

6-9 Contrast stratus clouds with cumulus and cirrus clouds.

There is a slight difference, usually, between fog and clouds. Fog is formed by low air temperature at or near the surface of the earth, as shown in Fig. 6-10 (page 144). It consists of water droplets condensed from and floating in the air near the surface of the earth. For example, one type of fog results from the nightly cooling of the earth. Can you explain this? Another condition which produces fog is the movement of warm, moist air over cold surfaces. Clouds consist of very tiny water droplets condensed from and floating in the air

6-10 How is fog formed?

6-11 What geometric pattern do most snowflakes take when they form?

well above the surface of the earth. Thus, we can say that fog, actually, is a cloud on the earth, and that a cloud is fog in the sky.

Different forms of precipitation depend upon the atmospheric conditions. You have learned that water vapor enters the air by evaporation and that the temperature of the air affects the amount of water vapor it can hold. In addition, warm air is less dense than cool air and, therefore, rises. As the warm air rises from the earth's surface, it is cooled. Its temperature reaches the dew point of the water vapor in the air. If the temperature of the air between the clouds and the earth is above freezing, it rains. If it is below freezing, we may have snow or sleet. Examine the beautiful snowflake patterns shown in Fig. 6-11. Can you give possible explanations for their various shapes?

Sometimes drops of rain will start to fall and then be carried up by rising air currents to where the temperature is below freezing. The drops freeze and then begin to fall again through warmer air. Here, they pick up another layer of moisture. If the air is moving violently, these drops may make several trips up and down before they finally reach the earth. Each time these frozen drops rise, another layer of moisture freezes on them. The process continues until they become too heavy to remain in the air. They fall to the earth in the form of *hail*. Sometimes, the process is repeated over and over, forming many layers of ice. These very large hailstones can cause great damage. If you broke open a hailstone, what would you see?

On the other hand, *sleet* is formed when rain freezes as it falls. This form of precipitation usually occurs during a temperature *inversion* (ihn-VUHR-zhuhn), when the air near the surface of the earth is colder than the upper air.

For accurate weather predictions, it is necessary to measure the amount of rain that falls in a given area in a certain length of time. For this purpose, weathermen use a *rain gauge*. The newest type of rain gauge measures the amount of rain electrically. The gauge records the amount of rain automatically on a moving chart.

REVIEW

1. Name the three main types of clouds.
2. What is the difference between fog and clouds?
3. Why are cirrus clouds made of ice crystals?
4. How are hailstones formed?
5. How does a weatherman use a rain gauge?

C/Why Does It Storm?

Air masses usually form at the tropical or polar regions. An *air mass* is a large volume of air over a certain part of the earth. They are either *warm* or *cold,* depending on where they form. In the United States we find that air masses are of four kinds: (1) Polar Maritime (*mP*), (2) Polar Continental (*cP*), (3) Tropical Maritime (*mT*), and (4) Tropical Continental (*cT*), as seen in Fig. 6-12. The polar air masses, formed at the poles, are always colder than the tropical air masses formed at the equator. Martime air masses are formed over oceans. They usually contain a great deal of moisture. Continental air masses are formed over land and are usually drier than maritime air masses.

When two air masses meet a front is formed. As the air masses move over the earth, two different masses may meet. When a warm air mass pushes into a colder air mass, a *warm*

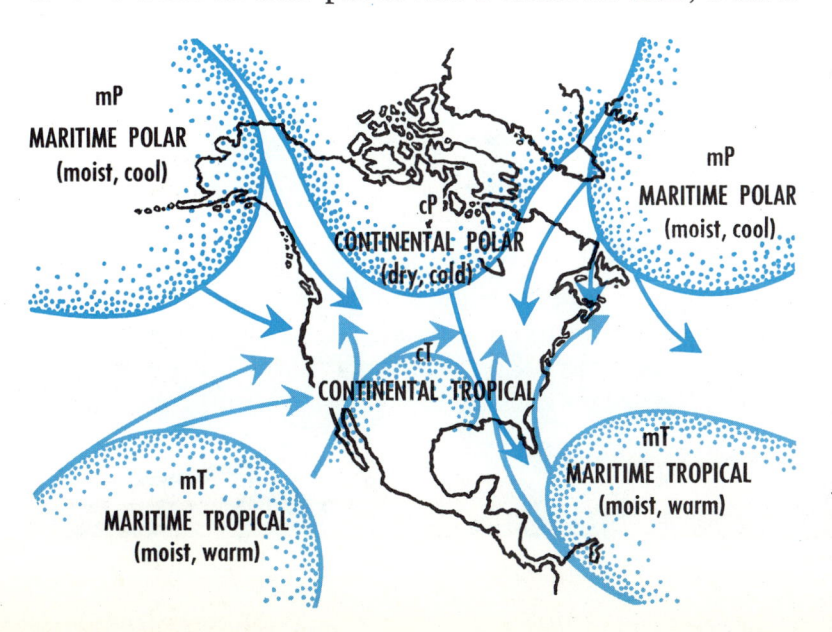

6-12 What weather does each of these air masses usually bring?

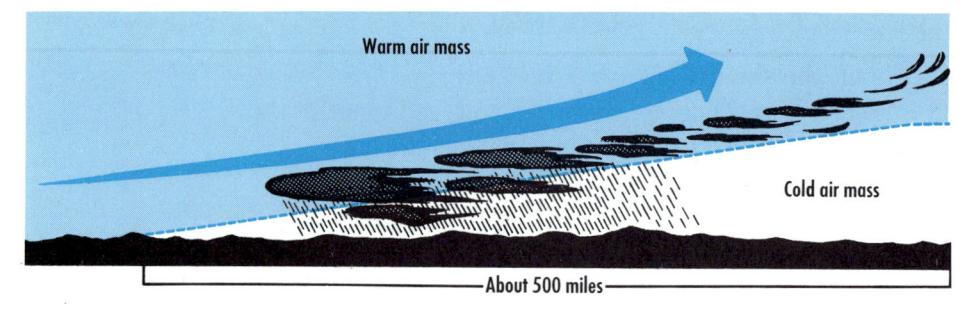

6-13 What kind of weather do we usually find along a warm front?

front occurs. Figure 6-13 shows the type of weather that develops at a warm front. High, thin clouds are the first indication of the approach of a warm front, which moves at speeds of 20 to 30 miles per hour. As the front comes closer, the clouds become lower and darker. Rain or snow usually begins to fall. A warm front may take a period of two or three days to move by.

A *cold front* is formed when a mass of cold air overtakes a mass of warm air. Figure 6-14 shows the type of weather produced by a cold front. Cold fronts usually move through an area more rapidly than warm fronts. Storms develop swiftly, accompanied by high winds and thunderclouds. A cold front is usually followed by weather which is clear and cooler.

6-14 What kind of weather do we usually find along a cold front?

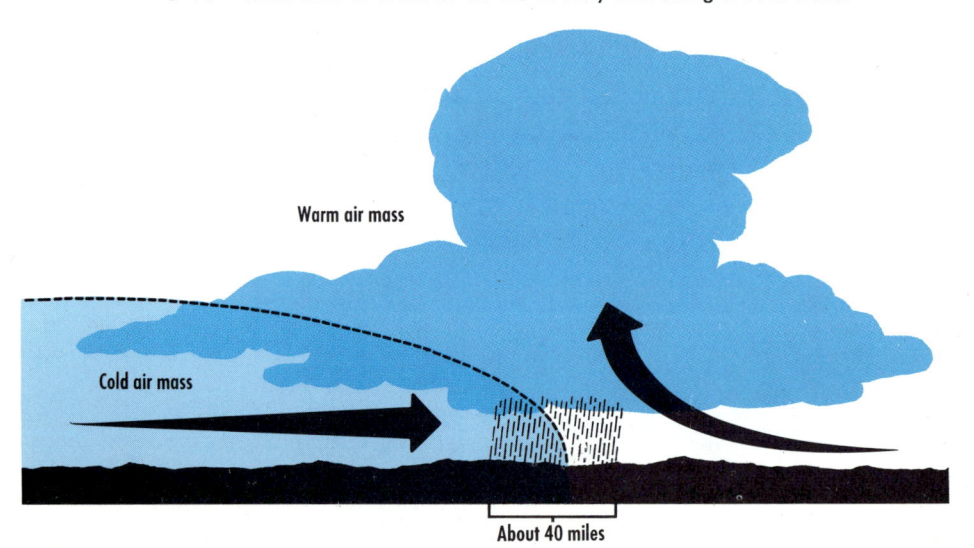

In naming air masses, the terms warm and cold are relative. A cold air mass in summer may be much warmer than a warm air mass in winter. The weather formed along a front depends on the difference in temperature between the two air masses and the amount of moisture present in the air. If the air masses remain over the same area for a few days, a *stationary front* will be the result. In a case such as this, no change in the weather may be expected for two or three days. You can observe the effect of cloud formations and air masses on the weather in the next activity.

OBSERVE

Watch the newspaper weather maps or follow the TV forecasts to find out when a warm front is moving into your area. Observe the cloud formations as the front moves in. What kind of high-altitude clouds form next? If the front produces rain, what kinds of clouds are present then? Make drawings to show the reason for these different cloud formations.

Warm, wet air masses are lighter than cold, dry air masses. If you were to weigh a certain volume of moist air and the same volume of dry air, you would find that the moist air is lighter. We have already learned that cold air is heavier than warm air. Therefore, an air mass that is both warm and moist will rise above a cold, dry air mass when the two meet. The rate of rise depends on the difference between the temperature and the moisture content of the two air masses.

As an air mass forms over a warm, wet area, it becomes less dense than an air mass forming over a cold, dry area. When two such masses meet, the cold air moves under the warm air. As the rising mass of warm, moist air is cooled at high altitudes, clouds and rain may be produced along the front.

Storm areas move across the country from west to east. You will remember that the prevailing winds in the United States are from west to east. The winds move the storm areas in the same direction, usually swinging them from the North Pacific, in a southeasterly direction, to the Mississippi Valley. The storms then move in a northeasterly direction out to sea along the New England states.

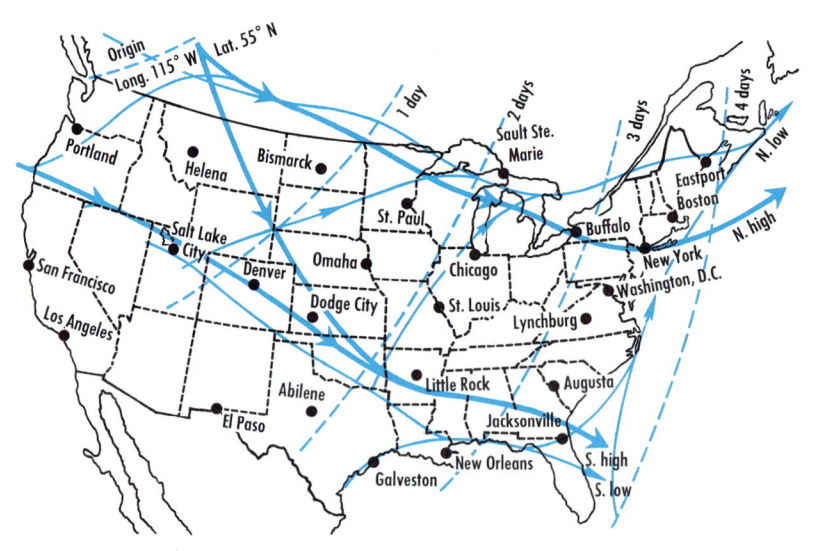

148 / UNIT 2 MAN EXPLORES HIS ENVIRONMENT

6-15 What is the direction of weather movement in the United States?

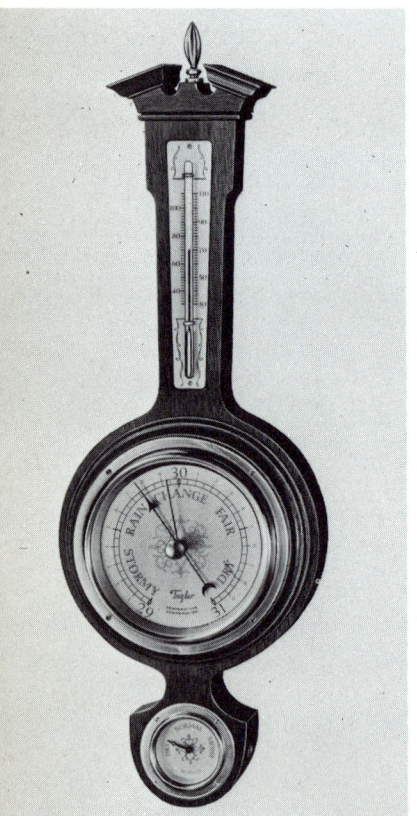

Local conditions such as differences in air pressure, mountains, and unequal heating of the air over deserts or large lakes may change the direction of movement of storm areas. The general directions and speeds of storm areas in this country are shown on the map in Fig. 6-15. Areas of stormy weather and fair weather, each 800 to 1000 miles in diameter, move across the country. Their pattern is fairly regular.

A barometer is used to measure air pressure. The common wall barometer, shown in Fig. 6-16, shows changes in *air pressure* as they occur. As you learned earlier, air pressure may be measured in inches of mercury. However, scientists who study weather and make forecasts, prefer to measure air pressure in international units called *millibars*. One inch of mercury is equal to a pressure of about 34 millibars. Therefore, a pressure of 30 inches is approximately equal to 1020 millibars. The weather map in Fig. 6-17 shows air pressure in millibars. Can you convert these readings into inches of mercury?

In preparing weather maps, weather forecasters receive readings of air pressure from many reporting stations. They plot these readings on the map. Then, they draw lines on it, connecting points that show the same atmospheric pressure.

6-16 What does this wall barometer record?

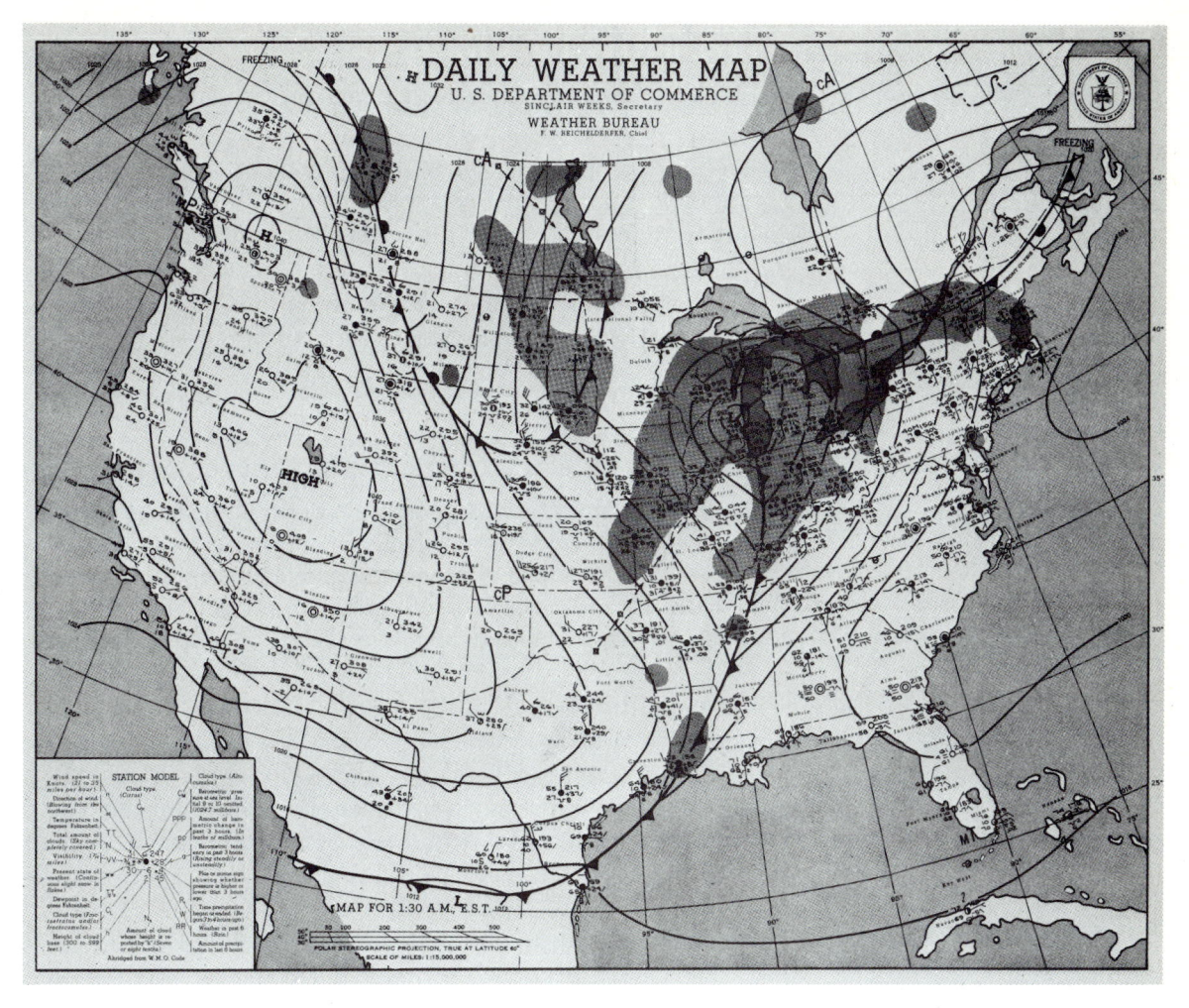

6-17 How is this weather map used to forecast weather?

These connecting lines, called *isobars* (EYE-so-bahrz), enable the meteorologists to identify the great areas of high and low pressure across the country.

Fair weather areas are usually those of high atmospheric pressure called *highs*. Areas of stormy weather are commonly called low pressure areas, or *lows*. The air pressure is high if cold, dry air is moving in, and we usually have fair weather. But if warm, moist air moves in, the pressure will be low. What happens to the moisture in the air? What kind of precipitation is likely to occur? By studying changes in barometer

6-18 What atmospheric conditions caused this tornado?

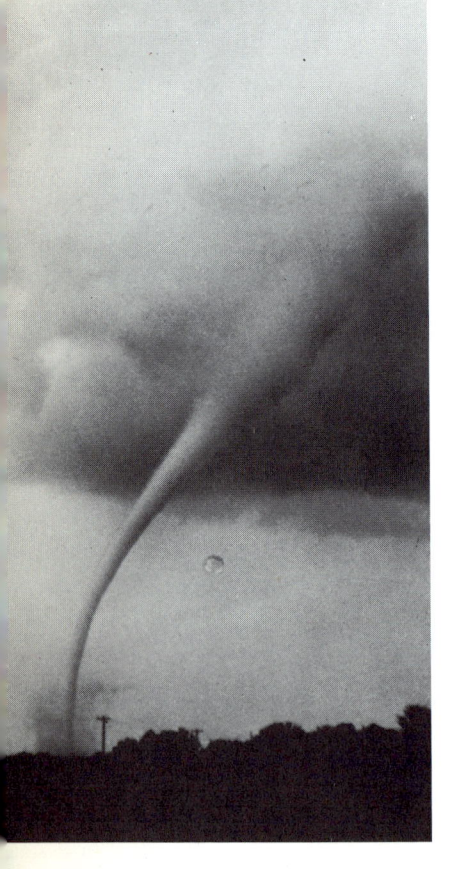

readings, a weather forecaster, generally, can predict the weather.

Since a barometer measures only air pressure, this information alone cannot tell us everything about the weather. There are other factors, like *wind direction*, *temperature*, and *humidity*, that must be considered. In fact, a series of readings have to be taken to find out whether the pressure is rising or falling. The next activity will help to illustrate the discussion for you.

RECORD

Keep a daily record of barometer readings for one month. Note the storms and rainfall in your area during the same period. Make a chart showing the daily barometric readings. Indicate the storms on the chart. Can you see any relation between changes in air pressure and formation of storms? Explain your answer.

A cyclone is a large whirlpool of air. Most people think of a cyclone as a destructive wind. Scientifically, this is not the correct use of the term. *Cyclones* (SYE-klonez) are large movements of air above the earth's surface. Bare areas on the earth, like deserts, and paved streets in cities absorb more heat than growing fields and forests. The air above these heated areas become warmer than the surrounding air and rises. When the warm air rises, it forms a low-pressure spot. Colder air flows in under it from the surrounding regions. The earth's rotation causes this flow of air to move in a spiral motion. This slow spiral of air under low pressure is a *cyclone* or *low*. Cyclones usually form along fronts where warm and cold air masses mix in a whirling motion. North of the equator, the spiral is counterclockwise. Can you explain why?

Sometimes the whirling action along a front becomes more rapid and develops into a *tornado* (tawr-NAY-do). A tornado may whirl with a velocity of 300 to 600 miles per hour. They cause severe damage. Tornadoes occur when air over a small area heats quickly, causing the pressure there to change rapidly. These dangerous storms are accompanied by thunder

CHAPTER 6   THE CHANGING WEATHER / 151

and lightning. The warm, moist air rises so fast that the whirling action forms a funnel-shaped cloud, as shown in Fig. 6-18. This whirling action can be illustrated in the following activity.

OBSERVE

Pour water into a narrow-mouth bottle. Turn the bottle upside down. As the water empties, whirl the bottle rapidly in a circle. Observe the way the water forms a funnel at the mouth of the bottle as it flows out. Compare the movement of air in a tornado with the action of the water in the bottle.

In a tornado funnel, the rapidly rising air causes a drop in pressure. Then, the surrounding air rushes in at great speed. Tornado paths vary in length from a few feet to hundreds of miles. Tornadoes move at speeds of 25 to 50 miles per hour. They can cause great damage in small areas where the funnel touches the surface. In an average year, between 500 and 600 tornadoes are reported in the United States. If a tornado forms at sea, the water is sucked up in the funnel to form what is known as a *waterspout*.

Hurricanes form in the same general way as tornadoes. However they cover a much larger area, often several hundred square miles. They usually originate over the ocean and may last from several days to weeks. If they move in over the land, hurricanes can cause extensive damage (Fig. 6-19). In the United States, for example, hurricanes may strike the Atlantic and Gulf coasts, sometimes wiping out entire towns. You have probably noticed that the weather bureau names the hurricanes each year by giving them feminine names in alphabetical order.

Thunderstorms are local storms usually occuring during the summer. They are caused by warm, moist air rising rapidly and cooling to form cumulus clouds (Fig. 6-20). Violent air movements cause the clouds to become electrically charged, causing flashes of lightning. The roll of thunder, strong winds, and heavy rains lasting for a short period of time, usually occur at the same time as thunderstorms.

6-19 In what regions are hurricanes likely to strike?

6-20 What causes a thunderstom to form?

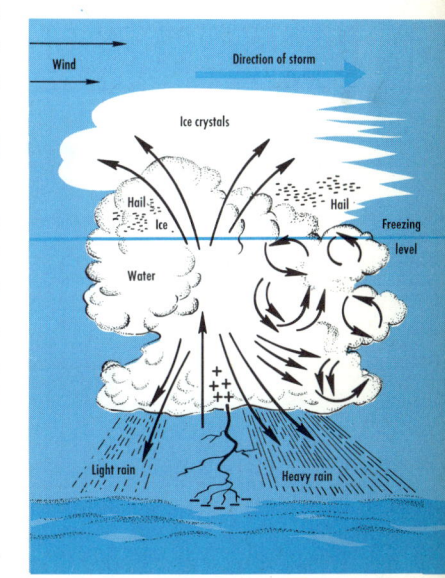

REVIEW

1. What two factors determine the kind of storm in an area?
2. Name the four main types of air masses found over the United States.
3. What two factors cause a warm air mass to have a lesser density than a cold air mass?
4. In what general direction do most storm areas move across the country?
5. What is meant by a: (*a*) cyclone, (*b*) tornado, (*c*) hurricane?

D/How Is Weather Forecast?

Weather stations report atmospheric conditions daily. To prepare the necessary maps, the weatherman must have accurate information from all parts of the country. Four times each day, over 2000 weather-observing stations report by telegraph, teletype, radio, and telephone to the United States Weather Bureau in Washington, D.C. Some stations send reports more frequently when the weather is changing rapidly.

The weather observers are stationed in all parts of this country, Canada, and the West Indies. They send reports of barometer readings, temperatures, wind speeds and directions, cloud cover, and precipitation. This information is charted, and the maps are sent out to all parts of the country. General short-range predictions based on weather reports are correct seven times out of eight. Since atmospheric conditions may change rapidly by the time the weather forecasts reach all parts of the country, this shows a high degree of accuracy. Examine the weather patterns in your community in the following activity.

RECORD

Divide a large sheet of paper into squares in the form of a calender. Keep track of the weather each day by placing symbols in each square to represent the weather conditions. Make up your own symbols for each type of weather condition, or use symbols from a daily newspaper weather map. At the end of the month, describe the weather changes that took place in your locality.

The Federal Aviation Agency also broadcasts weather reports every hour. Airplane pilots must know more about weather changes than the average person. The safety of a flight depends upon accurate knowledge of surface and upper-atmospheric conditions, temperatures at different altitudes, the height of cloud layers, or *ceiling*, and the possibility of storms. For aviation purposes, forecasts are made up to eight hours in advance.

Conditions in the upper atmosphere must be known to forecast weather. Weathermen use several instruments to study conditions in the upper atmosphere. Hydrogen-filled balloons carrying a thermometer, a barometer, and hygrometer, are sent up daily from many weather stations. These balloons are called *radiosondes* (RAY-dee-o-*sahndz*). The recording instruments are connected to a small radio transmitter. This sends back information about temperature, air pressure, and humidity as the radiosonde rises into the air. As the balloon rises, its path is observed with a small telescope called a *theodolite* (thee-AHD-o-lyet). This finds out the wind speed and direction. The radiosonde gives the weatherman a record of atmospheric conditions up to 100,000 feet above the earth's surface. Do you know how the instruments in these weather balloons are returned to the earth?

Weather forecasts are made from weather maps. Weather maps, like the one in Fig 6-17 (page 149), show the location of high- and low-pressure areas, temperatures, fronts, and wind conditions. Storm areas and fair-weather areas are charted from day to day to show their movement across the country. By knowing the speed and direction of movement of these areas, weathermen can predict the general weather several days in advance. Newly developed high-speed electronic instruments, which draw with a mechanical pen, enable forecasters to prepare weather maps in a short time (Fig. 6-21). The following activity with weather maps will illustrate the movement of high- and low-pressure areas.

6-21 Cloud formations from ESSA weather satellite (above) and weather map (below) are recorded.

CHECK

Collect the daily weather maps from newspapers on several successive days. Paste each in a notebook and keep a record of the dates. On a notebook-size map of the United States, show the movement of high- and low-pressure areas across the country. In which direction does most of the weather move? Where do most highs and lows start and how long do they take to move across the country?

Only in the last 20 to 30 years have weathermen been able to predict the weather accurately. They can predict the weather for a five-day period. Again, these forecasts are generally correct, although not as accurate as the daily forecasts. Five-day forecasts are made at the Extended Forecast Section of the United States Weather Bureau in Suitland, Maryland. Because of the thousands of observations needed to make a forecast, a high-speed electronic "brain" does all the routine figuring.

You can learn how forecasts are made by operating your own weather forecasting station in the next activity.

PREDICT

The operation of a model weather forecasting station is an interesting class project. One student should be assigned to read one instrument daily. The station will need a thermometer, barometer, hygrometer, rain gauge, anemometer, and wind vane. Make a chart (similar to the one shown below) to record each reading. From your observations try to predict the weather for the next day each time. **(Do not write in your book).**

Observations	Date	Date	Date	Date	Date
Temperature					
Pressure					
Humidity					
Wind speed					
Wind direction					
Amount of precipitation					
Cloud conditions					
Weather predictions for next day					

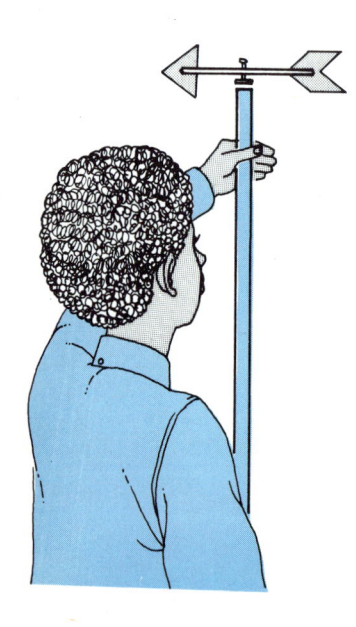

Rocket probes and satellites are useful in forecasting weather. Rocket probes sent hundreds of miles above the earth's surface have sent back valuable information about conditions that affect weather and climate. Photographs taken from high-altitude rockets and satellites are giving weathermen a new look at the earth's cloud patterns. Artificial satellites measure and send back information about solar radiations. These affect daily weather and long-term climate changes on the earth.

Weather-observing satellites, like the *Tiros, Nimbus* and the *ESSA* satellites are in orbit around the earth. These satellites, equipped with television cameras, are able to take pictures of the earth's surface. The pictures are recorded on magnetic tape. These pictures are broadcast later when the satellite passes over ground stations that are equipped to receive the pictures. At the same time, the ground stations now instruct each satellite when to take pictures during its next orbit.

The satellites furnish information for more accurate worldwide weather maps. Satellites do this by observing weather conditions in remote polar and ocean areas where dangerous storms might be forming. Such a satellite system is able to keep track of each important storm around the world. Weather forecasters know where and when future storms make their many appearances (Fig. 6-22).

Spotting storms from outer space saves the United States many millions of dollars yearly. Advance warnings of the approach of a severe storm allow those in its predicted path to take steps to prevent property damage and to save lives. A hurricane forecast, for example, shows the extent and path of the dangerous storm. Tornadoes, when photographed by a weather satellite, might show some special characteristics that could be used to spot them before the destructive funnels reach the earth.

The *wind-barometer table* (page 156), gives some general forecasting guides by uniting barometer readings with general observations of the wind and other weather factors. At the bottom of the table, there are two barometer readings of 29.8°. What accounts for the weather differences even though the barometer readings are similar?

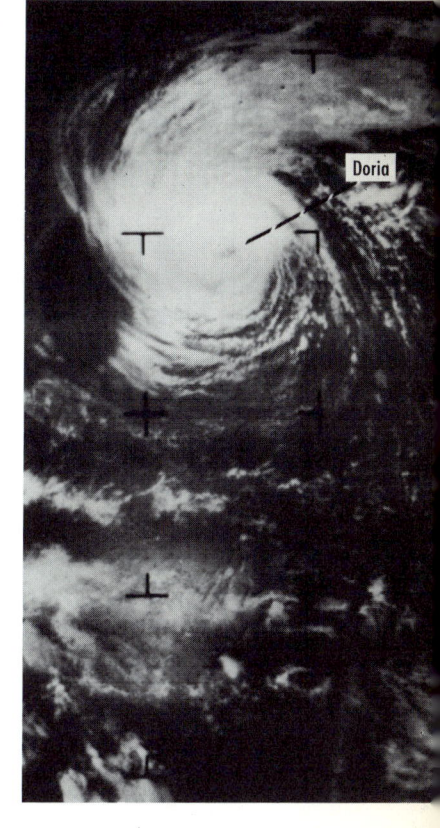

6-22 Hurricane Doria photographed by a Tiros satellite.

WIND-BAROMETER TABLE

Wind direction	Barometer reduced to sea level	Character of weather indicated
SW to NW	30.10 to 30.20 and steady	Fair, with slight temperature changes for 1 to 2 days.
SW to NW	30.10 to 30.20 and rising rapidly	Fair, followed within 2 days by rain.
SW to NW	30.20 and above and stationary	Continued fair, with no decided temperature change.
SW to NW	30.20 and above and falling slowly	Slowly rising temperature and fair for 2 days.
S to SE	30.10 to 30.20 and falling slowly	Rain within 24 hours.
S to SE	30.10 to 30.20 and falling rapidly	Wind increasing in force, with rain within 12 to 24 hours.
SE to NE	30.10 to 30.20 and falling slowly	Rain in 12 to 18 hours.
SE to NE	30.10 to 30.20 and falling rapidly	Increasing wind, and rain within 12 hours.
E to NE	30.10 and above and falling slowly	In summer, with light winds, rain may not fall for several days. In winter, rain within 24 hours.
SE to NE	30.00 or below and falling slowly	Rain will continue 1 to 2 days.
SE to NE	30.00 or below and falling rapidly	Rain, with high wind, followed within 36 hours by clearing, and in winter by colder temperatures.
S to SW	30.00 or below and rising slowly	Clearing within a few hours, and fair for several days.
S to E	29.80 or below and falling rapidly	Severe storm imminent, followed within 24 hours by clearing and in winter by colder temperatures.
E to N	29.80 or below and falling rapidly	Severe northeast gale and heavy precipitation; in winter, heavy snow, followed by cold wave.
Going to W	29.80 or below and rising rapidly	Clearing and colder.

CHAPTER 6 THE CHANGING WEATHER / 157

REVIEW

1. About how many weather-observing stations report daily to the Weather Bureau?
2. Name the five main atmospheric conditions reported.
3. What device is used to obtain a record of conditions in the upper atmosphere?
4. What weather information is recorded on weather maps?
5. Name three types of weather-observing satellites that are used in preparing weather forecasts in the United States.

E/What Determines the Climate?

Climate is an average of weather conditions over a long period of time. People often confuse the meaning of weather and climate. We have learned that certain atmospheric conditions, such as temperature, wind, and humidity affect the weather. These conditions also affect the *climate* in a particular region. In describing the climate, however, we consider the average rainfall and temperature. This means that over the years, the weather conditions of a region follow a certain pattern. This information is obtained by keeping daily weather records over a long period.

The average temperature helps determine the climate. As we have learned, the sun's rays strike the earth less directly as we move north or south from the equator. Hence, the amount of heat energy absorbed by the earth is less because the average temperature decreases with increase in distance. Thus, the distance north or south of the equator affects the average temperature of a region.

The average temperature of most areas also varies with the seasons. The temperature is usually higher in the summer because the hours of daylight are longer. Also, the tilt of the earth on its axis is such that the sun's rays strike the earth more perpendicularly during the summer. This means that the earth can absorb more heat energy and, in turn, further warm the air.

158 / UNIT 2   MAN EXPLORES HIS ENVIRONMENT

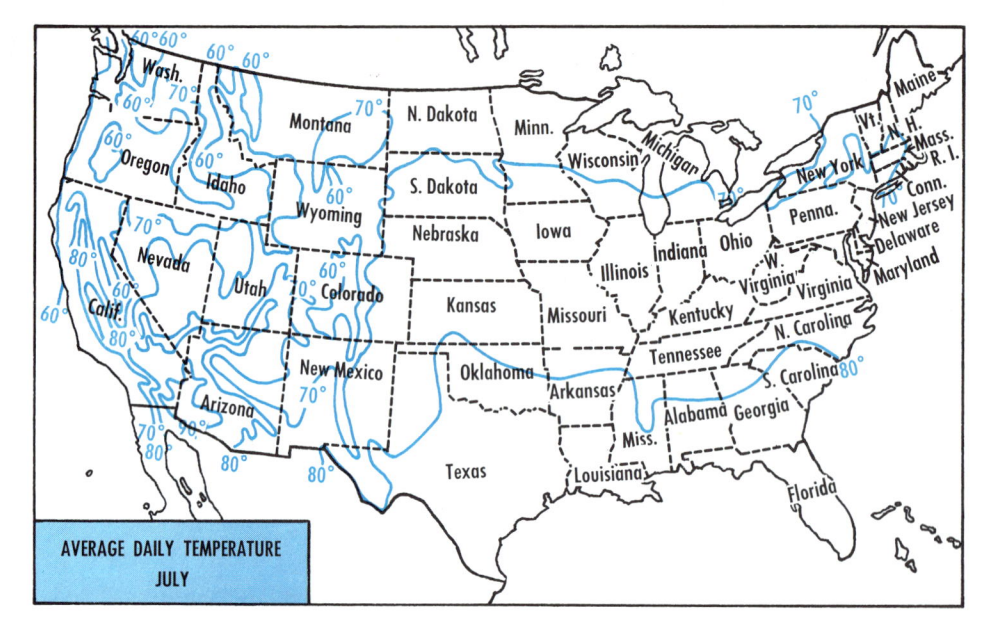

AVERAGE DAILY TEMPERATURE JULY

6-23   What do we call lines connecting points of equal temperature?

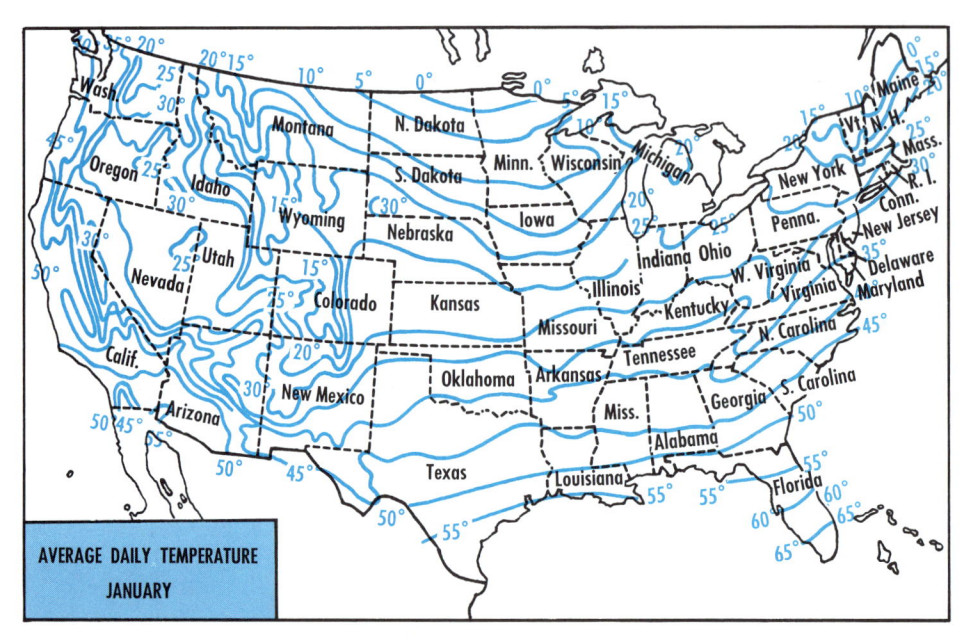

AVERAGE DAILY TEMPERATURE JANUARY

6-24   What change in temperature occurs as you travel southward?

Figure 6-23 shows the average temperatures in the United States for the month of July. Fig. 6-24 shows the average temperatures for the month of January. The lines drawn across each map connecting the places that have the same temperatures are known as *isotherms* (EYE-so-thuhrmz). You will note that the isotherms across the central and eastern parts of the United States show that the average temperature increases as you move south toward the equator. Can you explain why the isotherms over the western part of the country have a different shape?

The average direction of winds helps determine the climate. We know that wind direction seems to change from day to day. But when records are kept for a long period of time, we find that winds usually move in a definite pattern over the earth. If you look back at Fig. 6-1 (page 135), you will see the principal wind directions of the world. Since most of the United States lies in a region of westerly winds, the average wind direction is from west to east.

Since air masses move in the same general direction as the prevailing winds, the climate in various parts of the country is affected. For example, maritime air masses move in from the Pacific Ocean and the Gulf of Mexico. Since these air masses are usually warm and moist, the climate in the regions along the northwest coast and gulf coast is mild and wet. Why, then, is the climate in the central and northeastern parts of our country different? There are many other factors that also affect the climate.

Mountain ranges help determine the climate. As we have said, the air masses moving in from the west are warm and moist. As you saw in Fig. 6-5, these masses are forced to rise as they move across the mountains in the western United States. Warm air that rises to high altitudes is cooled. The moisture in the air condenses and drops out as rain or snow. Thus, warm air masses are "milked dry" when they cross over mountainous areas.

After the air mass has crossed the mountains, it moves down and becomes warmer. Because the air is now warmer, it can hold more moisture and absorb water from all available sources. Instead of bringing moisture to the east side of the

160 / UNIT 2   MAN EXPLORES HIS ENVIRONMENT

mountains, an air mass that has passed over the mountains actually absorbs water. Only when strong winds carry considerable amounts of water vapor over the top of the mountain can the eastern side of the mountain expect to have some rain or snow.

By looking at the annual rainfall map of the continental United States (Fig. 6-25), you will see that the areas west of the Sierra Nevada Mountains and the Rocky Mountains are moist and fertile. The eastern slopes of these mountains and plains are dry. Some regions east of the mountains in Washington, Oregon, Utah, and Nevada receive so little rainfall that they are deserts. In general, the climate on the western side of the mountains in our country is much wetter than it is on the eastern side. Note also the abundance of rainfall that occurs in the southeastern portion of the continental United States. In the world, the heaviest rainfall occurs on mountain slopes which lift moist, warm air from oceans that are near the equator.

6-25   The map shows annual rainfall in the continental United States. Why is there heavy rainfall west of the Rocky Mountains?

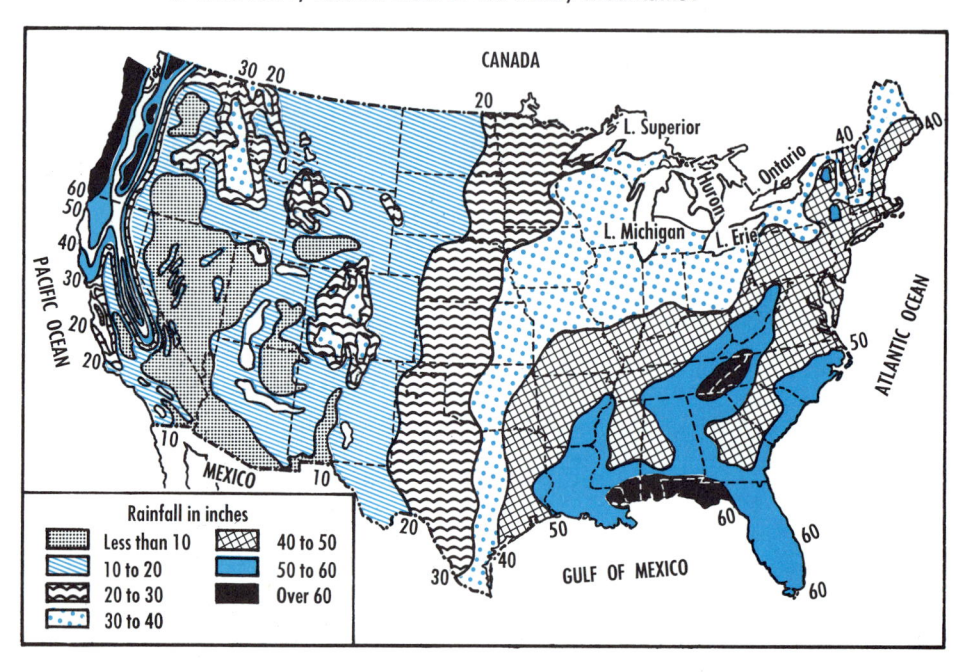

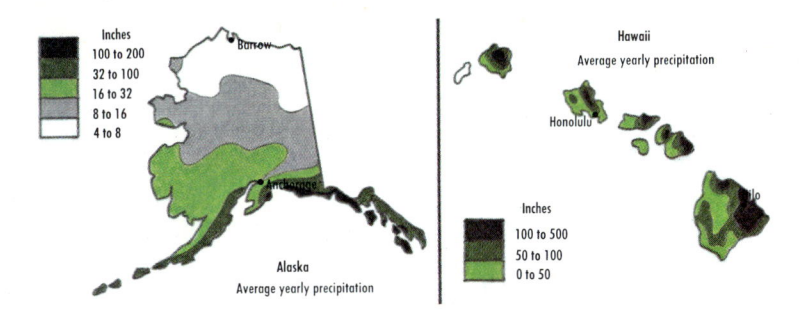

6-26 (right) Contrast the amounts of precipitation among the islands of Hawaii.

6-27 (left) What factors account for precipitation in Alaska?

Study Figs. 6-26 and 6-27 showing the average yearly precipitation of our newest states, Alaska and Hawaii. It is interesting to contrast Alaska, the 49th state and Hawaii, the 50th state. Hawaii has mountainous areas, as you probably know. How does this affect rainfall in those areas?

Surrounding water areas help determine the climate. As you have already learned, water does not absorb or lose heat

6-28 What are the factors mainly responsible for the climate that is found in your state?

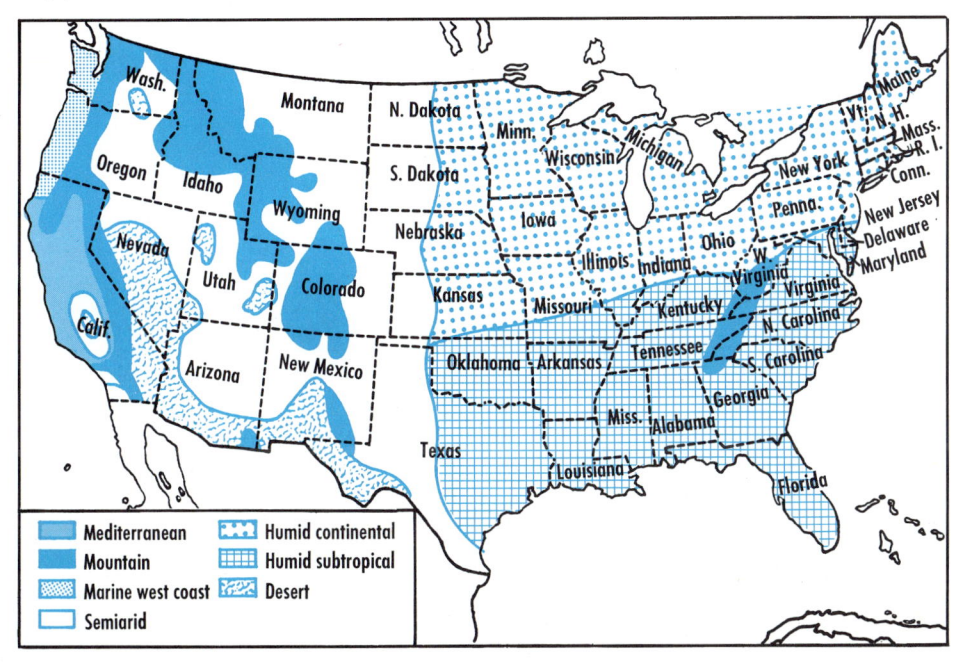

as rapidly as land. This affects the temperature of the region near large bodies of water. There are fewer sharp changes in temperature. In winter, the average temperature is warmer and in summer, the average temperature is cooler than that of inland areas. Thus, climates of areas near large bodies of water are usually milder than those of areas far from the oceans (Fig. 6-28). For example, Florida is almost surrounded by water and parts of the interior are swampy. The average temperature for January, the coldest month, is 60°F. For July, the warmest month, it is 80°F. The water helps keep Florida warmer in winter and cooler in summer.

Study the tables, which show monthly weather in two principal cities of the state of Alaska, and two principal cities in the state of Hawaii. Note the sharp temperature differences between the two states.

MONTHLY WEATHER IN ANCHORAGE AND BARROW (State of Alaska)

	JAN	FEB	MAR	APR	MAY	JUNE	JULY	AUG	SEPT	OCT	NOV	DEC	Average of:
ANCHORAGE	20	27	34	44	55	63	65	64	56	43	29	20	High Temperatures
	6	10	16	27	36	45	49	47	40	29	16	7	Low Temperatures
	7	6	7	4	5	7	11	15	15	11	8	7	Days of Precipitation
BARROW	−9	−11	−8	7	24	39	46	44	34	22	7	−4	High Temperatures
	−22	−24	−22	−8	13	29	33	33	27	12	−5	−17	Low Temperatures
	4	4	3	3	3	4	6	10	9	9	6	4	Days of Precipitation

Temperature in °F

MONTHLY WEATHER IN HONOLULU AND HILO (State of Hawaii)

	JAN	FEB	MAR	APR	MAY	JUNE	JULY	AUG	SEPT	OCT	NOV	DEC	Average of:
HONOLULU	77	77	77	78	80	81	82	83	83	82	80	78	High Temperatures
	67	67	68	69	71	72	74	74	74	73	71	69	Low Temperatures
	12	12	13	12	11	11	13	13	12	13	13	14	Days of Precipitation
HILO	78	79	79	79	81	83	83	83	83	82	80	79	High Temperatures
	62	62	63	64	65	66	67	68	68	67	66	64	Low Temperatures
	20	19	24	25	25	24	27	27	23	24	23	24	Days of Precipitation

Temperature in °F

Study the changes in climate along the parallel on which you live in the next activity.

EXPLAIN

On a large map of the United States, find the parallel (number of degrees north of the equator) on which your town is located. On a notebook-size map of the United States, draw in this parallel. Show such natural features as mountains, large lakes, rivers, and deserts. Describe typical climatic conditions along this parallel from west to east. Explain the reason for each of the different climates you listed.

Climates have changed over a long period of time. Climates change slowly over thousands of years, but how do scientists know this? Obviously, primitive man had no means to record daily temperatures, wind speeds, and precipitation. Scientists, however, have discovered many clues to the climate of the world in past ages.

For example, *fossils* found in the earth are remains of ancient plants and animals (Fig. 6-29). Workmen digging in the Midwest often find fossils of coral. Coral is known to grow only in warm oceans. Scientists conclude that this part of the country must have been covered by a warm ocean millions of years ago.

Other scientists have found fossilized remains of giant dinosaurs in Utah, Colorado, and Wyoming. These animals lived only in hot, swampy regions, much different from the climate in the western United States at present. The climate during the time these animals lived must have been very different from the cooler, dry climate found today.

Interestingly enough, microscopic pollen grains found in rocks give scientists a clue to the climate of past ages. *Pollen* is the colored, powdery substance produced by flowers. Each flower produces pollen grains of a specific shape. Pollen grains have hard covers, so they were preserved as fossils for millions of years. By identifying the plants from which the pollen came, scientists can tell what kind of climate each plant needed in order to thrive.

6-29 How have fossils helped us understand changes in climate over long periods of time?

For example, pollen buried in now dry New Mexico lake beds has been used to study the changes in climate during the Ice Ages. These Ice Ages covered North America with sheets of ice and snow, many thousands of years ago.

The climate of the world seems to change in cycles. Some scientists believe that over a period of thousands of years, the climate became colder. This resulted in an Ice Age. Then, the climate gradually became warmer again, producing more temperate conditions. For example, the interior of Greenland is almost completely covered with ice today. Yet early stories of the Norse people describe Greenland as a land of green, growing plants.

Scientists are studying the ice fields and glaciers in Greenland. An icecap, estimated to be more than 7000 feet thick in some places, may play an important part in influencing our weather. Scientists want to know if the icecap will continue to melt, as it has been doing in recent times, or whether it may now be growing in size. Further melting might be a sign of a milder climate in our part of the world. Growth in the size of the icecap would be an indication that the climate is gradually turning colder.

REVIEW

1. What three average weather conditions help determine the climate?
2. What factors influence the temperature in a particular region?
3. What is the general wind direction in the United States?
4. What effect does a large body of water have on the climate of land areas near it?
5. What clues are used by scientists to determine past climates on the earth?

THINKING WITH SCIENCE

A. *On a separate sheet of paper, write the numbers 1 to 15. Some of the following statements are true and some are false. Rewrite the statements, changing the terms in italics if necessary, to make them all true.* **Do not write in your book.**

1. Warm air masses formed over the equator are called *polar* air masses.

2. An instrument used to measure changes in air pressure is called an *anemometer*.

3. A whirling storm which is seen as a funnel-shaped cloud is a *tornado*.

4. Clouds made up of ice crystals high above the earth are known as *stratus* clouds.

5. Air that moves in from sea to land is a *land breeze*.

6. When a mass of warm air pushes into a mass of colder air, a *cold front* is formed.

7. An instrument used to measure the velocity of the wind is called a *barometer*.

8. Clouds that appear as huge dome-shaped masses piled up to a great height are known as *stratus* clouds.

9. The average of weather conditions over a long period of time produces the *climate*.

10. When a mass of cold air overtakes a mass of warmer air, a *warm front* is formed.

11. Uneven heating of the earth's surface produces *wind*.

12. When rain freezes as it falls *sleet* is formed.

13. Air pressure is measured in units called *isotherms.*

14. Areas of a weather map having the same air pressure are connected by lines called *millibars.*

15. Rain, snow, and hail are all forms of *precipitation.*

   B. *Write the answers to the following in your notebook. Be sure to use complete sentences and correct grammar and spelling.*

1. Explain why the winds in our part of the world are usually westerly winds.

2. How does the tilt of the earth on its axis affect the amount of heat received from the sun?

3. How does the unequal heating of land and water areas produce wind?

4. Why does precipitation occur in different forms such as rain, snow, or hail?

5. What kind of weather is usually produced at a warm front? at a cold front?

6. Why do storm areas usually move from west to east across the country?

7. In what ways have weather satellites extended man's knowledge of the weather?

8. What long-range factors affect the kind of climate found in a particular part of the world?

9. Why do the western sides of mountain ranges in the United States usually have more rain than the eastern sides?

10. What evidence has been found to show that climates change over long periods of time?

RESEARCH IN SCIENCE

1. Set up an experiment to show how fog can be formed in a large jar.

2. Look at a weather map for any given day. Describe the weather conditions that would be encountered in flying an airplane from coast to coast.

3. Analyze carefully some weather superstitions for scientific accuracy.

4. Experiment to see if you can produce artificial precipitation in a home freezer.

5. Make a report on the conditions which produce smog and on some of the methods of control.

READINGS IN SCIENCE

Bixby, William, *Skywatchers: The U.S. Weather Bureau in Action.* David McKay Co., 1962. The work of the U.S. Weather Bureau in forecasting and studying weather for the community, farmers, business, and the government, plus a glimpse of some of the basics of meteorology, the history of this science, and its future.

Blumenstock, David, *The Ocean of Air.* Rutgers University Press, 1961. One of the most comprehensive books on the topics in this unit.

Cantzlaar, George L., *Your Guide to the Weather.* Barnes & Noble, 1964. This book introduces the science of meteorology and discusses in detail modern methods of observing and forecasting the weather.

Dwiggins, Don, *Space and Weather.* Golden Gate Junior Books, 1968. Illustrates how satellite photographs have brought new knowledge about the atmosphere and the oceans. Shows potential uses of weather satellites in predicting world-wide weather.

Kovalik, Vladimir and Nada, *The Ocean World.* Holiday House, 1966. Chapters on what an oceanographer does and how to prepare for careers in oceanography are included. Imaginative writing deals with the sea floor, seawater, currents and life in the sea.

Longstreth, T. Morris, *Understanding the Weather.* Collier, 1962. Tells what signs to look for and what they mean so that one can observe and interpret weather signals and understand how they produce weather.

Middleton, W. E. Knowles, *The History of the Barometer.* The Johns Hopkins Press, 1964. The development of the modern barometer and related instruments is traced in this book.

Pilkington, Roger, *The Ways of Air.* Criterion Books, 1962. This is a comprehensive survey of the atmosphere, its composition, and its relation to heat, energy, rain, wind, clouds, and sea currents.

Rosenfeld, Sam, *Science Experiments with Water.* Harvey House, 1965. Experiments on water pressure, water and light, gases and water, the nature of solutions, among others, introduce basic laws in chemistry and physics for the young scientist. Materials are simple and the methods are clearly stated and illustrated.

Spar, Jerome, *The Way of the Weather,* Creative Educational Society, 1962. A comprehensive, well-illustrated study of weather and the science of meteorology with many black-and-white photographs and drawings.

Stambler, Irwin, *Weather Instruments: How They Work.* G. P. Putnam Sons, 1968. Descriptions and explanations of weather-observing instruments and their use are clearly given.

Weather Almanacs are a valuable source of weather information and give details about annual rainfall, temperatures, etc. Many different publications are available and the following are two sources: News Syndicate, 220 East 42nd St., New York City; New Information Bureau, 767 Second Avenue, New York City.

Woodbury, David O., *Fresh Water from Salty Seas.* Dodd, Mead and Co., 1967. The author clearly explains the technology of preparing fresh water from salt water. The book is a factual account of the history of man's attempt to secure pure water for drinking and industrial needs from salt water.

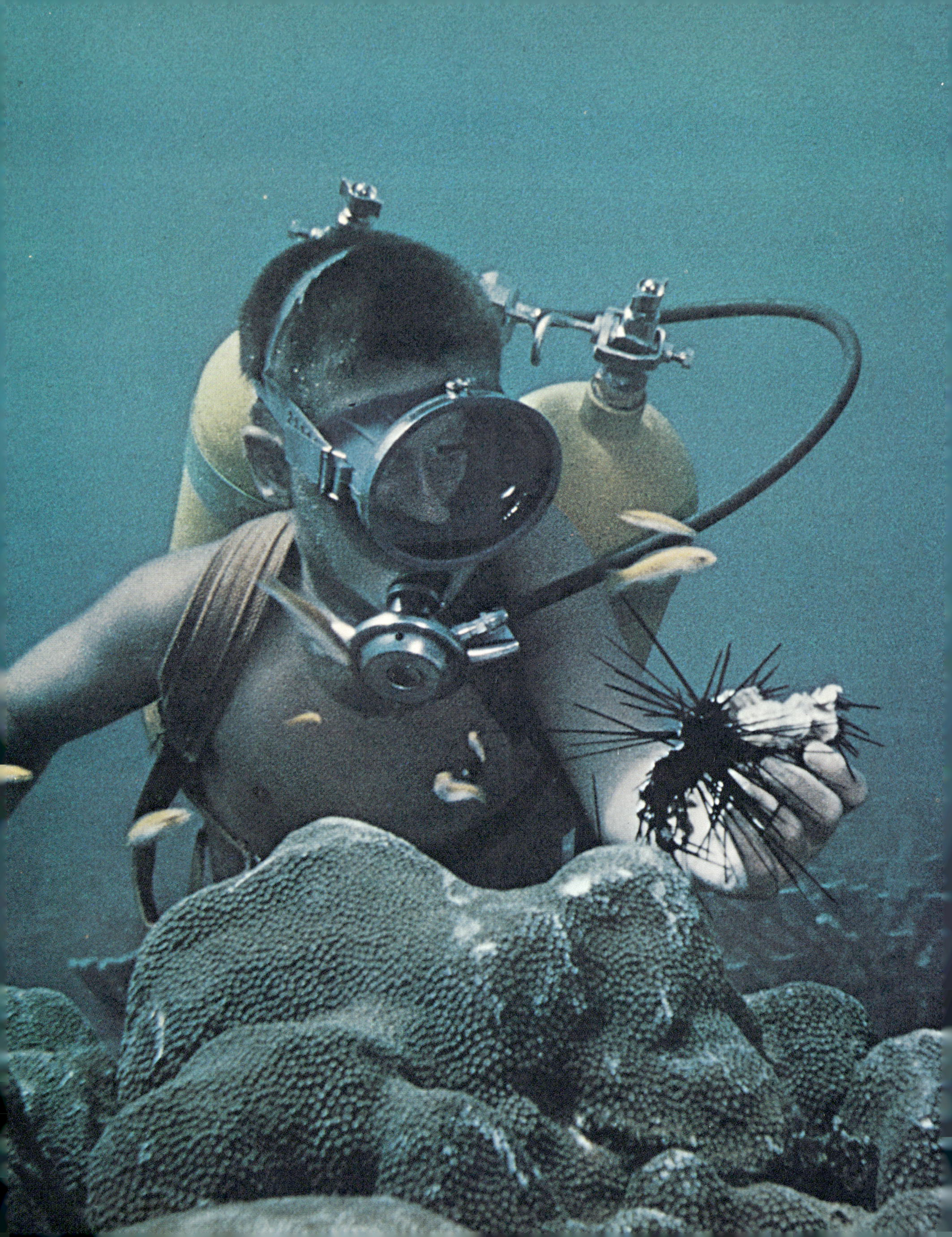

unit 3

Man Explores Living Organisms

Life is all around you in the form of plants, animals, and people. How did all these living creatures, with their great variety, first come into existance? In religions which are based upon the Holy Scripture, people believe that living creatures came into existance because God created them.

The early scientists had some theories about how life began. Frogs were thought to develop from mud, and flies from decaying meat. You may think that these are peculiar theories for intelligent men to believe. However, these theories were thought to be logical at that time. In 1668, *Francesco Redi (1626–1697)*, an Italian scientist, conducted experiments indicating that living organisms cannot be produced spontaneously. All scientists were not convinced. They continued to believe in the *theory of spontaneous generation of life,* in which living organisms were thought to develop from nonliving substances. Supporting Redi, however, was *Louis Pasteur (1822–1895),* the noted French scientist. He showed by brilliant experimentation that the theory of spontaneous generation was false.

The work of these scientists, however, did not answer the question of how life began. One modern theory developed by *Harold Urey (1893–   ),* an American scientist, states that the first living substance arose in the seas about two billion years ago. This theory has not yet been proved completely. Scientists do not believe that *protoplasm,* the living substance in cells, was formed from nonliving materials in a few simple steps. The theory proposes that the first living cell took many millions of years to come into being.

Chemical analysis shows that protoplasm is composed mainly of certain chemical *elements:* carbon, hydrogen, oxygen, nitrogen, sulfur, and phosphorus. Scientists have learned that these elements are linked in long chains with the carbon *atoms* attached to each other to form large *molecules.* Molecules of this type are found only in living matter. According to this theory, the first molecules of this kind were formed by the chemical combination of four gases: *methane* (carbon and hydrogen), *ammonia* (nitrogen and hydrogen), *hydrogen,* and *water vapor* (hydrogen and oxygen). It is believed that these four gases composed the ancient atmosphere of the earth.

Is there any scientific proof for this theory? *Stanley L. Miller,* a student working with Harold Urey at the University of Chicago, experimented with a mixture of methane, ammonia, hydrogen, and water vapor. He set up two flasks, one above the other, with connecting tubes. As the water was heated in the lower flask, steam rose into the upper flask and mixed with the other gases. Droplets of water containing the dissolved gases trickled back to the lower flask through a chamber. In the chamber, an electric spark jumped back and forth.

The water was heated continually so that the steam kept rising and trickling back over and over again. After the first day, the water in the lower flask turned pink and by the end of the week it was red. A chemical test showed that the water now contained various chemical substances made up of carbon, hydrogen, oxygen, and nitrogen. A few simple chains of molecules were also beginning to form. This experiment seems to offer proof that this theory may be partly true.

chapter 7
The Nature of Life

A/What Is a Living Organism?

An organism is a complete living thing. All *organisms* carry on basic life activities. Such activities as food-getting, digestion, respiration, removal of waste products, and reproduction are necessary for life. These activities are called the *life processes*. Both a simple organism, made up of a single cell, and man, made up of many billions of cells, carry on the same general life processes.

The process of taking food into the body is called *ingestion* (ihn-JES-chuhn). There are a number of methods by which organisms take in food (Fig. 7-1). Let us see how a simple animal, like a *hydra,* ingests food by doing the next activity.

7-1 How does a hydra capture its food?

COMPARE

Place a drop of water containing some hydras and water fleas (Daphnia) on a glass slide. (These can be found in stagnant pond water or purchased from biological supply houses.) Carefully place a cover glass on the drop of water. Use a hand magnifier or microprojector to see how the hydras capture the water fleas. Compare your observations with Fig. 7-1.

Food taken in by an organism is changed to usable soluble forms by the process of *digestion* before it can be absorbed by the organism. Almost all organisms produce digestive substances which help to change insoluble food materials into soluble forms.

After digestion, most substances are either absorbed and changed into living material or combined with oxygen to release energy. You will recall from your previous study that when oxygen combines with other substances, energy is released. The process in which an organism uses oxygen to release energy is called *respiration* (*res*-pih-RAY-shuhn). Do not confuse breathing and respiration. Respiration, which takes place in both plants and animals, includes both the process of supplying oxygen to the cells in the body, and the process of removing waste products from the cells. Breathing is only one part of respiration and is found only in more complex animal life. Breathing is the mechanical movement of air into and out of the lungs.

The waste materials produced by organisms are removed by the process of *excretion* (eks-KREE-shuhn). These waste materials may be the result of materials that were ingested but not changed into soluble substances, or they may be produced in the digestion and use of the food in the body. *Carbon dioxide* and *water* are the main waste materials resulting from starch and sugar digestion. *Nitrogen wastes* are the main products of protein digestion.

We can summarize some of the important life processes of living organisms in a table. Can you list more?

LIFE PROCESSES IN ORGANISMS

Process	Activity
Ingestion	The taking of food into the body.
Digestion	The changing of food into soluble forms.
Respiration	The use of oxygen in the body.
Excretion	The removing of wastes from the body.
Reproduction	The production of new organisms.
Absorption	The passing of essential materials into the cell

CHAPTER 7 THE NATURE OF LIFE / 175

All living organisms are made up of cells. The smallest basic unit of life is the *cell*. Most individual cells are not visible as separate units without using a microscope or microprojector (Fig. 7-2). *Robert Hooke (1635–1703)*, an English scientist, observed cells under magnification and recorded his observations. He examined cork, wood, and other plant specimens. *Anton van Leeuwenhoek (1632–1723)*, a Dutch biologist, explored the organisms in ponds and streams, using a simple microscope he invented.

It was not until 1838 that a clue to the true nature of the living cell was found. Scientists learned that both plant and animal cells are composed of *protoplasm* (PRO-to-*plaz*-uhm), the living material of the cell. After careful studies, scientists developed the *cell theory*:

(1) All living organisms are composed of cells.
(2) All cells come from previous living cells by cell reproduction.
(3) The cell is the unit of structure and function in all living organisms.

The microscope is an important scientific instrument. As you continue your study of science, you will find that the microscope and microprojector become very important tools (Fig. 7-3). From the crude magnifiers used by early scientists, modern microscopes with great magnifying power were de-

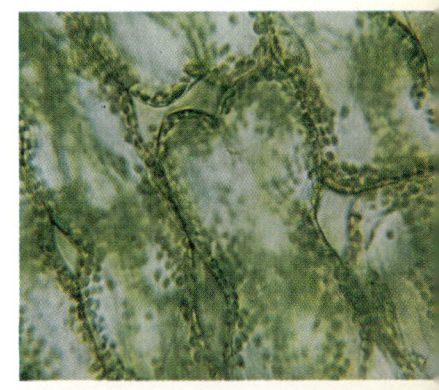

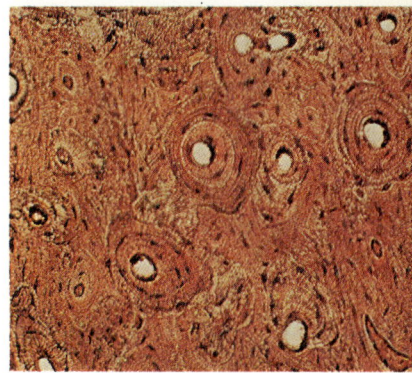

7-2 Compare the differences between plant cells (top) and animal cells (bottom).

7-3 What are the advantages of a microprojector to the students of a science class?

veloped. The most commonly used microscope in the science laboratory is the compound microscope. The magnification you obtain will depend on the power of the *objective* (ahb-JEK-tihv) lens and that of the *eyepiece* lens. By multiplying these two magnifications together, you can find the magnification of the field you are observing.

In order to use the microscope correctly, you should first learn its main parts and how to care for it. Study Fig. 7-4 carefully and compare the drawing with an actual microscope. To prevent damage to the microscope, keep the microscope covered against dust when not in use; use only lens paper to clean the lenses and mirror; do not remove the objectives or the eyepiece from the microscope; turn the adjustment knobs only as far as they go without forcing them; and always carry the microscope with one hand on the arm and the other on the base.

When you are familiar with the parts of the microscope and know how to take care of it, you are ready to learn the proper use of the microscope. Do the following activity which will teach you how to use a compound microscope correctly.

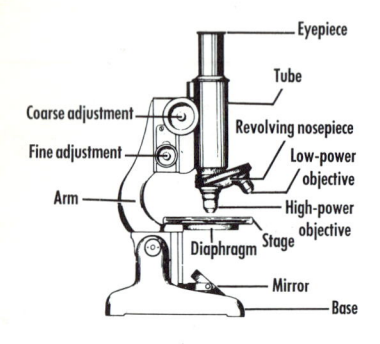

7-4 How is the magnification found in a compound microscope?

OBSERVE

Place the microscope on the table with the arm toward you and the back of the base about an inch from the edge of the table. Adjust your position so that you can look into the eyepiece comfortably. However, if you are going to use a slide with water on it, do not tilt the microscope. Wipe the top lens of the eyepiece, the lens of each objective, and the mirror with lens paper. Then turn the low-power objective so that it lines up with the body tube. Open the diaphragm to the largest opening so that the greatest amount of light is admitted. Place your eye to the eyepiece and turn the mirror toward a source of light, but **never directly toward the sun**. Adjust the mirror until a uniform circle of light without any shadow appears. This is the *field*. The microscope is now ready to be used.

Using a clean cloth, wipe a blank slide and cover glass clean and dry, and place a small drop of water in the center of the slide. (The water is a mounting medium and is necessary to obtain a clear image of the material.) Cut a single letter, *a, b,* or *c* from the smallest newsprint you can find and lay it right side up in the drop of water. Place the cover glass over it. Place the mounted slide on the stage, clip it into place, and move it with your thumb and forefinger until the letter is in the center.

Turn the low-power objective down as far as it will go, being careful not to touch the slide. Do not force the tube when it reaches the automatic stop. (Most microscopes are equipped with such a stop to prevent the low-power objective from hitting the slide.)

Place your eye to the eyepiece. As you watch the field, turn the coarse adjustment *slowly toward you*, raising the body tube. Watch for the material to appear in the field. When in proper focus, the low-power objective should be about half an inch above the slide. *The microscope should always be focused by raising the body tube.*

Bring the letter into sharp focus with the fine adjustment. As you examine the letter, turn the fine adjustment slowly back and forth, not more than half a turn. This will shift the focus to bring out details at different levels of an object. Compare the position of the letter as it is mounted on the slide with its appearance under the microscope. Has the microscope changed the position of the letter?

Now, move the letter to the exact center of the field. Check for sharp focus, and turn the high-power objective into position. The object should now be visible under high power. Correct the sharpness of focus with the fine adjustment. Is as much of the letter visible under high power as under low power?

Now that you understand how to use a microscope correctly, you should look at some real cells.

EXPLORE

Cut an onion in half and peel off the outer covering. Between the layers of the onion, you will find a transparent skin thinner than the finest tissue paper. Cut a small piece of this onion skin and spread it carefully on a clean microscope slide. Examine the unstained onion skin. What do you see? Put a drop of iodine solution on it and cover with a cover glass. Examine the onion skin first under *low power* and then under *high power* of the microscope. Compare with the photograph of onion cells.

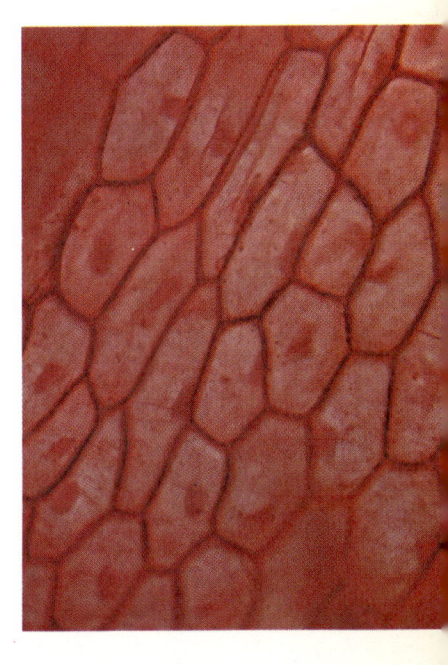

Why was the iodine solution added to the onion skin? Describe the appearance of the cells. Make a drawing of the cells that are seen under low power and under high power of the microscope.

The cell is a storehouse of highly-organized substances. All the material of the living cell taken together is called protoplasm. Although scientists have studied cells for many years, they still do not know everything that causes protoplasm to behave as it does. They have learned that it is a complex organization of substances which can be produced only by other living cells.

All cells have the same general parts. Figure 7-5 shows the general structure of a plant and animal cell. Compare their similarities and differences. The *cell membrane*, a living part of the cell, encloses or surrounds all of the cell material. It is thin and delicate and regulates the amount of liquids and gases which enter and leave the cell.

In the plant cell, shown in Fig. 7-5, notice that in addition to the cell membrane, there is a cell wall, which is quite thick by comparison. The largest and most distinct object inside the cell is the *nucleus* (NYOO-klee-uhs). The nucleus controls and regulates most of the cell's activity.

Cytoplasm (SYE-to-*plaz*-uhm) is the protoplasm lying outside of the nucleus of the cell. Within the cytoplasm, there are several other structures. *Vacuoles* (VAK-yoo-*olz*) are found in most older plant cells and in animal fat cells. Most vacuoles store dissolved food substances in the form of cell sap. *Plastids*

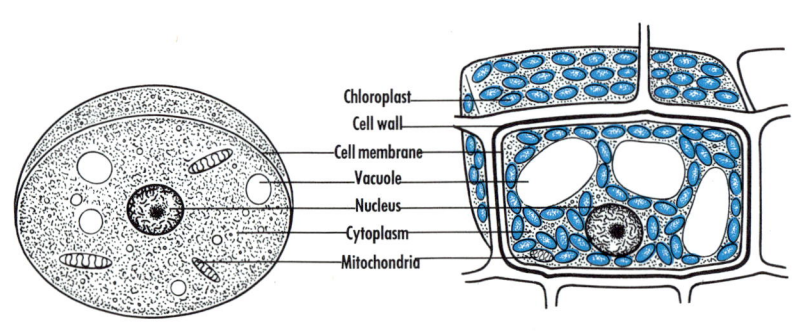

7-5 Compare the parts in the animal cell (left) with the plant cell (right).

(PLAS-tihdz), present in plant cells, are small bodies which store or form substances important in the cell's life processes. They make up the colors in plants. In some plastids, starch or fat is formed and stored. *Chlorophyll* (KLAW-ro-*fihl*), located in special plastids called *chloroplasts*, gives many plants their green color. It is also necessary for the making of their food.

Small bodies called *mitochondria* (mye-to-KAHN-dree-uh) are also found in the cytoplasm of cells. These structures are enclosed in a membrane. Mitochondria have a complex system of folded membranes within them. They are "food furnaces" because they contribute chemical substances needed in changing food to energy. It is believed that the mitochondria supply the energy used in making many of the chemical compounds necessary for the growth and development of the cell. They are sites of cell respiration. The table summarizes some principal parts of a cell.

PRINCIPAL PARTS OF A CELL

Living Substances	Nonliving Substances
Protoplasm Cytoplasm Nucleus Cell membrane Plastids Mitochondria	Cell wall Vacuoles Cell sap

There are two main types of organisms. We have mentioned organisms that consist of only a single cell, the *one-celled* organisms. We shall examine several one-celled organisms in the next section. Other organisms are made up of many cells. The cells in *many-celled* organisms must work together to carry on the life processes. Therefore, a many-celled organism usually has specialized cells that have separate functions. For example, certain cells help digest food, others carry oxygen, and others remove waste materials.

Thus, cells which are organized as a group of similar cells and perform a similar activity in the organism make up a *tissue*. Examples of tissues in plants are wood tissue and conducting tissue. Similarly we find muscle and nerve tissue in animals. Each cell in a tissue carries on the life processes necessary to stay alive and to carry on the special activities of the tissue.

As the organism becomes more complex in structure, simple tissues cannot carry on all the activities needed to keep the plant or animal functioning properly. Groups or tissues work together as *organs*. The leaf stem and flower are examples of plant organs. The heart, kidneys and stomach are examples of animal organs. A group of organs working together make up a *system*. For example, the mouth, stomach, and intestines are parts of the digestive system. This organization of cells into tissues, tissues into organs, and organs into systems makes the complex activities in living organisms possible.

REVIEW

1. Name the living substance that makes up all cells.
2. What are the functions of the: (*a*) cell membrane, (*b*) nucleus?
3. What is the main difference in structure between a plant cell and an animal cell?
4. Name three nonliving parts found in cells.
5. What is meant by: (*a*) a tissue, (*b*) an organ?

B/How Do Organisms Produce and Consume Food?

Green plants are the only important food producers. Green plants are capable of using the sun's energy along with carbon dioxide and water to build energy-rich compounds, such as *glucose* (GLOO-kose). Glucose is a sugar used by most living organisms for energy. Chlorophyll, the green pigment found in the chloroplasts within the cell, must be present so that a plant can manufacture its own food. However, the chlorophyll itself does not change into plant food. It is used by the plant over and over again. Chlorophyll functions by *absorbing the energy* in sunlight and making it available so that a chemical reaction between the carbon dioxide and water can take place within the chloroplasts.

Figure 7-6 shows a cross section of a green leaf. The upper layer allows the sunlight to reach the cells containing the chlorophyll. In most plants, the pores called *stomata* (STO-muh-tuh) are found on the underside of the leaf. These stomata allow gases from the air to enter the leaf. They open and close according to conditions surrounding the leaf. Can you think of some of these conditions? The opening and closing of the stomata are controlled by the action of the *guard cells* shown on the bottom of Fig. 7-6. The next activity shows one way of taking the green color out of leaves.

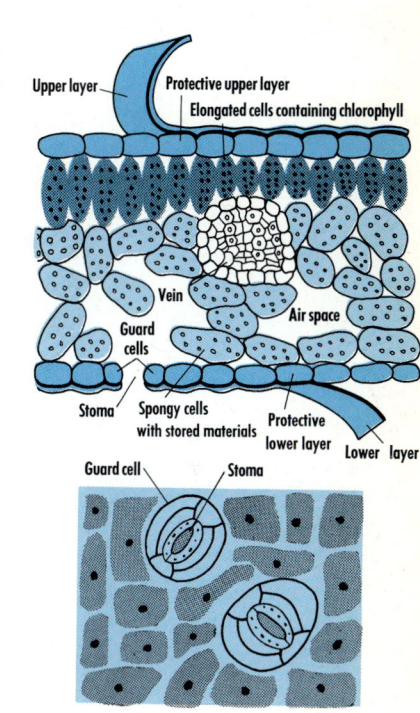

EXPLAIN

Place a green leaf in a beaker containing water. Boil the leaf for a few seconds. What does this do to the leaf? Remove the leaf and put it into a beaker of alcohol. Put the beaker in a pan of water and heat the pan so that the water boils. (**Caution: Do not heat the beaker of alcohol directly.**) What happens to the color of the alcohol? What is the color of the leaf when you remove it from the beaker? What is the green substance dissolved in the alcohol? Where did it come from? What is its purpose?

7-6 What is the function of the cells containing chlorophyll (top) and the stomata (bottom)?

Carbon dioxide and water provide the raw materials in the food-making process. *Photosynthesis* (fo-to-SIHN-thuh-sihs) is the food-making process which takes place in a green leaf. About 300 million tons of carbon are made into food materials by green plants each year. Of this amount, about 90 percent is from the activity of single-celled plants. You should begin to realize that without green plants almost all life on earth would die.

Jan B. Van Helmont (1577–1644), a Belgian scientist, first showed that plants make their own food material; they do not obtain it from the soil. It was not until the nineteenth century, however, that scientists discovered that water was needed in addition to carbon dioxide for green plants to make sugars.

Plants absorb most of the water they use through their roots. Carbon dioxide enters the plant through the stomata in the leaves. These two substances are the raw materials needed for the process of photosynthesis. However, since photosynthesis is a chemical process, an energy source (sunlight) is also needed. Although the chemistry of photosynthesis is complex, we can show the process in the equation:

$$\text{water} + \text{carbon dioxide} + \text{energy} \rightarrow \text{sugar} + \text{oxygen}$$

Observe that water and carbon dioxide are nonliving substances which are changed in this process into sugar (a carbon, hydrogen, and oxygen compound), which is a substance in living organisms. Photosynthesis is actually a "two-step" reaction. First, the light energy causes the water molecules to split. This frees the hydrogen and oxygen of which it is composed. In the next step, the free hydrogen combines with carbon dioxide to form sugar. The sugar may be used directly by the plant or it may be changed into starch, protein, or fat. The extra oxygen is given off by the leaves through the pores. This helps to increase the oxygen in the air. How does this action affect the balance of gases in our atmosphere? To find out if oxygen is produced by a green plant, let us do the following activity.

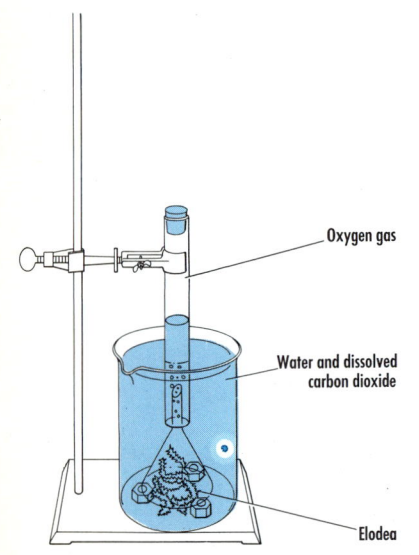

EXPLORE

Arrange the equipment as shown and place a few strands of Elodea under the funnel. With a sharp blade cut off the upper end of the strands while under the water. Fill the tube completely with water and put in a small pinch of sodium bicarbonate (baking soda). Place the jar in sunlight and note the bubbles of gas arising from the top of the plant.

What are the gas bubbles being given off by the plant? How does the rate with which the gas bubbles are given off show you the degree of photosynthesis carried on by the plant? Where did the gas come from?

Scientists are using chemical *tracers* which can be followed and identified as they are absorbed by plants. These tracers may be compounds containing *radioactive carbon*. The plant uses radioactive carbon in the same way it uses the ordinary form of carbon. In this way, scientists have discovered that the oxygen released in photosynthesis comes from water and not from carbon dioxide. In addition, they found that carbon in the sugar comes from carbon dioxide. Figure 7-7 shows scientists using tracers to study plant life processes.

Varying conditions affect the rate of photosynthesis. You have learned that carbon dioxide is needed in photosynthesis. It is possible to increase the rate at which photosynthesis takes place by increasing the amount of carbon dioxide present in the air. In fact, in some greenhouses, carbon dioxide is added to the air.

You can perform many activities to find the effects of light, temperature, and water on the rate of photosynthesis. Let us see how one factor, light, affects photosynthesis in a green leaf by doing the activity.

OBSERVE

Fasten some circles of cork to both sides of a green leaf so that part of the leaf cannot be exposed to light. Allow the plant to grow normally for a day in its usual conditions. Remove the pieces of cork. Dissolve the chlorophyll from the leaf as described in the activity on page 181. Place a few drops of the iodine stain on the bleached leaf. Does the whole surface change color? Which parts did not? Since this color change is a test for starch, what conclusion can you make about the food-making process and the role of light?

Plants produce several substances in photosynthesis. During the sunlight hours, the plant usually produces more sugar than it can use. The plant changes some of the extra sugar into starch and stores it for future use. If we were to test a leaf

7-7 Why do scientists use chemical tracers, like radioactive phosphorus, shown above?

for starch, as was done in the activity, we would find that in the afternoon the leaf would have the greatest amount of starch stored in it. The plant leaf also changes some sugar into fats and oils.

The green leaf can also make protein substances. It changes sugar into protein by the addition of *nitrogen*. A certain group of soil-living bacteria, known as *nitrogen-fixing bacteria,* take nitrogen from the air. They change the nitrogen in the soil into the form of *nitrates*. Plants use this nitrogen from these soil compounds to manufacture protein.

All of the processes related to photosynthesis take many steps and a great deal of energy change. Only recently have scientists begun to understand the complicated nature of these processes and to learn more about the complex chemical changes that take place within the tiny cells.

Respiration takes place in all organisms. Respiration involves (1) the taking in of oxygen, (2) the releasing of energy by the oxidation of food within the cell, and (3) the removal of the waste products which are produced in this food oxidation. All living cells carry on the process of respiration. The chemical reaction for respiration can be shown in this equation:

$$\text{sugar} + \text{oxygen} \rightarrow \text{carbon dioxide} + \text{water} + \text{energy}$$

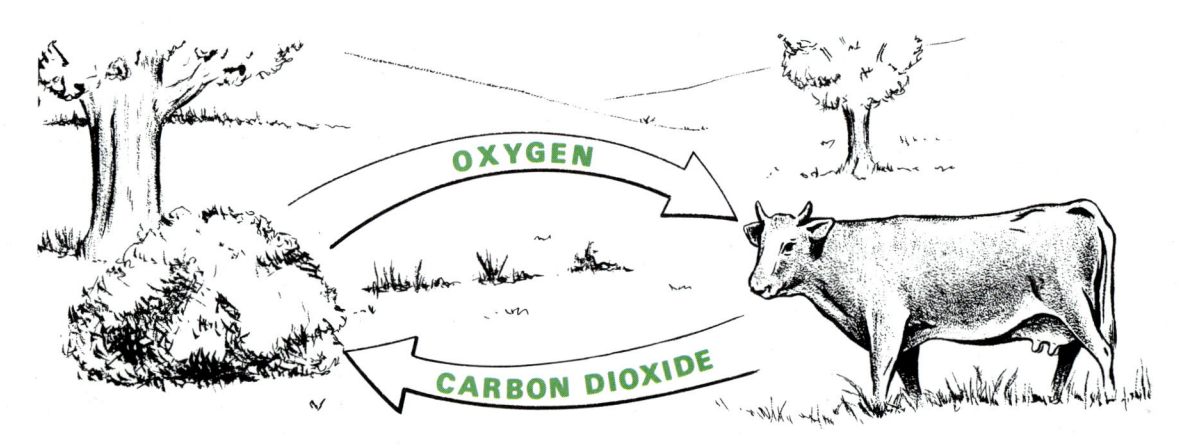

7-8 What part do plants and animals play in the carbon dioxide-oxygen cycle?

As you can see by comparing this equation with the one shown on page 182, the two processes are the opposite of each other. That is, in *photosynthesis, energy is being stored, while in respiration, energy is being released.* Do you know where all this energy comes from originally? Since oxygen and carbon dioxide change sides in these equations, we refer to this process as the *oxygen-carbon dioxide cycle* (Fig. 7-8).

COMPARISON OF PHOTOSYNTHESIS AND RESPIRATION

	Photosynthesis	Respiration
Occurrence	Only in cells containing chlorophyll exposed to light	In all living cells at all times
Raw materials	Carbon dioxide and water	Sugar, fats, some protein, and oxygen
Energy change	Energy stored	Energy released
Products formed	Sugar and oxygen	Carbon dioxide and water

Plants and animals carry on respiration 24 hours a day. Yet, green plants give off more oxygen during the daytime than they take in. This occurs because photosynthesis takes place at a much faster rate during the daytime than respiration does. In other words, much more oxygen is produced by the plant during daylight hours than is needed for its own use. Thus, the extra oxygen is given off by the plant. The table above compares the process of respiration with the process of photosynthesis.

Animals and nongreen plants are food consumers. Since green plants are the major source of food production, all other living organisms which cannot manufacture their own food can be called food consumers. The chemical breakdown of food material is similar in all living organisms.

Food substances taken in by an organism are digested, that is, broken down into small soluble molecules which can be absorbed by the cells. For example, sugars and starches are broken down into *glucose* (which is a simple sugar), proteins are changed into *amino* (uh-MEE-no) *acids,* and fats are converted into *fatty acids.*

REVIEW

1. What is meant by: (a) photosynthesis, (b) respiration?
2. What are the two main raw materials that are used in photosynthesis?
3. Name several conditions which affect the rate of photosynthesis.
4. What other substances does the plant produce in photosynthesis?
5. During what part of the day is the process of photosynthesis most active?

C/How Do Living Organisms Reproduce?

Reproduction is the process of producing new individuals of the same species. Prior to 1600, it was a common belief that young rats came from soiled rags, bees from the bodies of dead bulls, and flies from dew or decayed meat. Francesco Redi conducted the first reported experiment which tried to find out whether flies really came from decayed meat. Now you can perform a similar experiment.

EXPLAIN

Take three jars and place some beef in the bottom of each jar. Leave one jar open, cover the second with gauze, and cover the third with brown wrapping paper, as shown. Allow the jars to stand in an open place for a week or more where flies can get to them. What do you observe happening to the meat? Is it the same in all three jars? Are there living organisms on the meat in any jar? If so, how did they get there?

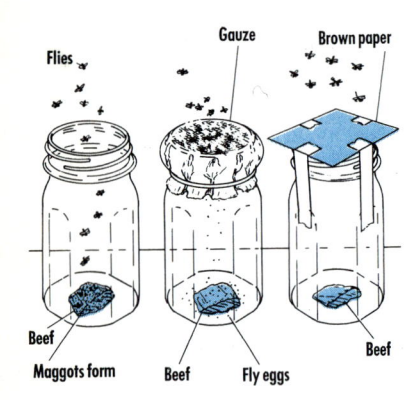

Redi concluded that flies, attracted to the meat odor, laid their eggs there. Since maggots appeared in the jar sealed with gauze, he guessed that eggs were small enough to slip through the gauze. No maggots appeared on the meat in the sealed jar. Based on this experiment, would you accept Redi's conclusion that flies do not come from decayed meat? Is there another explanation for the results?

The idea that living organisms arise from nonliving organisms is called *spontaneous* (spahn-TAY-nee-us) *generation.* This idea was not disproved until Louis Pasteur finally demonstrated that no decay and no living organisms appeared on meat or broth if they were first boiled in hot water and then were completely sealed from the air.

One type of reproduction involves only one parent. Scientists have found that certain organisms reproduce *asexually.* That is, a new plant or animal is produced from a single parent. Many one-celled organisms undergo a process of cell division called *fission* (FISH-uhn) when conditions are favorable. Fission also involves only one parent.

Yeast cells, which are plants, reproduce by *budding* or the division of a cell into two unequal-sized cells, as shown in Fig. 7-9 (top). Notice that the smaller cell, called the *daughter cell,* remains attached to the larger cell, known as the *mother cell,* for a period of time. So that you may see how rapidly yeast cells are capable of multiplying, perform the activity that follows.

DESCRIBE

Mix molasses and water together until the mixture is very thin. Add a small piece of baker's yeast and stir thoroughly. Allow the mixture to stand for about 30 minutes and look at a drop under the low power of a microscope or microprojector. Count the number of yeast cells seen. Allow the mixture to stand in a warm place overnight and examine another drop. Were more yeast cells seen the second day? At about what rate did the yeast cells seem to increase?

An animal, hydra, also reproduces by budding. The bud grows from the parent hydra, and when large enough separates from the parent. The bud continues to grow and becomes a separate animal (Fig. 7-9 bottom). Other animals, such as some sponges and worms, also reproduce asexually. We shall study this process further in the next chapters.

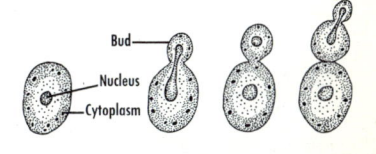

7-9 How does a yeast cell (top) and a hydra (bottom) reproduce?

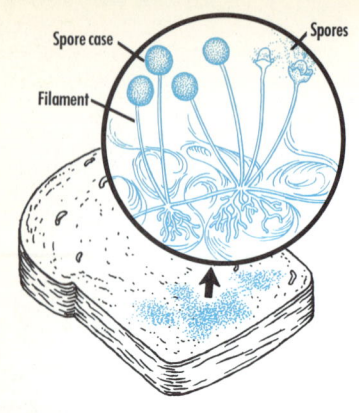

7-10 What type of reproduction is illustrated by mold plants?

7-11 What will a rhizome produce?

7-12 What does an "eye" represent?

188 / UNIT 3   MAN EXPLORES LIVING ORGANISMS

When conditions are unfavorable for growth, some organisms, such as yeast and other simple freshwater plants, produce *spores*. At this stage, the cell forms a hard coat around itself. When conditions become favorable, the spore case breaks open and the spores develop into new organisms.

Plants, such as molds, reproduce asexually by spores as their usual means of reproduction. Special threadlike strands which grow upward produce *spore cases* at the top of each strand (Fig. 7-10). Spores are formed inside this case and when fully grown are released and carried away by the wind. Each spore is able to produce a new plant if it reaches a condition favorable for growth. You can grow molds in the following activity.

OBSERVE

Take a clean coffee can and place a moist blotter on the bottom. Put a piece of bread inside. (The bread should be the kind that does not have a mold-preventing chemical added.) Allow the can to remain open in a dusty place for about 30 minutes. Then close it and place in a warm place, such as a closet. Do not open the can for about one week. After a week, open the lid. What do you observe? Is the growth all the same color? Compare the kinds of molds that grow on some canned pumpkin smeared on the bottom of a dish with those grown on bread. Place a thread of the mold under a microprojector or microscope and examine.

Another type of asexual reproduction involves a nonreproductive part of the organism. *Vegetative reproduction* is the process by which a plant or animal develops from some piece of the parent organism. For example, some plants send out underground stems which are called *rhizomes* (RYE-zomz), shown in Fig. 7-11. These are thickened stems which contain stored food. Trillium, Solomon's seal, and lily of the valley are all plants that produce rhizomes. Since only one parent is involved, will the new plant be exactly like its parent?

There are other underground stems used for reproduction. *Tubers*, found in the potato, are swollen stems containing small buds called "eyes" as well as stored food. Tubers can be used to produce new plants. Each piece of tuber containing an eye can be planted, as shown in Fig. 7-12. Under favorable conditions, a new plant will grow. *Bulbs,* which are underground leaves containing food, also are used for reproduction. Onions, tulips, and hyacinths are plants which reproduce by this method.

Certain plants send stems along the surface of the ground. Roots and leaves are produced and form a new plant. Strawberries and some grasses are all plants which produce by this method. These special reproductive stems are called *runners* (Fig. 7-13).

The upright stems of certain plants may be bent and covered by soil, permitting roots to grow from the stems. If conditions are favorable in this process of *layering* (Fig. 7-14), a complete new plant develops. Raspberry bushes and forsythia plants can be grown by this method.

In some cases, even leaves may be used to produce plants. The begonia leaf can be sliced into pieces and placed in moist sand, soil, or on blotting paper, and roots begin to grow out of the cut ends. Once roots are formed, you can plant the leaf and roots in the soil and a new plant might grow from every cutting. This type of reproduction is known as *leaf cutting.* The activity on vegetative reproduction shows how this is done.

7-13 How do runners reproduce?

OBSERVE

Take a leaf from a snake plant, or *Sansevieria* (san-sih-VEER-ee-uh), and cut it into inch-length sections. Place them in some moist soil. Cover the pot containing the leaf and soil to keep the moisture sealed in. Observe your cuttings daily. What changes take place? Where do the roots begin to grow? Transplant your cuttings only after the roots are well developed.

RASPBERRY

7-14 How is the raspberry plant reproducing?

Sometimes, roots can be used to produce new plants. A sweet potato, kept in a darkened room and partly submerged in water, soon begins to grow roots. The sweet potato is a swollen root containing stored food. Once the roots are well developed, the sweet potato can be placed in sandy soil in a sunny place, and new sweet potato plants will develop.

Man has invented artificial ways of developing new plants. *Budding* and *grafting* are two important methods of producing new kinds of plants. These two methods assure that the new plants will bear fruit exactly like the parent plant. Let us follow the process of grafting.

Suppose a farmer wants to develop a seedless orange grove. Since seedless oranges do not produce seeds, he would have to use a vegetative means to produce the grove. He buys one seedless orange tree for his grove. After the tree is growing well, he takes one branch from this tree and grafts it onto another orange tree, the stock, which produces oranges with seeds. The growing tissues of the seedless orange branch must be in contact with the growing layer of the stock, as shown in Fig. 7-15. The branch and stock must be a closely-related species. If the graft is successful, the rest of the tree will produce oranges with seeds while the grafted branch produces seedless oranges. In other words, *the grafted branch always breeds true to its parent.*

Budding is very similar to grafting but, instead of transferring a whole branch, a bud with some surrounding bark is grafted onto the stock. Peaches, plums, and some varieties of grapes are examples of fruit obtained in this way. It is important that the growing layers be in contact with each other, since the bud must develop into a branch. It must get its water and minerals from the stock, and therefore a close connection must be made between the bud and stock.

Many animals have the ability to grow missing parts. The ability of an organism to grow new parts to replace missing ones is called *regeneration* (ree-*jen*-uhr-AY-shuhn). If a piece of hydra is removed, in a short time the rest of the body will regenerate the missing portion. If a hydra is cut part way down the middle, a "two-headed" hydra may result.

You can observe an organism's ability to regenerate lost parts by experimenting with the *planarian* (pluh-NA-ree-uhn), a flatworm. It is larger than the hydra, making it easier to work with in the laboratory. Observe planarians in the next activity.

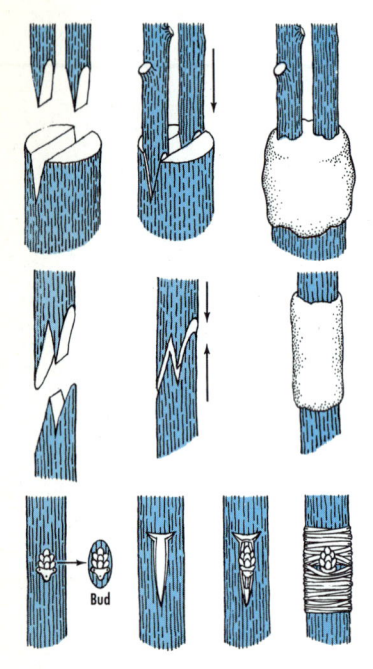

7-15 What is the reason for making plant grafts?

OBSERVE

Obtain a planarian and place it in the bottom part of a Petri dish with some water. Place this dish under the low power of a microscope. Cut the planarian in half with a sharp razor. Now place the Petri dish in a cool place. Cover it with the top half of the dish to prevent drying out. Observe the two pieces of planaria each day. What is happening to each half? Are both halves regenerating their lost parts?

You might want to cut the planarian in other ways, as shown. What are your results in these cases?

Crabs and lobsters often regenerate lost limbs. In fact, if a limb gets caught on an object, and the lobster cannot move, it is able to separate itself from the limb and in time grow a new one.

Perhaps you are familiar with the story about the oyster fisherman who once cut starfish apart and threw them back into the water to be rid of them because starfish eat oysters. However, since starfish have the ability to regenerate lost parts, more and more starfish were actually produced. Today starfish are placed on the beach where they are left to dry out and die.

The more complex the organism, the less it is able to regenerate lost parts. Human beings can grow new skin, new bone tissue, hair, toenails, and blood cells, but cannot grow new limbs or other body parts.

Sexual reproduction involves two parents. When organisms reproduce by means of special *sex cells* which join together to form a new individual, we have *sexual reproduction*. Most highly developed plants and animals reproduce sexually.

The sex cells are called *gametes* (GAM-eets). The union of two gametes is termed *fertilization*. In simple forms of living organisms, the sex cells are almost the same. However, in higher forms of plants and animals, there is a definite difference. The male gamete is called the *sperm cell*, the female gamete is the *egg cell*, and it is usually much larger than the sperm cell. The sperm cell has a whiplike tail and must swim to the egg cell in order to fertilize it.

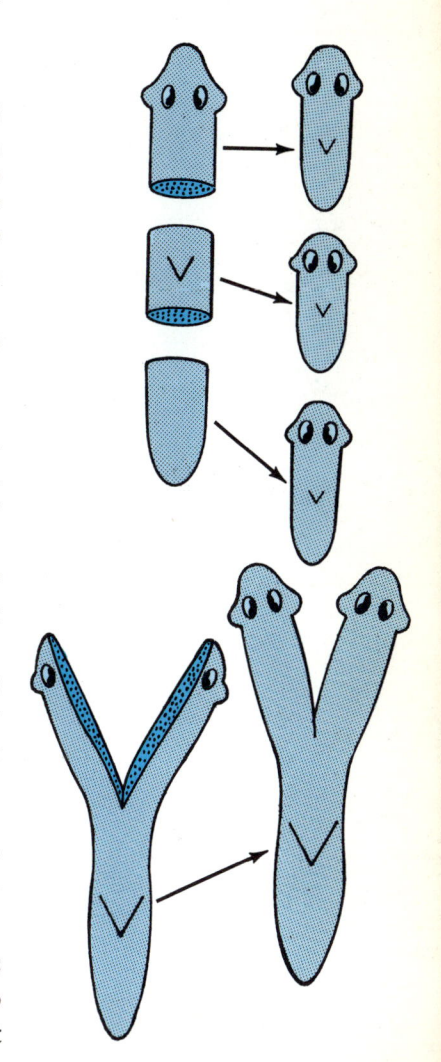

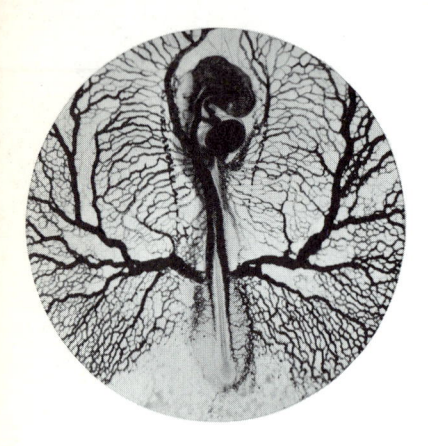

7-16 Where does the chick embryo obtain food?

You know that chickens, fish, and a number of other organisms lay eggs. However, let us not confuse the common use of the word, as in bird's eggs with the scientific term egg cell, a microscopic, single cell. An egg cell which is fertilized begins to grow into an *embryo* (EM-bree-o), the first stages in development of the new organism. Sometimes the embryo develops within the egg, which is protected by a shell, like the chicken or pigeon (Fig. 7-16). In other animals, like fish and frogs, the egg has no shell.

The more parental care given a developing embryo, the fewer eggs the female lays at any one time. For example, the female salmon may deposit as many as 3500 eggs. Since fertilization occurs in the water, can you see why the chances of a sperm uniting with an egg are not too great? Many of these eggs never are fertilized and soon die; other fertilized eggs are found and eaten by other organisms. Birds, however, which build nests and care for the eggs until they are hatched, produce fewer eggs but have a much higher percentage of eggs hatching.

Most mammals bear their young alive. Fertilization occurs inside the body of the female in mammals. Since this method makes sure that there will be a greater chance for fertilization to take place, most organisms that bear their young alive do not have many offspring. The hamster usually has a litter of ten or twelve, the cat or dog has a litter of about six, and the cow, elephant, and human being usually produce one offspring at a time. Organisms that bear live young give longer parental care to their offspring than those organisms which lay eggs.

REVIEW

1. What is meant by spontaneous generation?
2. Define asexual reproduction and give examples of two methods.
3. Define vegetative propagation and give examples of three methods.
4. What is the difference between budding and grafting?
5. What is the name of the: (*a*) male gamete, (*b*) female gamete?

D/How Are Organisms Classified?

Many different kinds of organisms are found in nature. There are about two million different kinds of organisms in the plant and animal world today. Because there are so many, a means of classifying organisms had to be found.

Carolus Linnaeus, (1707–1778), a Swedish scientist, developed our present classification system. In his book on classification of organisms, called *Systems of Nature,* he grouped all living organisms into two kingdoms, plants and animals. Each kingdom was then grouped into *phyla* (FYE-luh) [sing. *phylum*] and each phylum was grouped into *classes.* Each class was further grouped into *orders, families, genera* (JEN-uhr-uh [sing. *genus* (JEE-nuhs)], and *species* (SPEE-sheez).

Scientists use the last two groupings for scientific naming of organisms. The genus part of the name always begins with a capital letter and the species part, which follows, is always written with a small letter. No two organisms have the same name unless they are of the same species. For example, the scientific name of the common cat is *Felis* (FEE-lihs) *domestica* (domestic cats of all sizes), and the scientific name of the dog is *Canis* (KAY-nihs) *familiaris* (dogs of all sizes). Man also has a scientific name, *Homo sapiens* (SA-pee-enz).

In the table, the common grasshopper (*Schistocerca americana*) shown in Fig. 7-17, is used to illustrate how we classify an organism according to this system. (It is not necessary to learn the scientific names of the classification below.) Now, perform the activity on classification.

7-17 What is the scientific name of the grasshopper?

CLASSIFICATION OF THE GRASSHOPPER

Kingdom	Animal
Phylum	Arthropoda (joint footed animals)
Class	Insecta (body divided into three parts)
Order	Orthoptera (wings are straight)
Family	Acridiidae (locust group)
Genus	Schistocerca (cleft tail)
Species	americana (originally found in America)

194 / UNIT 3 MAN EXPLORES LIVING ORGANISMS

COMPARE

To understand the basis for classification, collect and examine a group of closely-related insects, such as groups of flies, beetles, or moths. Compare the types of wings, the shapes of the legs, the kinds of feelers and any other characteristics you can observe. Make a chart similar to the one below for each group of insects. **Do not write in your book.**

GROUPS OF INSECTS: _____			
Name of insect	Type of wings	Shape of legs	Kind of feelers

How are the insects in each group similar and how are they different? Try to find out the main characteristics used in placing each insect in a particular order.

Organisms with similar structures are related. Later, you will study the main plant and animal groups and learn that they are classified according to their similarities and differences in structure. Thus, all cats belong to the same genus, *Felis.* All dogs belong in the genus, *Canis.* But you know that wolves and foxes look like dogs (Fig. 7-18). Which group is more similar to the dog, and which is less similar? You can answer this question by examining the table below.

CLASSIFICATION OF DOGLIKE ANIMALS			
	Dog	Wolf	Fox
Kingdom	Animal	Animal	Animal
Phylum	Chordates	Chordates	Chordates
Class	Mammals	Mammals	Mammals
Order	Carnivores	Carnivores	Carnivores
Family	Canidae	Canidae	Canidae
Genus	Canis	Canis	Vulpis
Species	familiaris	lupus	fulva

7-18 Which animal is more closely related to the German shepherd (top), the timber wolf (center) or the red fox (bottom)?

As you can see, the dog and the wolf are so much alike that they are placed in the same genus. The fox, with a few more differences in structure, is placed in a different genus. However, these three animals are still closely related because they all belong in the same family of doglike animals, *Canidae*. Can you name other examples?

Scientists study the internal structure of organisms as well as external appearances, so that they can classify living organisms more exactly. It is interesting to note that a shark is classified as a fish, but the whale is called a mammal. They both live in water and share some common outward appearances. But the whale is a warm-blooded animal which breathes with lungs, and the shark is a cold-blooded animal that breathes with gills. Internally, these two animals of the open sea differ in many ways.

Since classifications are based on man's observations, scientists sometimes do not agree where a particular organism belongs in the classification scheme. For example, in the past, *Euglena* (yoo-GLEE-nuh), a simple one-celled organism, has been classified as both a plant and an animal. It contains certain characteristics common to plants, and yet it also has certain characteristics found in animals. We shall learn more about some of these simple forms of life in the next chapter.

Organisms are influenced by their environment. The branch of science dealing with the relationship of living organisms to their environment and to each other is called *ecology* (ee-KAHL-uh-jee). The environment affecting living things is made up of all the forces acting on the organism. This includes the climate, consisting of such factors as temperature, rainfall, wind, sunshine, and humidity. Since water is also affected by these conditions, even organisms living in the sea must react to changes in climate. Desert plants, seen in Fig. 7-19, cannot live in moist areas, while plants found in swamps cannot live in dry areas.

For each organism there is a lowest and highest temperature for survival, as well as a temperature at which it grows best. Figure 7-20 shows how two different species of trees react to changes in temperature. Warm-blooded animals, like birds and

7-19 What kind of environment do desert plants need?

7-20 How do the two species of trees react in winter?

7-21 How high can some birds fly?

mammals, have an almost constant internal body temperature, regardless of outside conditions. Yet, we know that warm-blooded organisms must change or modify their behavior when there is a temperature change. Many birds migrate south when it gets cold, and some mammals sleep through the winter. This might be due to a temperature change or to a food shortage.

Light is another factor that affects the growth of living organisms. Strawberries will yield large amounts of fruit when there are 14 to 15 hours of daylight, while chrysanthemums are short-day plants. The long days in Alaska help farmers there to raise many fine crops even though the ground is free from frost only 12 weeks out of the year.

It is known that above 9500 feet few trees grow in the Rocky Mountains. This elevation is called the *tree line* and varies in different parts of the world. Above this line, dwarf plants and relatively few insects, birds, and mammals are found. Exploration of the deep seas has indicated that life is rare much below one mile in depth. Observations of the habits of birds have shown that few birds are able to fly much above three miles from the earth (Fig. 7-21).

From studies in ecology, scientists feel that it is the proper combination of these different conditions which helps organisms survive. Each factor mentioned, as well as others you can name, are in a *critical balance*. An organism taken out of its natural environment does not usually survive. However, the ability of organisms to adapt to new surroundings has accounted for the many species of plants and animals living today that did not live in the past. You should now do the activity which deals with organisms and their environment.

OBSERVE

Make a list of the organisms in your locality which are adapted to live in a variety of environments. Look in ponds, under rocks, in trees and bushes and in open fields. Can you determine how each form of life you have seen is adapted to its environment? What conclusions can you draw about the animals you observe and their relation to their particular surroundings?

Organisms interact with their environment. In order to survive, organisms must be able to adapt themselves to changes which occur in their surroundings. If an organism lives in a pond, and the pond dries up, it must either have some other way of getting oxygen and food, or it will die. Organisms which cannot adapt to changing environmental conditions usually disappear from the earth and become *extinct*. The animals shown in Fig. 7-22 are examples of extinct organisms. Can you name others which are also extinct? Now do some research on birds and mammals as explained in the activity.

CHECK

Make a list of birds and mammals in your local area which seem to "disappear" during the winter months. Do library research to find out how they survive during this period. If they migrate, compare the climate where they go with that in your area. Find information about the behavior of birds before and during migration and their habits in these "summer" quarters. Perhaps you can keep records of when some of the migratory animals begin and end their migratory periods. Does this period of time vary?

The ability of a species to survive also depends on its ability to reproduce. Even if individual organisms adapt themselves to changes in the environment, the species will become extinct if its members cannot reproduce enough new generations to survive in the new environment.

Organisms need food to stay alive. Food is the source of energy in all living organisms. Green plants have the ability to make their own food. However, most other organisms must obtain food from other living organisms. One way we can classify living organisms is by what they eat.

Herbivorous (huhr-BIH-vuhr-uhs) animals, like cows and deer, eat only plants. *Carnivorous* (kahr-NIH-vuhr-uhs) animals, like cats and dogs, are the meat-eaters. *Omnivorous* (om-NIH-vuhr-uhs) animals are those animals which eat both meat and plants. Human beings, many birds, and a number of mammals are known as omnivores. Figure 7-23 (next page) is a diagram of a food chain.

7-22 Why are the dodo bird (top), brontosaurus (center), and the stegosaurus (bottom) now extinct?

A Food Chain

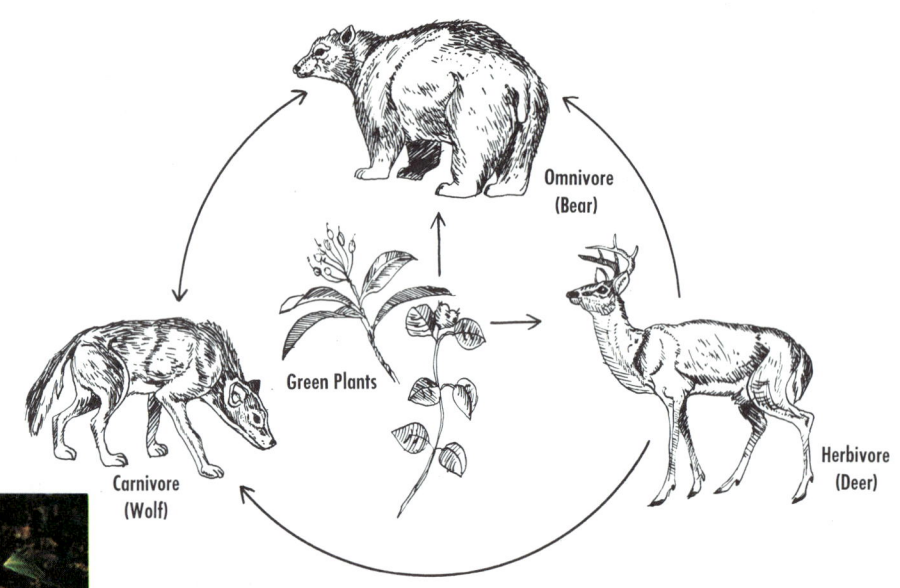

7-23 What is the original source of energy in this food chain?

7-24 Mushrooms (top) and mold (bottom). How do these organisms obtain food?

Many organisms depend on other organisms for their food. One group of living things, like mushrooms, bacteria, and molds which live and feed on dead organic materials are known as *saprophytes* (SAP-ro-*fites*) (Fig. 7-24). *Parasites* (PAR-uh-*sites*) live on other living organisms, known as their hosts, and usually have harmful effects on these organisms. Lice, fleas, and other disease-causing organisms are examples of this type of relationship. Another group, called *symbionts* (SIHM-bye-ahnts), consists of organisms which also live on or in other organisms but with both organisms benefitting from this type of relationship. For example, a *lichen* (LYE-ken) which is referred to as a single plant, is actually composed of two different plants (Fig. 7-25). One plant not capable of making its own food gets nourishment from the other which is able to make its own food. However, the plant which is capable of manufacturing plant food depends on the other plant for protection and moisture. Another example involves one-celled animals which live in the intestines of termites. They help termites to chemically break down the wood which they use for food. The activity will help you learn more about the food habits of organisms.

CHART

Make a chart similar to the one below. Consult biology books, and on each line list as many organisms as you can which belong in that group. **Do not write in your book.**

Food Habits of Organisms
Omnivores: _____
Herbivores: _____
Carnivores: _____
Saprophytes: _____
Parasites: _____
Symbionts: _____

REVIEW

1. List the main headings used in our scientific classification system.
2. What is: (*a*) ecology, (*b*) environment?
3. Name three food classifications by which organisms can be described.
4. What usually happens to a species if organisms are not able to reproduce?
5. Name five important life processes carried on by all living organisms.

7-25 Why is the lichen an example of a symbiont?

THINKING WITH SCIENCE

A. *On a separate sheet of paper write the numbers 1 to 15. After the number of each statement, write the letter of the term that correctly completes the statement.* **Do not write in your book.**

1. The scientific name of any organism consists of two parts, the: (*a*) family and species, (*b*) class and genus, (*c*) genus and species.

2. The combining of food and oxygen takes place in the process of: (*a*) digestion, (*b*) respiration, (*c*) excretion.

3. Animals which eat both meat and vegetation are known as: (*a*) herbivores, (*b*) carnivores, (*c*) omnivores.

4. An organism which lives with another organism for the mutual benefit of both is called a: (*a*) parasite, (*b*) symbiont, (*c*) saprophyte.

5. Two structures found in plant cells, but *not* in animal cells are: (*a*) cell wall and vacuoles, (*b*) vacuoles and chloroplasts, (*c*) cell wall and chloroplasts.

6. Four conditions which must be present for photosynthesis to occur are: (*a*) chlorophyll, oxygen, water, and sunlight; (*b*) sunlight, chlorophyll, soil, and carbon dioxide; (*c*) carbon dioxide, water, sunlight and chlorophyll.

7. The part of the cell that controls most of its activities is the: (*a*) plastid, (*b*) vacuole, (*c*) nucleus.

8. The changing of insoluble food materials into soluble materials takes place in the process of: (*a*) excretion, (*b*) digestion, (*c*) respiration.

9. An organism which lives on dead or decaying material is called a: (*a*) parasite, (*b*) saprophyte, (*c*) symbiont.

10. Animals which use only meat for food are known as: (a) carnivores, (b) herbivores, (c) omnivores.

11. The process of taking food into the body is called: (a) excretion, (b) ingestion, (c) digestion.

12. A sugar used by most living organisms is: (a) plasma, (b) glucose, (c) cytoplasm.

13. Most plant cells differ from animal cells in that they have a: (a) cell wall, (b) nucleus, (c) cell membrane.

14. The liquid portion of a cell is the: (a) cytoplasm, (b) plastid, (c) mitochondria.

15. The green matter in a plant cell is called: (a) mitochondria, (b) chlorophyll, (c) vacuole.

B. *Write the answers to the following in your notebook. Be sure to use complete sentences and correct grammar and spelling.*

1. What are some of the factors in the environment which affect organisms?

2. Explain the difference between a saprophyte and a parasite.

3. Describe the main life processes needed for organisms to survive.

4. What does the cell theory tell us?

5. Why are the cells in more complex organisms organized into tissues and organs?

6. Describe how a hydra obtains and digests its food.

7. Why are green plants the main source of food in the world?

8. Name several conditions which may affect the rate at which photosynthesis takes place.

9. Explain the relationship between photosynthesis and respiration.

10. Name three types of asexual reproduction and list examples.

RESEARCH IN SCIENCE

1. Set up an experiment to demonstrate that water is a necessary condition for photosynthesis.

2. Make some slides and examine the nitrogen-fixing bacteria found in the nodules on the roots of legume plants.

3. Use the iodine test to find out if starch is stored in other parts of the plant besides the leaf.

4. Make a collection of plants that do not contain chlorophyll and explain how each one obtains its food.

5. Collect pond water from several areas and make drawings of protozoans and algae found in the water.

6. Prepare a report describing recent advances in grafting involving both plants and animals.

chapter 8
The Relationships of Life

A/How Do Organisms Affect Each Other?

Life can exist under a variety of conditions. There is almost no place on our earth where some form of life, either plant or animal, does not exist. Different varieties of living organisms are found from the steaming jungles along the equator to the frozen wastelands of the Arctic. In this chapter, we shall study some of the different conditions under which living organisms exist. The natural environment in which a particular form of life is usually found is called its *habitat* (HAB-ih-tat). Examine habitats of insects in the activity below.

EXAMINE

The habitats of some insects are fairly easy to study because many insects remain in a small area unless disturbed. Examine such habitats as a rotting log, a patch of grass, an area under a rock, or the bark of a tree. Note the types of insects found in each habitat. Describe environmental conditions found in each habitat. Would insects from one habitat usually be found in one of the others? Can you explain the reason for this?

8-1 Explain the interdependence that is found among the organisms in this aquarium.

204 / UNIT 3 MAN EXPLORES LIVING ORGANISMS

Living organisms are affected by various factors of the environment, like temperature, soil, light, moisture, pressure, gases, water, and food. As we study living organisms further, we find that they also depend on other living organisms in their environment for survival. The *food producers*, the green plants, probably depend less on other forms of life than do the *consumers*. The consumers are those forms which do not make their own food. However, even the producers are dependent on other forms of life to supply them with carbon dioxide in the air and nitrogen compounds in the soil. We may conclude that *interdependence* (ihn-tuhr-dee-PEN-dens), the ways in which living organisms influence and depend upon each other, is an important part of the study of living organisms.

One of the best ways to understand what is meant by interdependence is to study a balanced aquarium in your science room (Fig. 8-1). This aquarium illustrates the interdependence of plants and animals in the oxygen-carbon dioxide cycle. A sealed aquarium, in which life processes are carried on while sealed off from the atmosphere, can be studied by doing the next activity.

EXPLORE

Fill each test tube about three-quarters full of clear pond water. Place a small snail in the first test tube, a small piece of *Elodea* in the second test tube, another piece of *Elodea* and a small snail in the third test tube, and no plants or animals in the fourth test tube. Add two drops of bromthymol blue solution to each test tube. Seal the tubes securely with the rubber stoppers. Place the test tubes in a rack in a lighted place, but not in direct sunlight. Examine the tubes daily for a week.

Bromthymol blue is a chemical substance that changes in color from blue to yellow in the presence of carbon dioxide dissolved in water. What happened to the color of the water in each of the sealed tubes? Explain the reason for any color changes observed in the tubes. Can plants live alone in a sealed aquarium? Can animals live alone in a sealed aquarium? What life cycle does this illustrate? What is the purpose of the fourth tube.

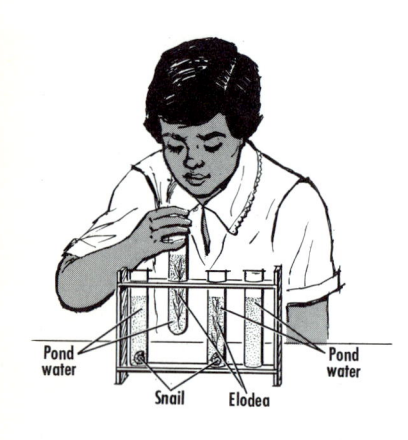

Perhaps you can begin to see why you cannot have too many fish in a small aquarium. The fish, or consumers, are dependent on the plants in the aquarium for oxygen and food. If there are too many consumers present, oxygen and food are used up at a faster rate than producers can supply them. As a result, animal life begins to die off. The plants, or producers, are also dependent on the animals for proper carbon dioxide levels and for waste materials which enrich the mineral content of the water. Therefore, as the fish die, the needs of the plants cannot be met. This results in a further imbalance.

Plants and animals in a balanced aquarium should be able to maintain life indefinitely if the conditions do not change. In most school aquariums, however, it is wise to keep more plant life than animal life and to feed the animals from time to time. Why is this precaution necessary?

Most areas support many different kinds of life. Living organisms can be classified into many species, as you have already learned in the previous chapter. However, as you study the different species of organisms in a given area, you find that they can be grouped into *populations*. Ordinarily, we think of a population as the number of people in a city, state, or country. Scientists, however, also use the word to describe the numbers of all plants or animals in a specific place. Thus we can speak of the rabbit population found in a field, the oak tree population found in a forest, or the trout population found in a stream. Let us examine a typical population in the following activity.

OBSERVE

Tree populations are fairly easy to study. A key describing the general appearance of the tree, the shape of the leaves, and the kind of bark will be helpful in identifying the different trees in your area. Make a rough map of a wooded area and locate different species of trees with drawings or symbols. To help in keeping a record of the different trees making up each population, collect a leaf from each tree. Dry the leaves between sheets of newspaper and press them between sheets of waxed paper with a hot iron to preserve them. Make a display of the leaves to represent the different populations.

Population density is affected by food and space. In any population of living things there is a struggle for survival. Can you explain what they mean by this? As a population increases in number, certain natural checks go into operation. Within a population, members are eaten by invaders, die for several reasons, or leave the community. Thus, a population may decrease or change over a period of time. This cycle of increase and decrease helps maintain a balance in terms of space and food needs for a given population. You can study typical animal and plant populations in the activity that is described below.

RECORD

Several teams of students should select different habitats, such as an open field, a tree-covered plot, a densely wooded area, or the bank of a stream, as areas for study. A small space, about a yard square, is measured off and a careful count of all living organisms on or in this piece of ground is made. Scratching the surface of the soil should be included to count the living organisms that are found there. You can write down the results in tables like the ones shown on pages 206–207. Your results can be summarized in class. **Do not write in your book.**

ANIMAL POPULATIONS

Populations studied	Type & numbers of birds	Type & numbers of mammals	Type & numbers of insects	Type & numbers of fish	Other animals
Season					
Open field					
Forest area					
Stream bank					
Other habitats					

CHAPTER 8 THE RELATIONSHIPS OF LIFE / 207

PLANT POPULATIONS

Populations studied	Type & numbers of trees	Type & numbers of shrubs	Type & numbers of grasses	Type & numbers of wild flowers	Other plants
Season					
Open field					
Forest area					
Stream bank					
Other habitats					

Each population is made up of a number of individual plants or animals. The number of each kind determines the density of the population for a given area. By making several population counts of this type at scattered places, can you estimate populations for a larger area?

Can a population in a given area change indefinitely, or are there certain factors which limit the density of a particular population of living organisms? Let us examine the changes that occur in a population of common one-celled animals in the next activity.

EXAMINE

Prepare a culture of one-celled animals by breaking some small pieces of hay into a fruit jar and then filling the jar with pond water. Add two tablets of brewer's yeast two days before putting in drops from an active culture. Now add a few drops from an active culture to the water. Keep the jar in a lighted place, but not in direct sunlight. Do not add any water or change any other conditions once you begin.

Starting with the first day, examine ten drops of the culture, one drop at a time, under the low power of a microscope or microprojector by moving the slide back and forth on the stage so that all organisms in each drop can be counted. Count the number of any one-celled animals seen in each drop and find the average number each day in one drop of culture.

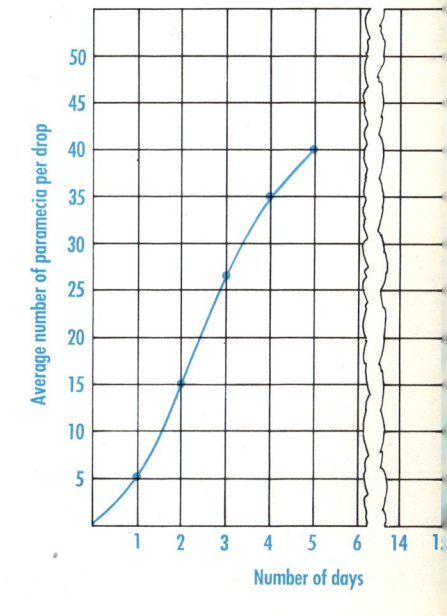

Make a graph of your results on a chart similar to the one shown (page 207). **Do not write in your book.** What happens to the number of animals for the first few days? after about a week? What do you note as to the population change after a period of about two weeks? Can you explain this change?

As organisms live and multiply, they use up the supply of food and oxygen at an increasing rate. In addition, they produce more and more waste materials. The increase in the population decreases the ability of the culture to carry on a balanced condition. Thus, the population begins to die off.

Do these same factors affect many-celled animal populations? A number of studies of mice populations have been made. It was found that food and space seem to be main factors controlling population size.

As the pens in which the mice were kept became more and more crowded, the animals began fighting among themselves and destroying the young. Thus, as the density of the population increased, the number in the population was reduced.

Animals have natural protection from their enemies. In an open environment, factors other than food and space may affect animal populations. For example, animals are consumers and feed upon one another in different ways. This factor affects the number of individuals in a population.

Many animals have special adaptations, known as *protective coloration*. This assists them in their struggle against being eaten by other animals. This characteristic takes several forms in different animals (Fig. 8-2). For example, the green and black areas on leopard frogs help these animals to blend with the plants growing around the ponds in which they are found. Some birds, like grouse, become almost invisible as they hide in a clump of plants. Some shore birds even have eggs that are protectively colored and which closely resemble their surroundings.

Fish have a slightly different form of protection coloration called *counter-shading*. In this case, the animal is darker on top to blend in with the dark colors on the bottom of the stream or lake when seen from above. The underside is lighter in

8-2 What special adaptations protect the chameleon?

8-3 What form of protective coloration do these fish exhibit?

color to blend in with the reflections of the sky in the water when seen from below (Fig. 8-3).

Another way in which animals are protected from their enemies is by means of *protective resemblance*. That is, the animal looks like something other than itself. Several kinds of butterflies look like brown leaves when their wings are folded. The walking stick looks like a dried twig on a branch (Fig. 8-4).

Mimicry (MIHM-ih-cree) is an unusual kind of protective resemblance found mostly in insects. In this case, the insect looks like another insect, rather than a part of the surroundings. To give examples, the viceroy butterfly looks very much like the monarch butterfly, a species that birds will not eat. Bee flies look like bumblebees, an insect that is left alone by most other organisms (Fig. 8-5).

It must be remembered that animals with protective coloration do not realize that they blend into their surroundings or mimic other species to stay alive. Rather, they are able to survive because they happen to have these protective adaptations. These adaptations for survival are the result of millions of years of development. During this time, those animals that were better able to escape their enemies survived in greater numbers than those forms which were not adapted in this way. Thus, we see one way in which a population of animals maintains itself by protection from its natural enemies.

Living organisms survive best in their natural environment. As we have learned, each living organism, plant or animal, carries on its life processes in relation to its total surroundings. The climate, food supply, the struggle with other forms of life, and other factors determine whether one form of life or another form can live, grow, and reproduce in an environment. These favorable conditions that make up the habitat in which the plant or animal organism lives are absolutely essential.

This habitat may be an area that is on land or in water. Some forms of life can exist better in forested areas and others in open grasslands. Still others live in swamps, deserts, lakes, rivers, or the ocean. Therefore, each form of plant and animal life is adapted to live in a habitat where it can carry on its life processes. Do the activity about habitats.

8-4 How did the walking stick get its name?

8-5 How does the bee fly (above) mimic the bumblebee (below)?

210 / UNIT 3 MAN EXPLORES LIVING ORGANISMS

EXAMINE

Make a map of the school grounds, or a nearby park, and locate different kinds of habitats. (An untended area would be even better.) Some parts of the ground may be covered with bushes, other parts with grass, and other areas may be dry and desertlike. Label these habitats as shady woodland, open woodland, prairies, deserts, or cultivated land. Show the main types of plants and animals found in each of these habitats.

You have probably noticed that certain organisms can live in a wide range of physical conditions. Others must live in a rather limited area. What are some examples of limiting factors? No organism can live in an area where it cannot get food or water. No animal species can survive in an area where all of its eggs are eaten. No animal can live for long in an area where all other animals eat the same food. Thus, over many years, plants and animals have become organized into small groups called *communities*. Figure 8-6 is an illustration of one kind of a seashore community.

No plant or animal lives by itself. Each is a member of its own species population and, at the same time, a part of a larger habitat. The way in which different plants and animals come to depend on each other and to live with each other is called a *biotic* (bye-AH-tihk) *community*; or simply, community for short.

Actually, no biotic community exists by itself in nature. Boundaries of different communities overlap. Plants and animals may move from one community to another. As some organisms move out, others move in. Thus, the living members of a community develop through a slow series of changes and adaptations.

These changes also are brought about by nonliving conditions as well as the growth of new plants and animals, and the moving in of other forms of life. When a biotic community, after a period of change and development, finally continues with little change for a long period of time, it has reached a state of balance. The community now is called a *climax community*. Can you give some examples of climax communities shown by plants or animals?

8-6 On what does the success of this community depend?

REVIEW

1. What substances do plants contribute to animal life in a balanced aquarium?
2. What is meant by a population of living organisms?
3. What factors can affect the density of a one-celled population?
4. Name two types of protective coloration that are found in insects.
5. What is meant by a biotic community?

B/What Forms of Life Are Found on Land?

There are broad zones of similar climates around the earth. As you may remember, the average amount of sunlight, temperature, precipitation, and many other factors determine the climate of an area. Thus, the land areas of the earth may be mapped according to zones of similar climates. Each main type of climate usually supports a particular type of plant life. Therefore, we find zones with different kinds of climax plants. For example, very wet and continually warm climates support tropical *rain forests*. Temperate climates, such as those found in eastern North America, support *deciduous* (duh-SIHD-yoo-uhs) *forests* (trees that shed their leaves seasonally). Partly dry climates support *grasslands* or prairies.

Different forms of animals are found in the main climatic zones. Arctic foxes would not be able to live successfully in a tropical rain forest. Squirrels would probably not live long in a hot, treeless desert. In other words, both the climate and the plant life determine the types of animals that are found in each zone.

Scientists refer to this relationship between plant life, animal life, and the climate as a *biome* (BYE-ome). The biomes are the main climatic groupings of plants and animals corresponding to the major conditions of life on land. Of course, biomes are not sharply divided. Plant and animal communities may change slowly and overlap. However, to make it simpler to study, scientists have divided the major worldwide biomes into the following classifications: (1) *tropical forests,* (2) *temperate forests,* (3) *grasslands,* (4) *deserts,* (5) *tundra,* and (6) *mountain areas.* Now do the activity concerning biomes.

OBSERVE

Contact the Weather Bureau in your area and find out the average temperature and precipitation in your community for the past year, month by month. Make a temperature-precipitation chart, similar to the one shown, for your area. **Do not write in your book.** In what type of biome do you live? How is your climate different from that found in a tundra or tropical forest?

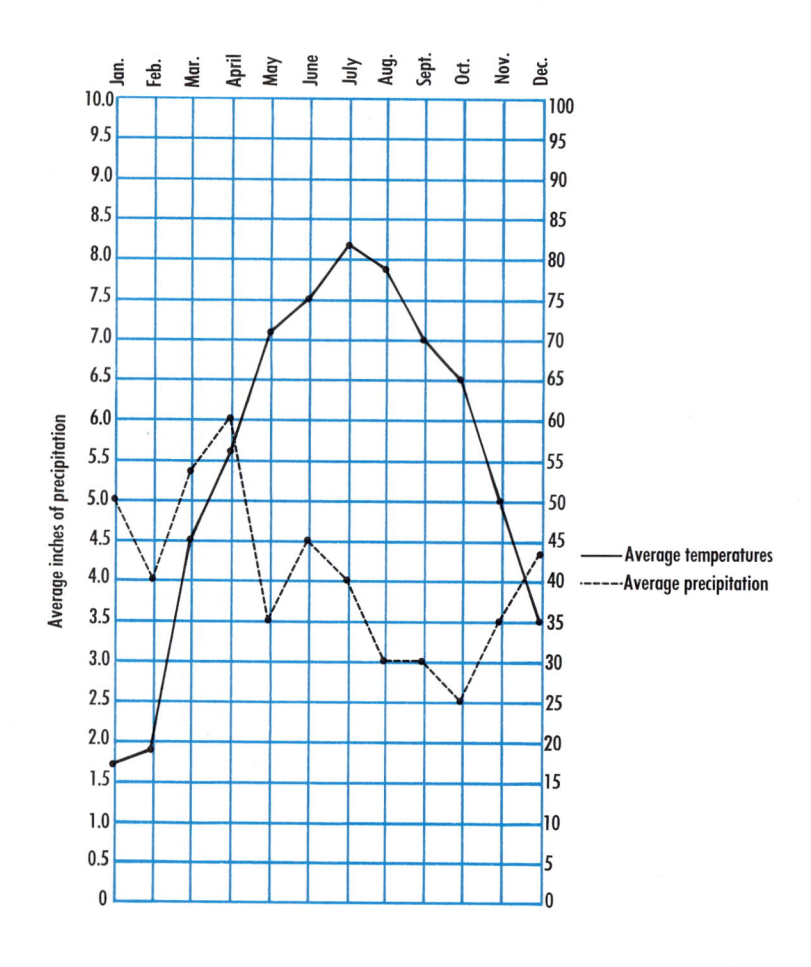

CHAPTER 8   THE RELATIONSHIPS OF LIFE / 213

Tropical lands have the most rainfall. As you might expect, the most complex forest communities develop where the rainfall is most abundant and the average temperature is high throughout the year. The area which best meets these conditions lies along the equator. The rain forests cover most of Central America, central Africa, southern Asia, and northern South America.

In these forests, there are hundreds of species of trees. The taller trees reach a height of approximately 200 feet and form a canopy, or cover, over the rest of the forest area, as seen in Fig. 8-7. If you visited this area, you would find only a very few species of green plants growing on the forest floor, since the canopy cuts off almost all of the sunlight. However, the humidity is high since evaporation from the forest floor is difficult.

Forest trees serve as support for other plants. Climbing vines, rooted in the soil, climb the trees to a point where they can get enough sunlight. Several groups of plants depend on the trees for support and grow well above the forest floor. Many of these plants have roots which are not anchored in the soil, making it difficult for them to get water. The orchid is adapted to live in this environment by having roots which grow upward instead of down. Now do the activity as directed.

OBSERVE

Examine the bark of trees growing on or near the school grounds very carefully. Collect samples of various forms of plant life growing on the bark for study in the laboratory. Shelflike and mushroomlike growths are fungus forms. A grayish scale that is often found is probably a lichen. A thin, greenish layer that may be located on the north side of the tree is a simple form of alga. What does each form of plant studied receive from the tree? Which ones are parasites?

8-7 How do tall trees affect the environment of the forest areas?

Most animals in a rain forest live in trees. During the day, the forest is rather silent except for the calls of monkeys and birds. At night, however, there is much activity as the tree

8-8 Trace the food cycle illustrated by the pyramid.

8-9 What are some characteristics of a deciduous forest?

frogs, lizards, and various rodents search for food under the cover of darkness. It is interesting to note that the mammals and reptiles are not as abundant in tropical forests as the other forms of animal life.

The *food chain* of the tropical forest follows a pattern similar to that in other biomes. The green plants are the food producers. They are eaten by the herbivores like rodents, antelopes, and wild pigs. These animals, in turn, are eaten by the larger carnivores. This food cycle may be illustrated by means of a food pyramid, shown in Fig. 8-8. In the intermediate stages are the numerous termites, flies, butterflies, beetles, and birds. Decay is rather rapid in this moist, warm area. This causes organic material to be decomposed and reused by the green plants very quickly.

The temperate forests have sufficient rain but wide temperature variation. The eastern portion of the United States was formerly covered by the temperate *deciduous forest*, made up mainly of trees which shed their leaves in winter. The British Isles and central Europe were also covered by the deciduous forests. The characteristics of this forest are a rapid growing season which is followed by a *dormant* or inactive period. During the spring, as the temperature increases and the days lengthen, the forest seems to come to life. The beeches, maples, oaks, elms, hickories, and birches are some of the varieties found in this type of forest. (See Fig. 8-9).

A thick leaf cover close to the ground enables such mammals as rodents and moles to live comfortably. The trees are not as tall as those found in the tropical forests, nor quite as dense. Therefore, more vegetation is found close to the ground.

The typical herbivores of this area are the deer. These animals are kept under control by game laws as well as by their natural enemies. There are some tree-living animals, mainly the squirrels, but not nearly as many as in the tropical forests. Mountain lions, wolves, and foxes are the common carnivores of this habitat.

At one time, many birds, such as turkeys, eagles, owls, and ravens inhabited this region. Can you name some reasons why they are now rather scarce? The scarcity of these birds has helped the smaller songbirds to flourish. Some of these forest dwellers are pictured in Fig. 8-10. You can construct a woodland habitat in the activity that follows on page 216.

8-10 Muskrat (top left), owl (top right), woodchuck (center left), porcupine (center right), turkey (bottom left), deer (bottom right).

CONSTRUCT

A woodland habitat can be made if materials are collected in the fall or spring. Place about one inch of washed gravel in the bottom of an aquarium. Cover this with a layer of crushed charcoal briquets, used in outdoor cooking. Cover the briquet pieces with one or two inches of sandy garden soil. Set a pan of water that is flush with the top of the soil to form a pond and moisten the air.

Collect small woodland plants, like evergreen seedlings and ferns. Transplant them into the habitat. Cover the soil between the plants with leaf mold or peat moss to keep the moisture from evaporating rapidly. Sprinkle well. Place a glass cover carefully on the aquarium so that some air can circulate through it. This habitat will support a few forest animals like snails, insects, and frogs. Report on conditions that are needed for life in a woodland habitat.

8-11 What kinds of animal life are supported by a grassland biome?

The grasslands are found where rainfall is low and irregular. Partly dry areas are located both in tropical and temperate zones. In the tropics, tall trees are found near water. There are vast areas, however, where the rainfall is not sufficient to support the growth of large trees. However, it is enough for the growth of grasses and small herbs. Called by various names, such as *prairies* in our West, *steppes* in Europe, *pampas* in Argentina, and *velds* in South Africa, these are all *grasslands*. Each area has its own characteristics, depending upon the soil conditions and the amount and frequency of rainfall. Usually the more rainfall in a particular region, the taller will be the few trees located near the streams and rivers.

You are all familiar with stories about the grasslands of our Great Plains. At one time, great herds of bison, pronghorn antelopes, and other herbivores roamed here (Fig. 8-11). Today, the grasslands have been plowed under. They have become a rich agricultural area where our corn and wheat belts are located.

Grasslands support numerous animal communities. For example, large herds of zebras, gazelles, antelopes, and giraffes are still found in the African grasslands. These are all herbivores which eat the plentiful grasses. Their natural enemies include lions, leopards, cheetahs, and hyenas. Rodents, burrowing animals, insects, locusts, and grasshoppers are also common in the grasslands.

Deserts result from the continual lack of water. It is difficult to determine where semidesert areas end and true *desert* areas begin. We have moved directly from the grassland to the desert because these usually lie in more direct contact than desert and forest. Some deserts, like our Salt Lake Desert in Utah, contain large amounts of salt. Others, like Death Valley (Fig. 8-12), are dry because of the geographic settings.

Evaporation is usually high in desert areas due to great heat and high winds. The absence of moisture allows the earth to cool very fast. Thus, the days are usually hot and the nights are cold.

Desert plants have certain characteristics in common, such as dense woody structures, thorns, enlarged green stems, and the absence of leaves. The large green stems and lack of leaves help the plants to save water. Their roots are usually very numerous and not very deep. This helps them to absorb some water even when only very little falls. The Joshua trees, shown in Fig. 8-13, and cacti are two plants that are associated with a desert habitat.

There are many animals adapted to live in this dry area. Lizards, kangaroo rats, snakes, and camels, some of which are shown in Fig. 8-14, are examples. Desert animals are adapted to the lack of water and the high heat. For example, the kangaroo rat is adapted to conserve water by producing urine which contains so little water that it appears as a moist solid. This animal also avoids the heat by remaining in a burrow during the day and coming out at night when the desert is coolest.

How does a camel get along without water for long periods of time? Some thought that it stored water in one of its stomachs, and others said that the camel stored water in its humps. The truth seems to be that it does not have any special organ to store water, but that its body is adapted to saving water. Camels excrete urine the way the kangaroo rat does.

8-12 What factors contribute to the formation of a desert?

8-13 What characteristics are common among desert plants?

218 / UNIT 3 MAN EXPLORES LIVING ORGANISMS

In addition, their body temperatures during the night drop to 34°C. The lower the temperature, the slower the body processes function. Thus, in the morning, the body is working slowly; this results in a saving of water.

It has also been found that the hair covering the camel aids in keeping the moisture inside and the animal cool. The camel does not begin to perspire until his body temperature reaches about 41°C. The hair usually prevents the camel's body from reaching this high temperature until late afternoon.

Thus far, we have been studying biomes in terms of their water supply. The following areas are affected more by temperature and elevation.

Special adaptations are needed for life in the tundra. The *tundra* region is a very large, treeless plain encircling the Arctic Ocean. It has an arctic climate, or very cold temperatures with a long, dark winter and a cool summer with continuous daylight. Most of the precipitation occurs in the form of snow.

The ground is permanently frozen except for the top layer during the short growing season. During the summer, the top foot or so of soil thaws and this results in a great deal of moisture in the ground. Since the growing season of about two months may have one or more frosts, any organism living in this area must be able to withstand sudden temperature changes. Dry winds increase the process of evaporation, making this a semidry region.

The plants found growing in a tundra are usually dwarfed in size. The main forms of plant life are mosses and lichens. The willows and birches that do grow remain close to the ground. Flowering plants, like herbs, are also present, and at times the tundra is colorful. Grasses found in this habitat are used for grazing.

During the summer, there is more animal life in the tundra. Waterfowl and other birds come to this area to nest. The climatic conditions help the birds to gather insects easily.

While most of the organisms migrate to the tundra, there are a few permanent residents such as the *ptarmigan* (TAHR-mih-guhn), snowshoe rabbit, and the Arctic fox. The Arctic fox eats animals such as the Arctic rabbit and a ratlike rodent

8-14 How are the lizard (top), kangaroo rat (center), and camel (bottom) adapted for desert life?

called the *lemming*. All these animals are brown in the summer but turn white in the winter. Why is this adaptation useful? (See Fig. 8-15.)

Since the tundra is completely covered by snow during the winter, the animals must either hibernate, migrate, or find other ways to get food.

Conditions in high mountains are similar to the tundra. Man has been surprised to learn that life exists up to elevations of 22,000 feet. We have learned a great deal about the areas above the tree line of high mountains. Up to 17,000 feet, the cover is similar to that which you studied earlier in this section. Dwarfed grasses and sage brush predominate in this area. Insects, like beetles, ants, wasps, moths, and even butterflies, seem to be common up to heights of 16,000 feet. Most of the flying insects find shelter in the cracks in rocks during the cold night.

At about 20,000 feet, the number of organisms thins out. Those species which do survive find a spot where submerged water runs off. This occurs near the base of rocks. Since rocks heat more rapidly than the surrounding area covered by snow, the snow resting on the rocks melts first and therefore provides water necessary for plant growth. The limit of flowering plants seems to be about 20,000 feet.

The food chain at this height depends upon the plants as food producers. Insects eat the plants and serve as food sources for other larger animals. Decay as a result of bacterial action occurs as elsewhere but much more slowly.

As you might expect, the farther one moves away from the poles, the longer the growing season and the larger the plants. In this region, called the *taiga* (TYE-guh), the forest area consists of spruce, birch trees, and juniper thickets. The forest area is known as *coniferous* (ko-NIH-fuh-ruhs) or cone-bearing. The many evergreen trees, including fir and pine, belong in this group.

The ground cover consists largely of small plants. Many animals which live on the surface spend all of their time in this habitat. Moose, elk, grizzly bear, wolverine, lynx, weasel, and mink are examples of animals living in the taiga and coniferous North American forest regions.

8-15 Ptarmigan (top), rabbit (center) and Arctic fox (bottom). Why do they turn white in winter?

Among the birds found here are chickadees, jays, crossbills, and woodpeckers. In the trees, we can also find mites, beetles, ants, and spiders. Life, therefore, is varied in the taiga region (Fig. 8-16).

8-16 Bear (top left), fox (top center), bluejay (top right), moose (bottom left), elk (bottom center) and chicadee (bottom right).

REVIEW

1. What are the main biomes that are found on the earth's surface?
2. Why is more growth found on the floor of the temperate forest than in the tropical forest?
3. How are the following organisms adapted to live in the desert: (*a*) a cactus, (*b*) a camel?
4. How do animals that are living in the tundra survive in winter?
5. How do plants living at 20,000 feet obtain water?

C/What Forms of Life Are Found in Inland Waters?

Inland water living conditions are different from those on land. All the land biomes, and even some of the desert areas, contain bodies of water of various sizes. Some water areas remain throughout the year, but others last from a few days to several months. Water areas include those found in and around inland waters, such as lakes, ponds, rivers, streams, and swamps, as well as the regions made up of the oceans. In this section, we shall turn our attention to the forms of life found in inland or fresh water.

It is possible to think of the life of inland waters as a part of the surrounding land biomes. However, although the waters are dependent in many ways on the land areas, they also have many characteristics that are quite different. The environmental conditions in a water area are more closely related to the nature of the body of water than to the surrounding biome. For example, a pond in the United States might have the same plant and animal life in it as a similar pond in France. A clear mountain stream that is located in South America will have probably the same fish population as a similar stream that is located in North America. Inland water life is considered separately from land biomes.

Water habitats may undergo change. Life in some of the water habitats faces many conditions not found on land. For example, some bodies of water are subject to wide differences in temperature. In fact, many water bodies dry up completely in the summer. At the other extreme, many bodies of water are subject to freezing temperatures from time to time. Some of these bodies of water may be frozen solid during the winter season. The amount of sunlight on the surface waters is another factor. Rivers and streams have currents which vary in swiftness with different seasons of the year. Thus, survival often depends on how well the organisms can adapt to these changing conditions. Do the activity on the next page which deals with temperature changes.

OBSERVE

The effect of temperature changes on life in a water environment can be studied by using a balanced aquarium in the classroom. Make two balanced aquariums by placing Elodea or similar water plants and some small goldfish in the clear water. As a general guide, about 20 square inches of water surface should be available for each one-inch length of fish. Do you know the reason for this? Place both aquariums in a north window so that they are lighted, but not in direct sunlight. Fasten a thermometer in each aquarium and take daily readings of the water temperature.

After about a week in this location, move one aquarium to a sunny exposure and record the temperatures for one week. Compare the types of growth in the aquarium under the two different conditions. Describe the results of the experiment. Why were two aquariums needed?

You have probably had the experience of being in water where there was a strong current. Do you remember how difficult it was to remain in one place? Organisms living in fast-moving water must be good swimmers or else be able to anchor themselves. For example, certain fish have suction cups on their undersides which help them to stick to rocks. Others are streamlined in shape to present the least resistance to the flowing water. Figure 8-17 shows several important types of fish and their adaptations.

8-17 Notice the adaptations of fin structure and body shape presented by each fish.

The oxygen content of running water is usually high. The water in streams is constantly "turning over" as it flows downstream. Small air bubbles are trapped in the running water and carried along underneath the surface. Another reason for this oxygen increase is that little decay takes place on the bottom. How does this affect the oxygen content?

The temperature in streams limits the organisms that live there because streams vary in temperature. Cold water has been found to support more life than warm water. One reason for this is that cool water can carry more dissolved gases. Can you name other reasons?

There is usually vegetation growing along the banks, as seen in Fig. 8-18. This is used as shelter for fish, air-breathing insects and larvae, and frogs. Since a stream usually has rapids and pools, as well as still-water areas, certain animals are found in other parts.

Common river animals include fish, turtle, mussels, and water insects. Shore animals are usually numerous, although many just come to the river to drink. On the surface, one may find ducks, gulls, and other water birds. Microscopic organisms are usually found where streams enter the river because the water movement is slower at that point. Why can a river support such a variety of organisms?

The brook trout is an interesting fish found in many of our freshwater streams. You are probably familiar with trout fly-fishing. This popular sport gets its name from the way in which the trout catches insects. The trout sees the insect, provided by the fisherman, on or just above the water. The trout jumps out of the water to catch it. The trout moves so quickly that it is usually successful.

Since it has a strong tail which helps it to move the back fin, the trout also is able to swim against the current. Even when it is at rest on the stream bed, it faces the current.

The bullhead is a good example of a fish modified to live in running water. Let us look at the bullhead, shown in Fig. 8-19, to see how it is adapted to live in a water habitat.

The fish has a flattened front end which reduces the amount of resistance in flowing water. Since it is flat, the water tends to flow over its surface rather smoothly. The long rear fin of

8-18 A freshwater stream supports what kinds of living organisms?

8-19 How is the bullhead adapted to live in a water habitat?

the bullhead is quite far back, which helps it to swim rapidly.

In addition to its eyes, the bullhead has three sets of feelers which help it to sense if anything is below, above, or in front of it. Since the fish faces the current, smaller organisms and dead organic matter pass by it. As they do, the bullhead opens its mouth and swallows them. It changes its direction depending upon its sight as well as its touch.

The bullhead is a *scavenger* (skav-en-juher), eating dead organic matter as well as living material. The fish does not have any scales and blends in well with its environment because it is mud-colored. In winter the bullhead digs deep into the mud where it remains until spring. Do the activity in which you observe a fish.

OBSERVE

Obtain a small bullhead (or catfish) and place it in an aquarium jar. Now add some plant life. What does the fish do? Where does it move in relation to the plants? Now add some gravel and mud to the bottom. How does this affect the behavior of the fish?

A lake is a large water community. Lakes resemble oceans in their features, but of course are much smaller. They are the most stable of the freshwater habitats. Because of this, biologists have studied them more than any other habitat. Types of organisms present in a lake depend upon the depth, mineral content, temperature variation, and elevation. The Great Salt Lake in Utah, because of its high salt content, supports very little life. However, a small species of shrimp does live in the water. Other forms of life are usually found near the mouth of freshwater streams which empty into the Great Salt Lake.

Organisms living in lakes are able to survive in a wide temperature range. Chemical processes in their bodies give them the ability to hibernate during the winter and reduce the rate at which their bodies function. Thus, the organism uses very little food and thus needs very little energy.

Most organisms living in lakes are *cold-blooded,* meaning that they take on the temperature of their environment. The colder temperatures slow down the rate at which organisms function. These animals have a built-in "thermometer," and as the water gets colder, the chemical processes slow down the rate of body activity.

The amount of microscopic life present will determine the number of food consumers the lake can feed. Since one-celled green plants are food producers, their number depends upon the mineral and gas content, as well as the depth of the water. Depending upon the type of shore, you might find beetles, ants, spiders, snails, frogs, toads, lizards, and even grasshoppers. If there are rocks close by, two birds, the kingfisher and swallow, might be present. You might also find gulls, crows, and blackbirds here.

The water lily (Fig. 8-20) represents an organism which has become adapted to a lake environment. The water lily must have still water, not too deep, and with a mud bottom. It is usually found in shallow areas in ponds and lakes. The root of the water lily is rather weak and just buries itself loosely in the mud. Its stem is flexible, yet strong. It contains hollow tubes which make it light enough to float. The leaf is modified by having stomata and guard cells on its upper surface. What would happen if the stomata were on the underside of the leaf? The leaf is rather broad, circular, and capable of floating. It is even waterproof, which prevents it from becoming too heavy.

8-20 Describe the habitat of the water lily.

The water lily is really a community in itself. Herbivorous organisms, such as beetles, eat part of the leaf. Young insects also enjoy it as a food source. The lily serves as a platform for the adult insect which emerges from the resting stage of its life cycle. Many insect eggs and tiny organisms cover its surface.

A pond is a small body of water which can support life. A pond is not very different from a lake, but it is usually not as deep. A typical pond community is shown in Fig. 8-21 (page 226). Let us study the life in a pond in your area by doing the activity that follows on page 227.

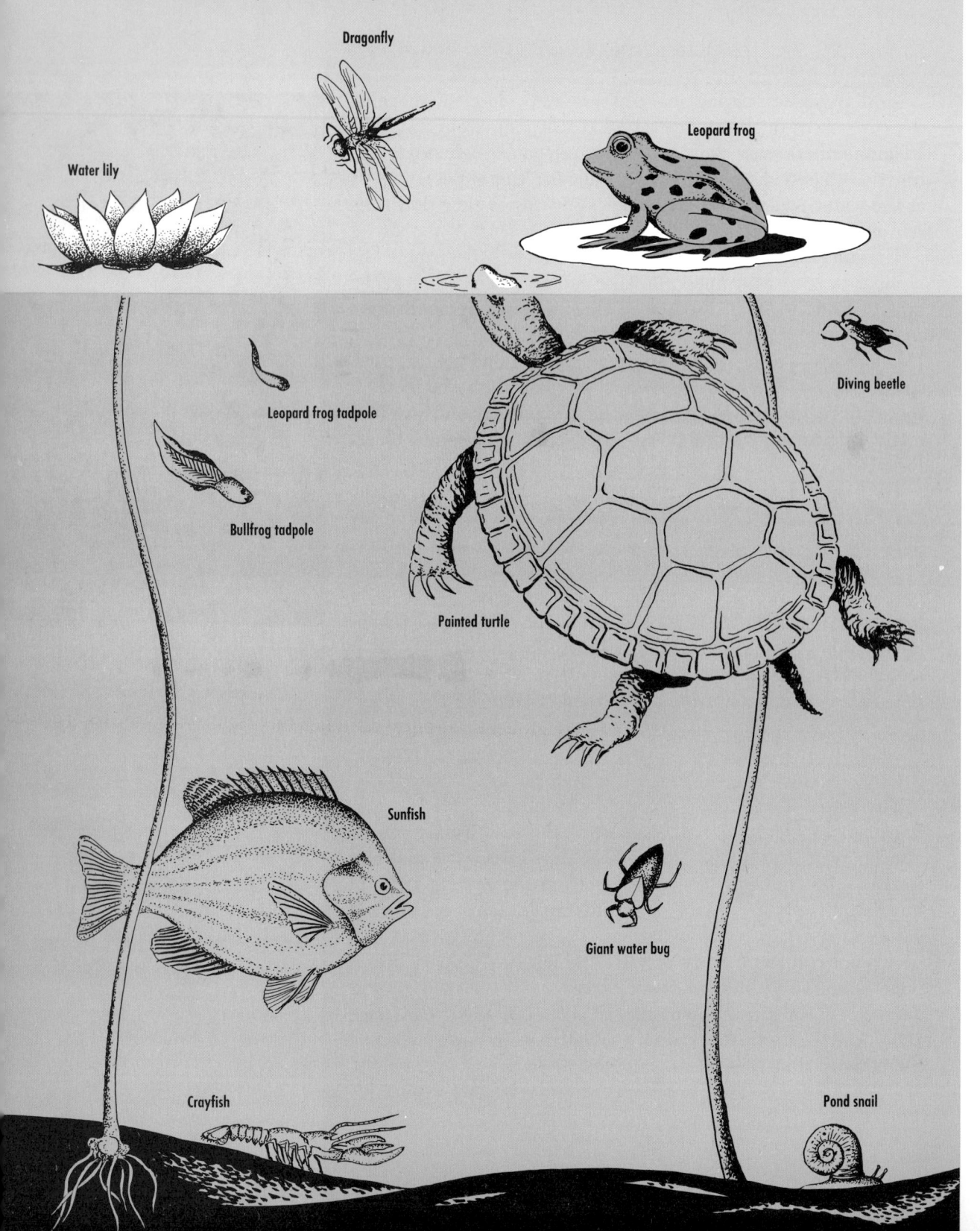

8-21 A pond community has plant and animal life above and below the water.

EXAMINE

Plan to visit a nearby pond as a class activity. Take some small jars or plastic bags into which you can put some specimens. You should also take a small net, a notebook, and a hand lens. Collect one or two members of each plant and animal group you find. Why should you not collect too many? Allow the water to remain for a few days. Then collect a sample of the water for microscopic examination. Carefully observe the many forms of living organisms in the pond.

Use a field guide to help you identify the different plants and animals. Summarize your observations on the relationships among the living organisms found in this community in your own table similar to the one below. **Do not write in your book.**

\ POPULATIONS IN A POND \ \			
Food producers	Furnish food for	Food consumers	Consumers feed on

You might think that the green plants you observed on the pond's surface or along its edge are the primary food producers in the pond. However, this is not true. Most of the food, you recall, is produced by the microscopic one-celled plants living in the water. Since different plants live in different types of ponds, depending upon the oxygen and dissolved gases in the water, it is difficult to predict which you will see when examined under a microscope. You can study pond life by doing the next activity.

OBSERVE

Take a drop of pond water from your sample (which has been standing for a few days) and place it on a microscope slide. Carefully place a cover glass on the slide to avoid air bubbles. Using the low power of a microprojector or microscope, adjust

228 / UNIT 3 MAN EXPLORES LIVING ORGANISMS

the light and observe carefully. Can you list the plant and animal organisms present in your drops? If you do not know the names, draw a sketch and look in biology books to see if you can identify the main types.

You probably saw many tiny organisms swimming by on your slide. Many of these get their food by eating the small plants and animals. They, in turn, serve as food for larger organisms which are later eaten by still larger ones. Usually, in a pond, there are no more than three or four steps within a food chain. The larger the size of the species, the fewer the numbers present in the community.

Pond insects also make up an important part of the consumer relationship. One cannot visit a pond without seeing the damsel fly or dragonfly. The dragonfly, shown in Fig. 8-22, is an important aid to man, eating the mosquitoes in the air while their young eat the mosquito larvae.

The dragonfly lays eggs which sink to the bottom of the pond. The emerging young dragonfly, called a *nymph*, does not resemble its parent. It usually remains hidden in the pond and eats mosquito larvae. Its legs have hooks which help to catch the passing larvae. It cannot fly in this stage. However, by successive sheddings of its outer skeleton, it grows and changes. It finally emerges from the water ready to fly.

The bottom of the pond supports many small worms. An examination of the sand taken from the bottom of a pond will reveal many interesting "homes." These little homes are occupied by the *caddis* (KA-dihs) *worm*. You can study these worms in the next activity.

8-22 In what way is the dragonfly helpful to man?

OBSERVE

Take a dip net or long-handled strainer and move it carefully along the bottom of the pond. Place your contents in a pail half full of water and bring it back to your classroom. Pour the water in a clean aquarium and observe the action of the caddis worm. How many different kinds of worm cases did you find? Of what help is the worm case to the worm? In what way is the worm case not helpful?

The worm is just one stage in the life cycle of the caddis fly. It begins as an egg and hatches as a larva. As it matures, it enters a resting stage by closing the doors of its "house" and spinning a silken material to keep it shut. Later, the adult fly, which is able to reproduce, emerges.

Fish are the second or third order consumers. The smaller fish, like sunfish, are found near the pond edges. They usually eat the smaller organisms, but they too have their natural enemies, like the pickerel and bass. These larger fish also have enemies like the hawk, kingfish, and man.

Bacteria decay the dead organic matter. In every food cycle, one must return to the starting point. Carbon dioxide, water, and certain minerals, like nitrogen, phosphorus, and calcium, represent the raw materials which plants use to build their food. Organic material must therefore be decomposed or broken down, in order to return the minerals to the bottom. Bacteria are, therefore, the last consumers or *decomposers* because they break down the material.

REVIEW

1. List five examples of inland water communities.
2. Name two adaptations of fish living in rapidly-flowing streams.
3. What are some typical animals that are found in a river community?
4. What factors influence the organisms that are found in a lake community?

D/What Forms of Life Are Found in the Oceans?

The oceans contain a wide variety of organisms. Up to this point, we have been studying the living organisms on land and in the related water communities. The oceans, however, cover about 75 percent of the earth's surface. They have many important habitats in which living organisms are found.

230 / UNIT 3 MAN EXPLORES LIVING ORGANISMS

The species of organisms on land are more numerous than those found in the oceans because insects and flowering plants account for the greatest number of species known. Species, however, differ from each other only in detail. The different forms of life in the oceans, on the other hand, from microscopic plants to giant whales, are very numerous. It is believed that the main forms of life found on land and in inland waters originally came from the sea.

Physical oceanography is the study of the relationship between salt content, temperature, motion of the water, the nature of the bottom and the brightness of the light in the oceans. As a dwelling place for animal life, this habitat offers conditions that change very little, because the gaseous and mineral content of the water change very little. A body of water warms and cools more slowly than land. Thus, the physical conditions which affect life in the oceans vary less than those on land.

In any community the plant which carries on photosynthesis at the greatest rate is called the *dominant species*. Within the oceans, the most dominant species are the *diatoms* (DYE-uh-tahmz) and *dinoflagellates* (dye-no-FLAJ-uh-layts). Many people think that seaweed is the most important food source. It is not because it is usually found close to shore and in a few warm spots. The tiny plants and animals which make up the main food source in oceans are called *plankton* (PLANK-tuhn). Different varieties are shown in Fig. 8-23. The total mass of plankton is far greater than the mass of all the other animals in the sea combined.

There is a food chain present in the oceans. It is always important to understand the food chain so that we do not upset its balance. Since you know that the energy within food material comes from the sun, we shall start our study with the diatoms and dinoflagellates, one-celled organisms capable of producing their own food. These, in turn, are eaten by small organisms called *copepods* (KO-puh-*pahdz*). Copepods are sometimes called the key industry animals of the ocean community since they eat the largest amount of plankton.

The copepods are in turn eaten by the smaller animals in the ocean, such as the jellyfish. These are then eaten by larger

8-23 Young jellyfish (top), copepod (center), and dinoflagellates (bottom) are food sources for larger animals.

CHAPTER 8 THE RELATIONSHIPS OF LIFE / 231

fish which are, in turn, eaten by still larger ones. A diagram of the food chain is shown in Fig. 8-24.

At this point, you might ask, "How long is this food chain?" Usually, it does not have more than five links. The chain is completed after the death of the largest member of the chain. When it dies, it sinks to the bottom where scavengers and bacteria decompose it back to basic minerals. These may remain at the bottom, or rise where they are again used by plants.

By using new instruments, scientists have been able to find the depth, light intensity, temperatures, and content of ocean bottoms. Underwater cameras have been able to take pictures of the ocean floor and some of the life living there.

Many organisms live in the shallow part of the sea. The *littoral* (LIHT-o-ruhl) *zone* is the name given to the shallow portions of the sea. It extends from the shore line to a point where the water is about 800 feet deep. This area provides three main types of habitats: rock, sand, and mud. Within the rock habitat, most animals move away from the light. Animals such as the prawn, small starfish, and snail are usually found living among the rocks. The snail, oddly enough, moves toward the light when exposed to the low tide. The *sea anemone* (uh-NEM-uh-nee), shown in Fig. 8-25, is attracted to the light.

Since light penetrates to about 150 feet, there is usually a large amount of "seaweed" to this depth. These plants belong in a group of *algae* (AL-jee) and are the food producers. The brown algae are the most dominant in the northern waters. Certain brown algae, called *kelp*, may reach a length of 50 feet or more. Figure 8-25 shows another kind, the red algae.

On the rocky shore, plants and animals must fasten themselves securely or swim freely. Otherwise, they would be washed up on the beach. They also must be capable of withstanding various degrees of drying when the tide is out. Temperature variation is greatest at the shore line. Snails and barnacles of various species are usually found close to the shore. Since they have protective shells, they can withstand the exposure to dryness and temperature change better than the starfish, crayfish, or worms which are found farther out

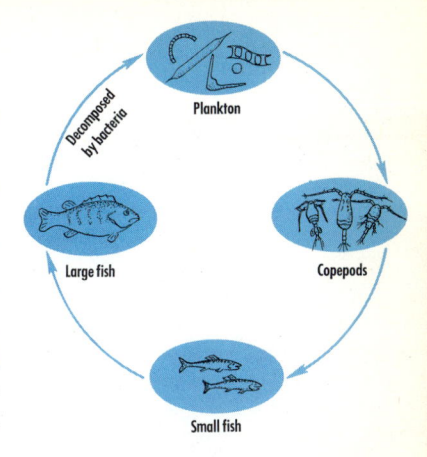

8-24 What link begins the food chain shown above?

8-25 Describe the habitat of the sea anemone.

8-26 What is the value of red algae, shown above?

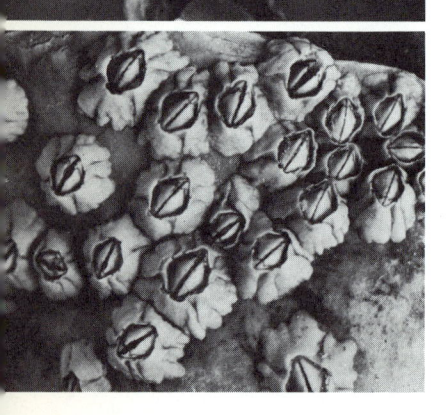

232 / UNIT 3 MAN EXPLORES LIVING ORGANISMS

in the water. (See Fig. 8-27.)

Sandy beaches have fewer organisms than rocky ones, since sand has no solid bottom for organisms to hold onto. Because the sand would bury them, only certain organisms capable of burrowing into the sand usually live in this habitat. Worms, snails, and some crabs might be found along these beaches. Examine some of these small organisms by doing the activity that follows.

EXAMINE

If you live near the shore, obtain a sample of the sand just at the water's edge. Place it in a container with some of the sea water. Examine the sand under the microscope or microprojector. Do you see any small organisms with hard shells and jointed legs? These are the copepods. What other microscopic organisms can you find in your sample?

Muddy shores pose similar problems to organisms as do the sandy shores. The mud crab, worms, and clams are most common along these beaches.

Many shore animals have protective coloration which enables them to blend in with their surroundings. Such an animal is the lobster, shown in Fig. 8-28, which has a bluish-black color similar to its environment, making the animal extremely difficult to see.

Warm tropical seas contain the coral reefs. The corals are largely found in warm seas where the temperature of the water is about 20°C. A coral reef is built up through the activities of several different organisms living together. Certain plants and small animals are binders and shapers.

It has been found that the color of many coral reefs is caused by the presence of red algae. Tropical sea bottoms may look like gardens, but many of the "flowers" are not really plants but colorful animal forms.

The number and kinds of individuals decrease with depth. Since all the food is produced near the surface, the only primary food available in deep water is that which drifts down from the surface. Thus, there is a great food shortage

8-27 Why can snails (above) and barnacles (below) withstand weather changes?

8-28 Why is a lobster difficult to see in its natural environment?

in deep waters. Those organisms living in the ocean depths have special adaptations for obtaining food. For example, many creatures have their own "light source," to attract other organisms which they then eat. Some deep-sea creatures have very large mouths which help them to swallow larger organisms (Fig. 8-29).

The Antarctic Ocean supports one of the richest food chains. You might think that little life would be found near the ice-capped waters of the Antarctic. It seems logical that waters always close to 0°C might not be able to support life. Yet, a small shrimplike organism, called *krill*, is a vital factor in the food chain. The krill is capable of eating one-celled plants of the sea and the krill, in turn, serves as food for fish. Penguins and birds feed on the fish and are thus indirectly dependent upon the krill. In fact, even seals and whales feed directly upon enormous quantities of krill.

It is of interest to scientists that organisms living in the Antarctic waters have longer life spans than those found in warmer waters. It is believed that their body rate slows down, thus slowing down their aging process. The number of different species living in these waters is much less than elsewhere, but the number of each species is very large. Seals and whales are found in large numbers here.

Some marine birds and reptiles can drink salt water. Man is not able to drink salt water to quench his thirst. You have probably read stories of sailors dying from lack of water while they were at sea.

Recently, scientists have found that certain marine birds and turtles have an organ which removes the salt from the water, thus helping them to drink salt water. The herring gull, penguin, blue heron, and brown pelican are examples of birds that are adapted to drink salt water (Fig. 8-30). The tears secreted by a marine turtle are produced by its salt glands which are located just behind the eyes. Thus, these animals eliminate salt which is present in the water that is taken into their bodies.

Much more research and study is necessary before we can learn the many secrets of the deep. Oceanographers and marine biologists are spending much time both at sea and in the

8-29 How are these fish (above) adapted to life at great depths?

8-30 Pelicans (top), heron (below). Why can they drink salt water?

laboratory trying to uncover the mysteries of the sea. Every day new information appears as a result of the work of these research scientists.

REVIEW

1. About what percentage of the earth's surface is covered by the oceans?
2. What are the two dominant plant species in the oceans?
3. What forms of life are: (*a*) plankton, (*b*) copepods, (*c*) krill?
4. What are some forms of animal life found in the littoral zone?
5. Name several birds which are able to drink salt water and survive.

THINKING WITH SCIENCE

A. *On a separate sheet of paper write the numbers 1 to 15. After the number of each question, write the term that correctly completes each statement.* **Do not write in your book.**

1. The shallow, off-shore part of the sea is the _____ zone.

2. The natural environment of an organism is called its _____.

3. The vast treeless plain around the Arctic Ocean is called a _____.

4. The way in which different plants and animals depend on each other and live together is known as a _____.

5. The one-celled organisms making up the greatest mass of food in the oceans are called _____.

6. Color adaptations that help animals avoid enemies are called _____.

7. The complex forest communities which are found near the equator of the earth are called _____.

8. The number of all species of a particular plant or animal found in a specific area is called the _____.

9. The final community that continues with little change over a period of time is called a _____.

10. The relationship between plant life, animal life, and the climate is known as a _____.

11. Animals which do not make their own food are called _____.

12. Insects resembling other objects to escape enemies illustrate _____.

13. Trees that shed their leaves each year are the dominant plant in the _____ biome.

14. Pampas, velds, steppes, and prairies are all types of _____.

15. The study of the relationships which control life in the oceans is known as _____.

B. *Write the answers to the following in your notebook. Be sure to use complete sentences and correct spelling and grammar.*

1. How does a balanced aquarium illustrate interdependence in plants and animals?

2. What is the difference between a population and a community?

3. Explain why density limits a population.

4. Compare the living conditions in a temperate forest biome with those in a desert biome.

5. Describe the food cycle in a tropical forest.

6. How are animals that live in the tundra able to survice in that environment?

7. Explain why the oxygen content of running streams is usually higher than in stagnant ponds.

8. Describe the food cycle in the oceans.

9. What kind of organisms are found in littoral zones?

10. Explain how a coral reef is formed.

RESEARCH IN SCIENCE

1. Set up several habitats by obtaining soil, water, and plants from as many of these areas as you can: seashore, lake, stream, pond, desert, and moist swamp. Study the various organisms present in each habitat and list the adaptations necessary for their survival.

2. Find a shaded area and a sunny spot on or near your school grounds. Mark off one square foot of ground in each location. Compare the quantity and type of life present in each.

3. Prepare a report on one of the following topics:
   (a) New developments in oceanography.
   (b) Algae as a food source for man.
   (c) The effect of water pollution on plant and animal life.

4. Make a table of the common native animals of your region. Include their habitat, diet, method of survival in winter, special adaptations, and other facts about their importance.

chapter 9
The Diversity of Life

A/What Are the Main Plant Groups?

There are many different groups of plants. In this section, we shall start by studying simpler plants and gradually work up to the more complex types. To help you with this study, look over the table on page 241 which summarizes the main plant groups. Characteristics of each group are given with common examples of each listed.

Bacteria are one-celled organisms. Like all plant cells, *bacteria* have a cell wall and are made of protoplasm. They are so small that it is difficult to see them even under the high power of a compound microscope. Bacteria are found everywhere and can live under many conditions. They are believed to be the most widespread forms of life. Bacteria belong to the group of plants which includes all plants lacking true tissues, like roots or stems. The *fungi* (FUHN-jye) and *algae* (AL-jee), which are the two forms of life we shall study next, are also members of this group.

As you know, bacteria do not contain chlorophyll and are decomposers. That is, they survive by breaking down living or dead materials into simpler substances. Therefore, most bacteria are parasites or saprophytes.

CHAPTER 9 THE DIVERSITY OF LIFE / 239

There are three basic forms of bacteria. *Coccus* (KAH-kuhs) forms are round, *bacillus* (buh-SIHL-uhs) are rod-shaped, and *spirillum* (spye-RIHL-uhm) are shaped like small corkscrews (Fig. 9-1). Most bacteria are found as groups of cells or *colonies*. Researchers prepare *cultures* which are used to grow the bacteria they are studying in the laboratory. You can find out how bacteria are grown in the following activity.

EXPLORE

Wash four fruit jars and lids thoroughly in hot, soapy water and rinse. Boil a potato in water until it is soft. While you are waiting for this, place the jars and lids in another pan of boiling water and boil for 30 minutes. Soak the test-tube holder, forceps, and knife in alcohol for 15 minutes. Remove the glass jars and lids from the hot water with the forceps. Cover the jars loosely with the lids. Be sure you do not touch the *inside* of the jars or lids with your fingers. Replace the forceps in alcohol.

Slice pieces of potato into each jar as shown and cover each one with the lid at once. Using the forceps, place a strand of hair on the slice of potato in one jar and a small piece of paper that you have rubbed with your fingers in the second. Remove the lid from the third jar and cough into it. Do not open the fourth jar or place anything in it. Screw the lid on each jar as soon as you have done this. Label the jars and leave them in a warm, dark place for several days.

Why were the jars and lids boiled in hot water? Why were the instruments used soaked in alcohol? What is the purpose of the fourth jar? Describe the appearance of the potato slices in each jar after several days. What conditions are needed by bacteria to grow and multiply?

Bacteria multiply by cell division. Bacteria reproduce by *fission* (FIHSH-uhn), a process in which the cell divides in two equal parts. Each new cell then grows to full size and may then redivide. Bacteria may divide as often as every 20 minutes. In an hour, one cell could produce eight cells, and in two hours there would be 64 cells. How many cells are there in four hours?

Bacteria can survive harsh changes in their environment by forming *spores*. In this process, the bacteria are able to form

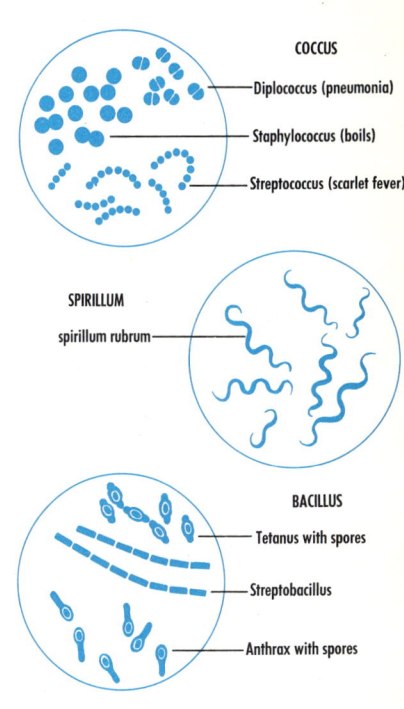

9-1 How are bacteria classified?

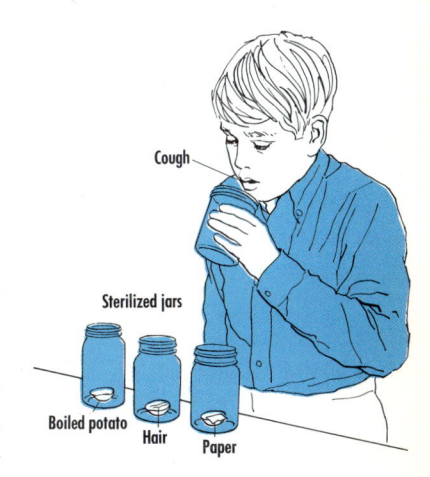

hard walls around themselves. They remain in this resting stage until conditions become favorable again for their growth. This condition permits the cells to survive in extremely hot, cold, or dry conditions.

Many bacteria are helpful to man. They are essential to the dairy industry because it is their action on milk which changes it into cheese. Bacterial action is also responsible for the production of vinegar from alcohol. Bacteria are important in the tanning of hides into leather. Some bacteria are extremely valuable in taking nitrogen from the air to change it, putting it back into the soil as nitrates. Bacteria also aid in decaying organic matter.

Although most bacteria are helpful, or at least not harmful, there are some forms that do cause disease in plants or animals. They may cause diseases by producing chemical substances called *toxins*. Toxins act as harmful poisons in an organism, or they may cause a fatal disease by destroying plant or animal tissues.

Fungi are another group of simple plants. *Fungi* [sing. *fungus* (FUHNG-guhs)] include many common organisms. Some of these organisms are single-celled plants and others are many-celled plants. *Yeast* is a one-celled fungus which is very important to both the baking and brewing industries. The breakdown of sugar by yeast produces two important products: *carbon dioxide* and *alcohol*. This process is called *fermentation* (*fuhr*-men-TAY-shuhn). You can learn more about yeast by doing the next activity.

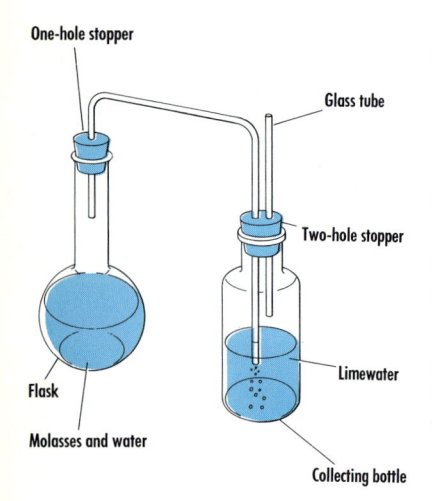

EXPLORE

Mix some molasses and water in the flask. Add some yeast and stir thoroughly. Pour clear limewater into the collecting bottle. Arrange the equipment as shown and allow the yeast to act overnight.

What kind of action do you observe in the flask? What happens to the limewater in the collecting bottle? For what gas is limewater used as a test? Describe the odor of the mixture in the flask after it has been standing a few days.

MAIN GROUPS OF PLANTS

Group	Characteristics	Examples
Bacteria	Microscopic single cells lacking in chlorophyll; reproduce by fission	Coccus, bacillus, spirillum
Algae	Single cells, filaments, or sheets of cells; have chlorophyll; reproduce by fission or spores	Spirogyra, Chlorella, diatoms
Fungi	Single cells; most occur as groups of threadlike cells; do not have chlorophyll; reproduce by spores	Yeasts, molds, mushrooms, bracket fungi
Bryophytes	Many-celled green plants living on land; reproduce by spores and gametes	Mosses and liverworts
Ferns	Many-celled green plants; have roots, stems (rhizomes), and leaves; reproduce by spores and gametes	Ferns, horsetails, club mosses
Seed-producing plants	Many-celled green plants; have true roots, stems, leaves, and seeds; reproduce by seeds	Conifers and flowering plants

Fungi do not make their own food. Thus, yeast, which is a member of this group, is a saprophyte. It must obtain its food from other food materials. The yeast cells secrete a digestive substance into the food material, and digestion takes place outside the cell. The digested substances are then absorbed by the cell. Most of the saprophytes obtain their food supply in this way.

Since fungi do not contain chlorophyll, they cannot make their own food. They are parasites or saprophytes. Although some species are useful, many others are harmful.

Higher fungi, like mushrooms and shelf fungi, have the main part of the plant below the surface. That is, the plant grows in the bark of a tree or in dead organic material. Digestion occurs outside the body. Thin, threadlike cells produce digestive juices. These digestive juices change the insoluble material so that it can be absorbed by the fungus. The visible portion you see is usually the *spore case*. Many fungi are poisonous.

Molds are composed of threadlike structures. Another group of plants belonging to the fungi are the *molds*. Molds resemble neither the yeast nor the mushroom. They are made up of tiny, threadlike cells or filaments called *hyphae* (HYE-fee). Molds are extremely helpful to man in many ways. Both *penicillin* (*pen*-ih-SIHL-ihn) and *terramycin* (*ter*-uh-MYE-sihn), disease-fighting drugs, are produced from molds. Molds are also used by man to flavor certain cheeses like Roquefort and blue cheese.

Algae are one-celled plants containing chlorophyll. Algae [sing. *alga* (AL-guh)] are found in the sea, in fresh water, in the soil, and on moist stones. Although these plants have no roots, stems, or leaves, they are able to make their own food by means of photosynthesis. Algae are very important to the food cycle in nature. They are a source of food for many water organisms.

Spirogyra (*spye*-ruh-JYE-ruh) is a common alga found as long, slimy strands floating on top of pond water. Each strand is called a *filament*. It is composed of a number of cells joined end to end (Fig. 9-2). Yet, each cell in the filament is independent, carrying on its own life processes. Each Spirogyra cell contains one or two ribbonlike chloroplasts in which the chlorophyll is located. As in all plant cells, each cell also contains vacuoles, plastids, a nucleus, and cytoplasm. You can study Spirogyra in the next activity.

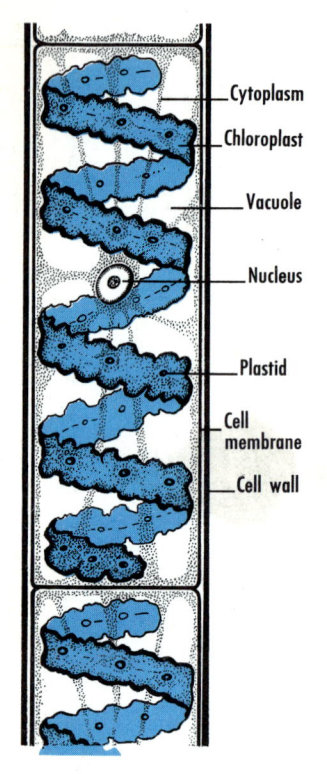

9-2 Where is the chlorophyll in a Spirogyra cell?

EXAMINE

Place a Spirogyra filament on a clean microscope slide and cover it with a cover glass. Examine the strand under the low power of a microscope or microprojector. Locate the cell wall and the chloroplasts. Under high power you might be able to see the nucleus. Compare it with the illustration (Fig. 9–2).

Spirogyra reproduces by two methods. As in all one-celled organisms, Spirogyra reproduces by fission. However, when a cell divides, it does so across the width of the cell and remains as a part of the filament. When a filament reaches a certain length, it may separate into two strands.

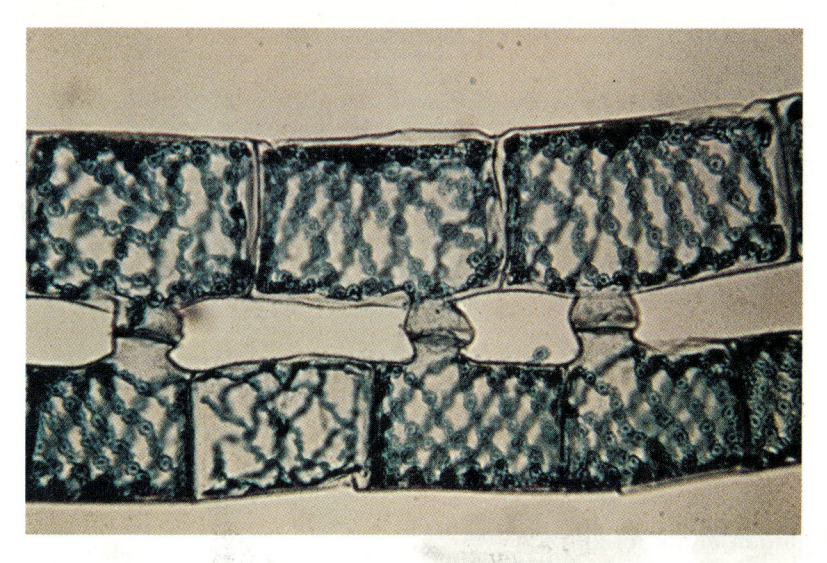

9-3 Describe the reproductive process of the Spirogyra cells above.

Spirogyra also reproduces by *conjugation* (kahn-joo-GAY-shuhn). During this process, two filaments line up next to each other (Fig. 9-3). The cells begin to form extensions or "bridges" of protoplasm between each other. The contents from one cell flow into the other cell, and the two nuclei unite. A hard wall then forms around this "fused" cell, forming a reproductive spore which can survive unfavorable environmental conditions for a long time. The "empty" cell breaks up and disappears. When pond conditions become favorable, the spore breaks open and a new Spirogyra grows.

Now let us look at *Chlorella* (klo-REL-uh), another species of algae, which is diagrammed in Fig. 9-4. It has a chloroplast which contains the chlorophyll, a cell membrane, cell wall, nucleus, and cytoplasm. It is usually found on the surface of ponds. Chlorella has been studied by scientists as a possible source of food and oxygen for man. Study algae in the next activity.

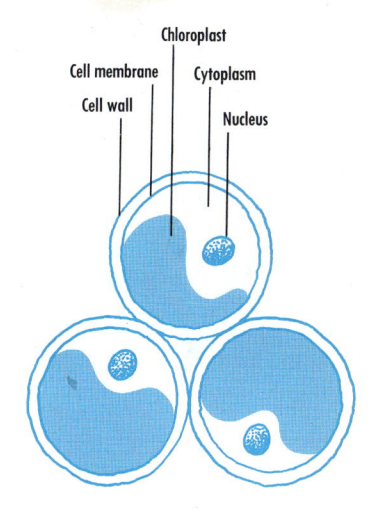

9-4 How does the size of chloroplasts in Chlorella cells compare with those found in other algae?

OBSERVE

Obtain some pond scum containing Chlorella. If you have difficulty, locate a similar species which grows like a green film on trees in moist areas. Bring in a piece of bark with these algae.

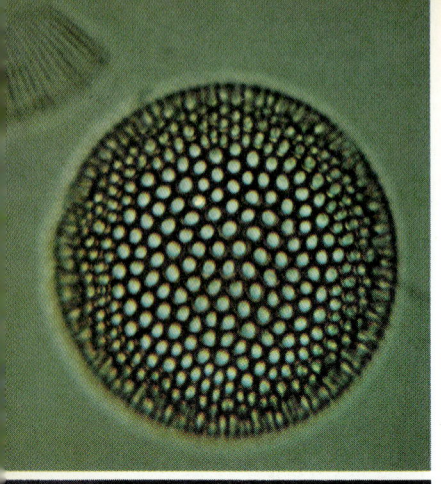

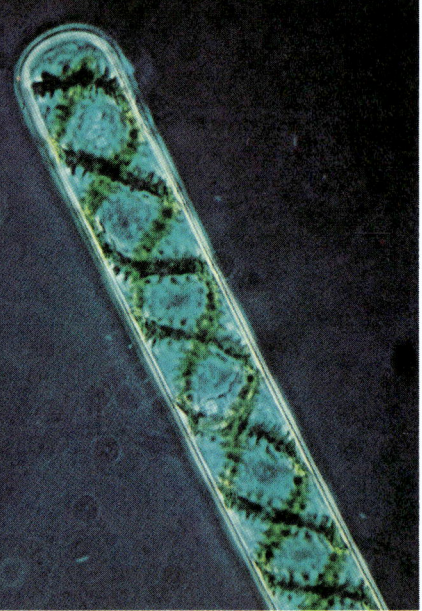

244 / UNIT 3 MAN EXPLORES LIVING ORGANISMS

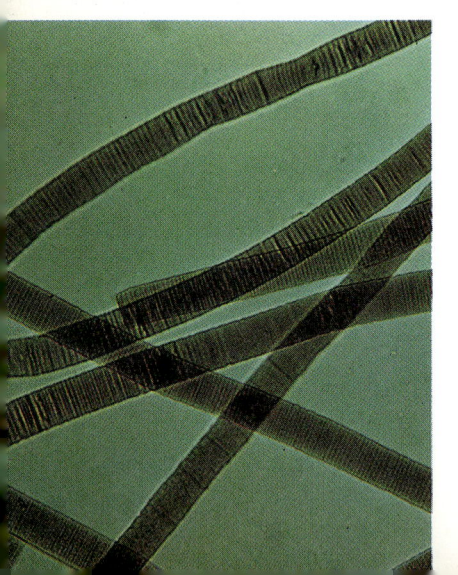

Scrape off some of these cells and place them on a slide. Add a drop of water and cover with a cover glass. Examine the slide under the high power of a microscope or microprojector. Locate the different cell parts and compare what you see with the illustration. Draw some of the cells.

In addition to green chlorophyll, algae contain many other pigments which are located in the cytoplasm. These pigments may give the algae different colors. As a result, there are classifications of red algae, brown algae, blue-green algae, and golden algae. Figure 9-5 illustrates some of the smaller forms of plants within this group. You may wish to collect some of these types from pond water and study them more closely under a microscope.

Mosses and liverworts are adapted for life on land. *Mosses* are small green plants which do not have well developed roots, stems, or leaves. They are found in moist shaded areas or in patches in deep woods. They grow only about six to eight inches in height.

An important moss is *sphagnum* (SFAG-nuhm), which grows in swampy places. This plant provides us with peat, which is used as fuel in some countries. Dried sphagnum is used by florists as a packing material because it retains large amounts of water.

Liverworts (LIH-vurh-*wuhrts*), simpler in structure than mosses, are not as common as mosses. Their leathery leaves are spread out flat on the ground. They are found clinging to damp soil and stones. Usually, liverworts grow only one or two inches in height.

9-5 Some varieties (above and below) in the algae group.

REVIEW

1. List the three main shapes of bacteria.
2. By what method do bacteria reproduce?
3. What is the main difference found between the groups, algae and fungi?
4. In what ways do Spirogyra reproduce?
5. Why do some forms of algae produce spores?

CHAPTER 9 THE DIVERSITY OF LIFE / 245

B/What Are the Characteristics of Higher Plants?

Ferns have well developed roots, stems, and leaves. There are about 9000 species of *ferns* on earth today. Millions and millions of years ago, there were many fern tree forests which covered the wet, marshy land common at the time. These ancient plants are the source of present coal deposits.

The ferns, shown in Fig. 9-6 (top), grow in wooded areas. They are common in flower displays. An examination of the fern leaf will help you to see small clusters of *spore cases* which are located on the underside of the leaf. These spore cases produce spores. The following activity will help you understand the structure of ferns and spores.

OBSERVE

Obtain a fern leaf and study its structure with a hand lens. Locate the spore cases. Shake some of the spores onto a slide. Add a drop of water and cover with a cover glass. Observe these spores under the low power of a microscope or microprojector. How are they different from bread mold spores?

The *horsetail*, also seen in Fig. 9-6 (bottom), is related to the fern. These, along with the *club mosses*, existed in great numbers many years ago. Today, few remain for us to see. However, they can still be found in lowlands and in shallow water near the shores of lakes. Club mosses are used during the Christmas season as decorations. The "clubs" on top of the moss are really spore-producing cones.

The majority of common plants are seed producers. Seed-producing plants are divided into two main groups: the *flowering plants* and the *conifers* (KAHN-is-fuhrz). The conifers have no flowers but bear their seeds in specially-adapted *cones*. The pine, cedar, fir, and redwood are all evergreen trees and are conifers. Their needlelike leaves are able to survive great temperature changes. The next activity will let you compare evergreen plants.

9-6 Ferns (top), horsetails (below). To what group do they belong?

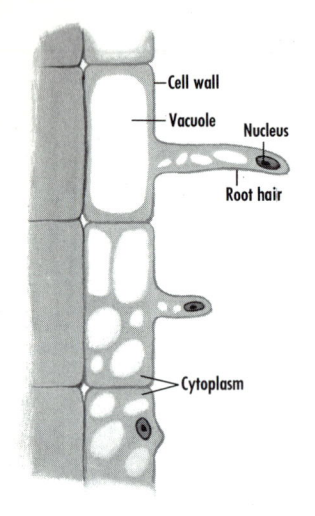

9-7 Are the root hairs one-celled or many-celled?

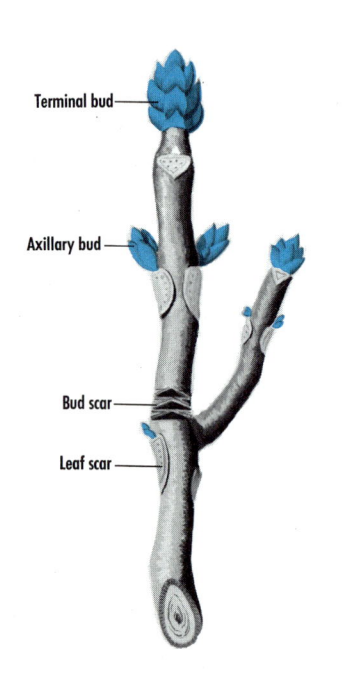

IDENTIFY

Bring in samples of leaves from evergreen trees growing in your areas. Can you identify the various species from their leaves? Are all the needles the same shape? Cut one in half and look at the inside of the needle with a hand lens. Note the thickness of the outer layer. Why is this thick layer helpful to the plant?

The flowering plants are adapted to carry on complex life processes. This group consists of more than 250,000 species. They are found in almost all of the habitats we have studied. Flowering plants range from those which live for only one season to those which live for hundreds of years. They represent the highest form of plant development and adaptation. Almost all these plants have well developed roots, stems, and leaves. Their most common characteristic is the ability to produce flowers and seeds.

The *roots* of these plants serve as anchors. Their main function, however, is to absorb water and minerals from the soil. *Root hairs*, shown in Fig. 9-7, are small one-celled extensions from the root. They increase the root's ability to absorb materials by providing a greater absorbing area. Because root hairs are so small, they can work their way in between the fine particles making up the soil.

The plant *stem* serves as a support for the leaves and as a pathway for the flow of substances from the roots to the leaves and back again. Some stems are soft and green, like the stems of the tomato, corn, and grass plants. Other stems are woody, as in trees, that can grow in length and increase in diameter. Stems grow in length by forming new tissue at their tips. You can observe buds, in the next activity, by examining a twig removed from a broad-leaf tree when the seasonal growth has stopped.

COMPARE

Take a twig from a tree during early winter. Compare the twig with the illustration and locate the terminal bud, leaf scars, bud scars, and axillary buds. Determine the role of each part.

The diameter of the stem increases because of the growing region located just under the bark. This region is called the *cambium* (KAM-bee-uhm) *layer*. During the spring and summer, the cambium layer is very active. Its cells rapidly produce new cells which help the tree to grow in diameter. If conditions for growth are favorable, the growth in diameter is quite large. If there is little rainfall or if it is very cold, however, the growth will be small. By examining cross sections of woody stems, scientists can determine the growth conditions of a particular year. They can also tell the age of the stem by counting the *annual rings* of wood as seen in Fig. 9-8. You can observe the growth of some common plants from seeds in the next activity.

9-8 A cross section of tree showing its annual rings. Why are some rings wider than others?

OBSERVE

Plant some bean or corn seeds in two flower pots. When the young plants are about two inches high, place one pot under a box with a window cut in it, so that no light can get in except through the hole. Leave the other plant where it gets light from all sides. Water both pots as usual, but do not change the position of the plants in the box during the experiment. Describe any differences in the growth of the plants. What seems to be the effect of light on stem growth?

The leaf contains chlorophyll which is needed for photosynthesis. Let us observe the parts of the leaf to see which adaptations help them to carry on this important process.

A typical leaf consists of a *blade* which is strengthened by many *veins*. These veins contain vessels made of specialized tissue which carry water and minerals from the stem to the leaf. They also carry food from the leaf to other parts of the plant. Each kind of leaf has a definite arrangement of veins as can be seen in Fig. 9-9. The blade of a leaf is attached to the stem by a stalk which is called a *petiole* (PET-ee-*ole*).

You will recall that a leaf consists of several layers (see Fig. 9-10). The *upper epidermis* (ep-ih-DUHR-mihs) usually has a single, clear cellular layer which allows sunlight to pass

9-9 Different vein arrangements in four types of leaves. Describe each type.

White oak

Maple

Pea

Strawberry

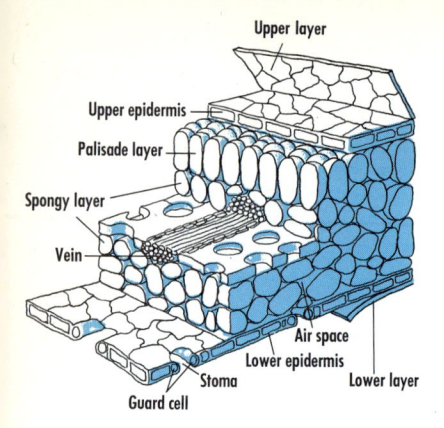

9-10 A cross section of a leaf. In which layer are the stomata found?

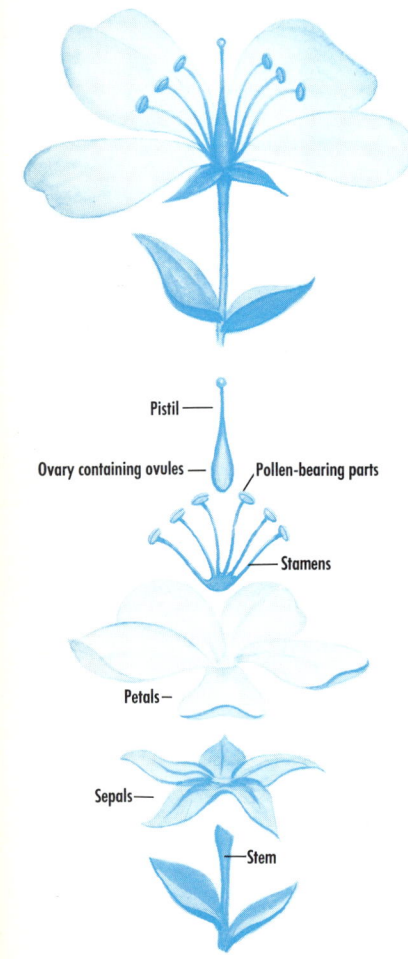

9-11 A simple flower. Identify the male and female parts.

248 / UNIT 3   MAN EXPLORES LIVING ORGANISMS

through. The layer just below, called the *palisade* (pal-ih-SAYD) *layer,* consists of long and narrow cells. They contain the chloroplasts which contain chlorophyll. A *spongy layer* of cells below these consists of thin-walled cells surrounded by air spaces. The cells of this layer store foods made in the palisade cells. In addition, the air spaces allow the carbon dioxide and other gases to reach the palisade cells. The *lower epidermis* usually consists of a one-celled layer which contains the *stomata* and *guard cells.* You will recall that the stomata regulate the amount of material passing into and out of the leaf, while the guard cells control the size of the stomata.

The flower is the reproductive organ of a plant. The main parts of a simple flower are shown in Fig. 9-11. The *stamen* (STAY-men), the male sex organ, produces *pollen grains.* The *pistil,* the female sex organ, contains the *ovary.* The sperm cells are produced in the pollen grains, and the egg cells, or *ovules* (o-vyoolz) are produced in the ovary. The colorful petals are used to attract insects. Insects aid in the process of *pollination* (*pahl-ih-*NAY-shuhn), the transfer of pollen from the stamen to the pistil. Study the various parts of a flower in the next activity.

EXAMINE

Bring in a simple flower and identify its parts. Remove the parts of the flower and lay them out on a sheet of paper. Count the number of petals, stamens, and pistils. Shake some pollen from a flower onto a slide. Then, add a drop of water and a cover glass. Examine under the low power of a microscope or microprojector. Draw some grains.

When the mature pollen escapes from the stamen and lands on the pistil of the same species, a tube begins to develop from the pollen grain. This tube grows down the center of the pistil until it reaches one of the ovules. One of the sperm cells, which is inside the pollen tube, unites with the egg cell to form a fertilized egg. The fertilized egg then develops into a tiny, young plant called the *embryo* (EM-bree-o). Around

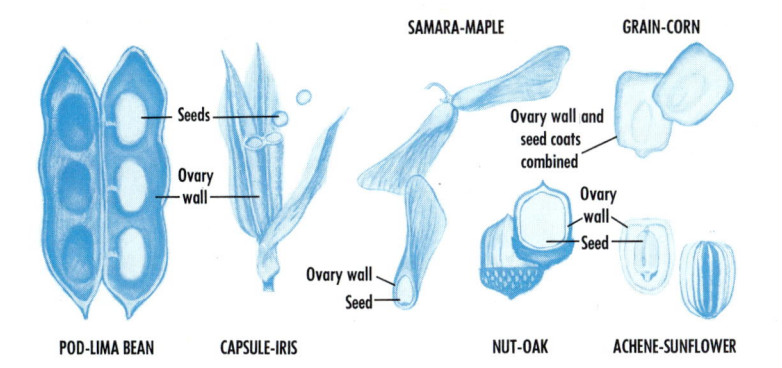

9-12 What parts make up a typical seed?

the embryo, we find stored food surrounded by a protective coat. Thus, a *seed contains a tiny embryo, stored food, and a protective seed coat.* When the seeds are ripe, the flower withers and dies, leaving the ripened and enlarged ovary. The ripened ovary containing the seeds is called a *fruit*. The fleshy fruit tissue protects the seeds and aids in scattering the seeds as you will learn later (see Figs. 9-12 and 9-13). Compare the structure of several fruits in the next activity.

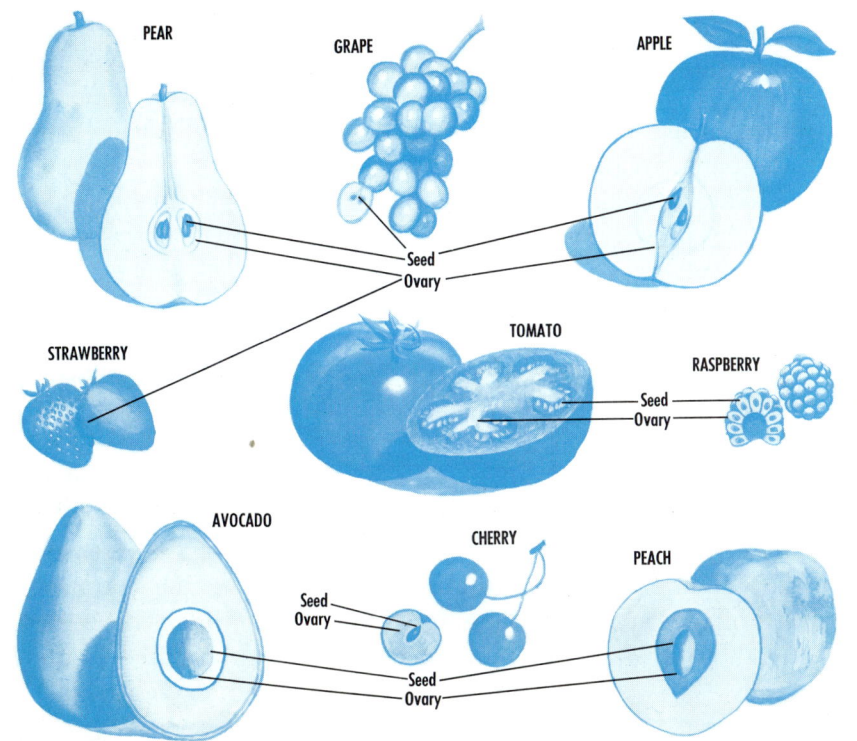

9-13 What is the purpose of fruit tissue surrounding seeds?

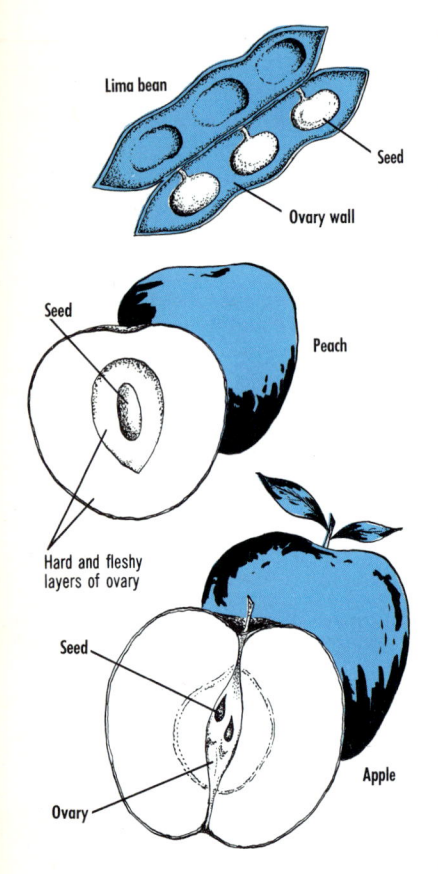

COMPARE

Take an apple, peach and lima bean and cut them open as shown. Identify the various parts. Observe the similarities and differences between them. Now carefully remove one bean seed and open the seed coats. Can you locate the embryo and stored food? Is there a relationship between the type of fruit and the amount of stored food in the seed?

Seeds are scattered in a variety of ways. If all the seeds that are produced by a parent plant fell to the ground near the parent, these seeds would be so crowded that only a few would be able to develop. Plants, however, have many ways of scattering seeds. Sometimes, only the seeds are carried away, and at other times, the fruit containing the seeds is also carried away some distance.

The wind carries many fruits and seeds. The fruits of the maple, elm, and dandelion can be carried many miles by the wind because they are shaped like tiny wings, light in weight.

Some seeds are exploded or shot out from their fruits. The pod fruits, like beans and peas, dry up when the seeds are ripe. As they ripen, the pods begin to curl. The springy action of the curling pod shoots the seeds some distance away from the parent plant.

Water also aids in seed scattering. The coconut palm, which usually grows along a seashore, drops its fruit into the water. The fruit does not sink. Floating, it is carried along by the water until it washes up on a beach, miles away. The seed that is inside the fruit then begins to grow, pushing through the protective cover of the fruit and taking root on the distant shore.

Animals also aid in carrying seeds considerable distances. Many fruits, like needle grass and wild carrot, have *stickers*. As the animal rubs against these plants, the fruits or seeds of these plants attach themselves to the animal's fur. You may have found some "stick-tights" or burdocks attached to your clothing. Animals also eat fruit but cannot digest the seeds. As they pass their waste material, the seeds drop to the ground. They may begin to grow in the ground if the conditions are favorable for growth.

CHAPTER 9 THE DIVERSITY OF LIFE / 251

REVIEW

1. How do ferns reproduce?
2. What are the two main groups of seed-producing plants?
3. What is the function of the: (*a*) root, (*b*) stem, (*c*) leaf, (*d*) flower?
4. How can you tell the age of a woody stem?
5. Give the main ways in which plants are able to scatter their seeds.

C/What Are the Characteristics of Single-celled Animals?

Animals are divided into two large main types. All animals, composed of only one cell, are members of the *protozoan* (pro-tuh-zo-uhn) group. All other animals which are made up of more than one cell are known as many-celled animals or *metazoans* (*met*-uh zo-uhnz.) Almost all of the common forms of animal life with which you are familiar belong in the metazoan group of animals.

The *ameba* (uh-MEE-buh), seen in Fig. 9-14, is a one-celled organism found in fresh water. It performs all the main life processes, even though it is a single cell made up of a mass of protoplasm. The ameba is thought to pull itself along. The flowing protoplasm contracts at the tip of the "false foot" or *pseudopodium* (*soo*-do-PO-dee-uhm), pulling the cell towards this point.

An ameba ingests a food particle by flowing around it and over it. Once the food is in the cell, a *food vacuole* is formed and digestive substances are produced in the vacuole. These secreted substances help change the insoluble material. The digested food can now be used as an energy source by the ameba.

The soluble material passes into the surrounding material and enters the cytoplasm and nucleus where it is used. The undigested material which remains in the food vacuole is eventually excreted through the cell membrane or through a *contractile* (kuhn-TRAK-tuhl) *vacuole*. The contractile vacuole,

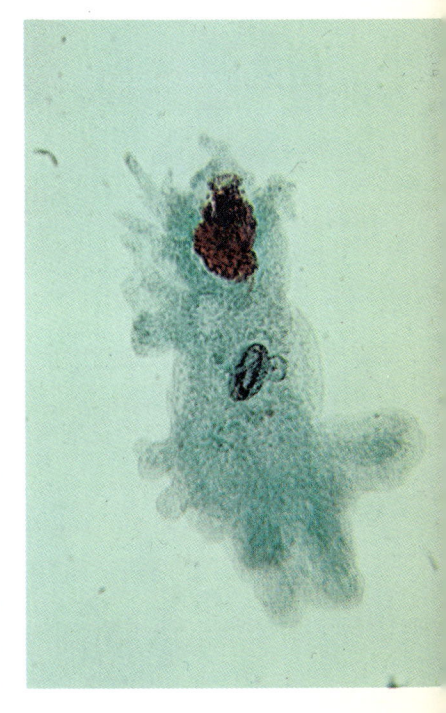

9-14 How does an ameba move from place to place?

by expanding and contracting, regulates the amount of fluid in the cell. You can learn more about this one-celled animal in the next activity.

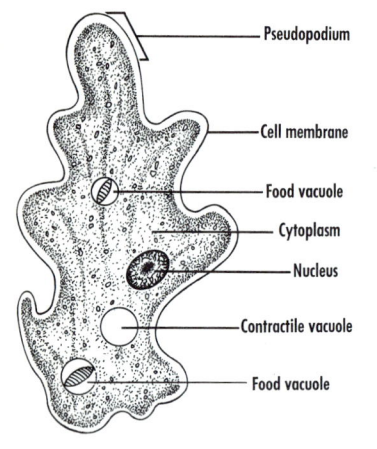

OBSERVE

Place two or three amebas in a shallow dish with water. With a hand lens, observe their normal behavior. Shine a light on one side of the dish. Do the amebas move toward the light or away from it? Now hold a lighted match on the bottom of one side of the shallow dish. Do the amebas move toward the heated water or away from it?

Place an ameba on a microscope slide with a drop of water and cover the drop with a cover glass. Observe the ameba carefully through the low power of a microscope or microprojector. You may have to adjust the light so that you can observe the internal structures clearly. Locate the contractile vacuole. How many times does it fill up and shrink during a two-minute interval? Now, add a drop of salt water to the microscope slide. How does the salt water affect the rate of contraction in the vacuole?

Set up your own experiment to find out how amebas react to other chemicals. Compare your results with those of other members of the class to be more certain of your conclusions.

Gases and liquids are exchanged through cell membranes. There is a greater amount of oxygen dissolved in the surrounding water than there is in the cytoplasm of the one-celled animals. Thus, the oxygen tends to pass into the ameba until there is a balance. In a similar way, because waste gases and liquids are present in greater amounts in the cell than in the surrounding water, they tend to pass through the membrane out of the cell to again reach a balance.

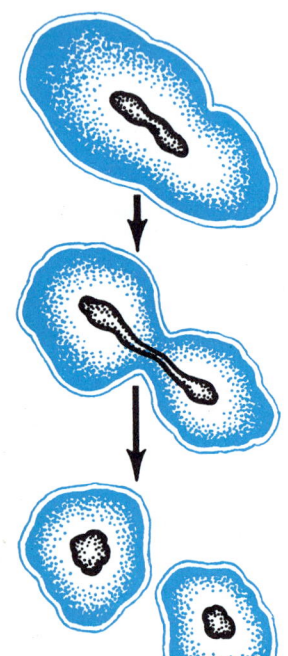

The ameba reproduces by cell division, or fission. When an ameba reaches its full size, and the conditions are favorable, the nucleus and the cytoplasm begin to divide in half. The complex process ends when two small amebas form, separate, and go their independent ways (Fig. 9-15).

We know that living organisms grow old and finally die. However, this is not the case with most protozoans. Instead

9-15 Describe the process of fission, illustrated above.

of dying, the cell divides to form two new cells, which later repeat the process of division all over again. Thus, in a sense, you might say that these organisms can live forever. Assuming this is so, why isn't the world overrun with these one-celled protozoans?

The paramecium is a more complex protozoan. The *paramecium* (*par*-uh-MEE-see-uhm) is a more complex one-celled organism than the ameba. It has specialized areas within its single cell where essential life processes take place. It has a more definite body shape, and its behavior is more complex than that of the ameba. You can study the habits of paramecia in the next activity.

INVESTIGATE

Break some hay into small pieces and place in a dish. In a second dish, place ten grains of soaked rice. Pour a little pond water in each dish. Add two tablets of brewers' yeast to each dish and cover. Leave the dishes in a warm place, but not in direct sunlight, for several days. When a film forms on the water, the protozoans are ready to be studied.

Using a medicine dropper, put a drop of the filmy water on a clean microscope slide and add a cover glass. Examine the drop under the low power of a microscope or microprojector to see what kinds of protozoans are moving in the water. Locate some good-sized paramecia. Since these animals move rather rapidly in the water, you can slow them down by adding a drop of methyl cellulose to the slide. You can identify other kinds of protozoans by comparing the ones you see with the one shown in the diagram.

Study one paramecium under the low power of the microscope or microprojector and compare its structure with the illustration.

A paramecium moves by means of tiny, hairlike structures called *cilia* (SIHL-ee-uh). These cilia grow out of the thickened cell membrane, which gives the paramecium its cigar shape, as shown in the diagram. Cilia are used not only for movement, but also for obtaining food. You can observe the action of cilia in paramecia by doing the next activity.

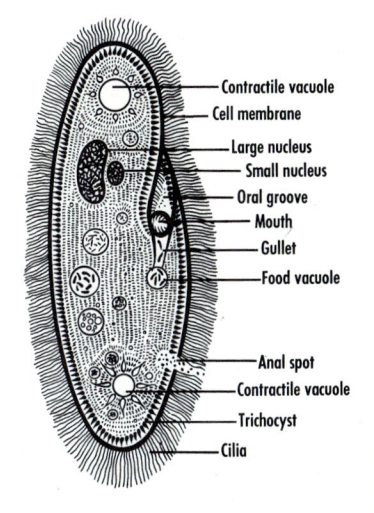

254 / UNIT 3 MAN EXPLORES LIVING ORGANISMS

DESCRIBE

Add one drop of carmine dye to a ten percent solution of methyl cellulose. Place one drop of this mixture on a slide containing some active paramecia. Cover with a cover glass and examine under the low power of a microscope or microprojector. Describe how the cilia beat. Can you locate the oral groove? How many food vacuoles do you see? Make a summary of all the observations you have made about the movement of the paramecium cilia.

The cilia sweep food particles into the *oral groove* and into the *mouth cavity*. The food particles then pass into the *gullet* where a *food vacuole* is formed in the cell. As in the ameba, digestion takes place in the food vacuoles. Oxygen enters the paramecium through the cell membrane. Two contractile vacuoles are present. These remove waste products and help regulate the amount of fluid contained in the cell. The vacuoles are at either end of the cell.

Under the cell membrane, you will note a group of special structures called *trichocysts* (TRIHK-o-*sihsts*). These may be shot out and used as protection against enemies and as a way of capturing other smaller forms of life for food.

Paramecia reproduce in two ways. Two *nuclei* are seen in each paramecium cell. The large nucleus is believed to regulate the everyday activities of the cell while the smaller one is believed to function only during reproduction. A single paramecium reproduces by fission, like the ameba. However, these animals also undergo a more complex process called *conjugation* (*kahn*-joo-GAY-shuhn), which involves two organisms. Some scientists believe that conjugation is not a true reproductive process, because, although two paramecia are involved, no new organism is formed. Therefore, conjugation is considered to be only a part of the reproductive process in which the cells are given new vigor and energy (Fig. 9-16).

There are about 20,000 different species of protozoans with a wide range of variation. Most protozoans are important food sources for larger organisms. They also live in the intestines of larger animals to help in the digestive process. Many protozoans secrete hard protective walls around themselves

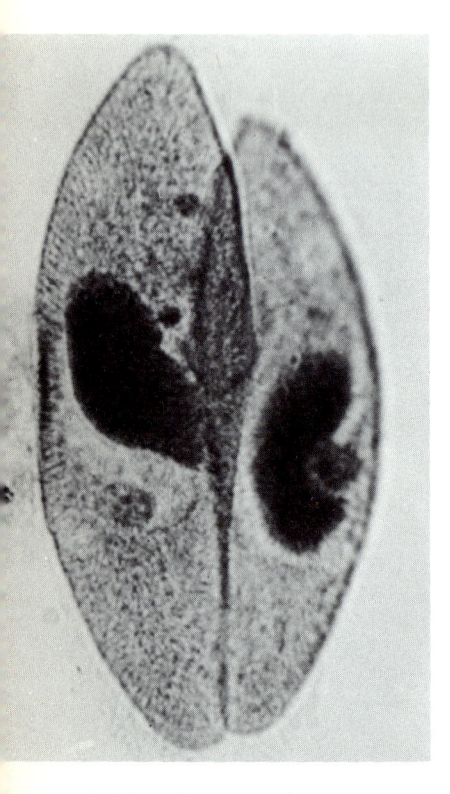

9-16 Why is conjugation not a separate reproductive process?

which remain as mineral deposits in many parts of the world. It would be interesting for you to do further library research on the economic importance of different kinds of protozoans.

REVIEW

1. Name the two main groups of animals.
2. How does an ameba move?
3. How does a paramecium obtain its food?
4. By what method do most one-celled animals reproduce?
5. What is meant by the following terms: (*a*) contractile vacuole, (*b*) food vacuole, (*c*) trichocyst?

D/What Are the Characteristics of Invertebrate Animals?

Metazoans have two main types of body organization. Many-celled animals are generally divided into two main groups, *invertebrates* (ihn-VUHR-tuh-brayts) and *vertebrates*. The structural difference between these two main animal groups is easily seen. Vertebrates have a backbone, but invertebrates do not. The invertebrates make up about 95 percent of all members of the animal kingdom. Figure 9-17 shows the way scientists believe the major groups of animals in these two main types of development are related.

Sponges are simple metazoan animals. When we mention *sponges,* you probably think of sponges used to wash dishes or your car. Most of the "sponges" used today are made of plastic materials and have no relation to living sponges. The living sponge, a simple metazoan, is a colony of cells. The various cell groups must work together to carry on the life processes. Sponges are usually found attached to solid objects in water. While they are found mostly in warm salt water, there are also freshwater species.

One group of invertebrate animals has a saclike body arrangement. This group of animals, called the *coelenterates* (sih-LEN-tuhr-ayts), includes the jellyfish, hydra, sea anemones,

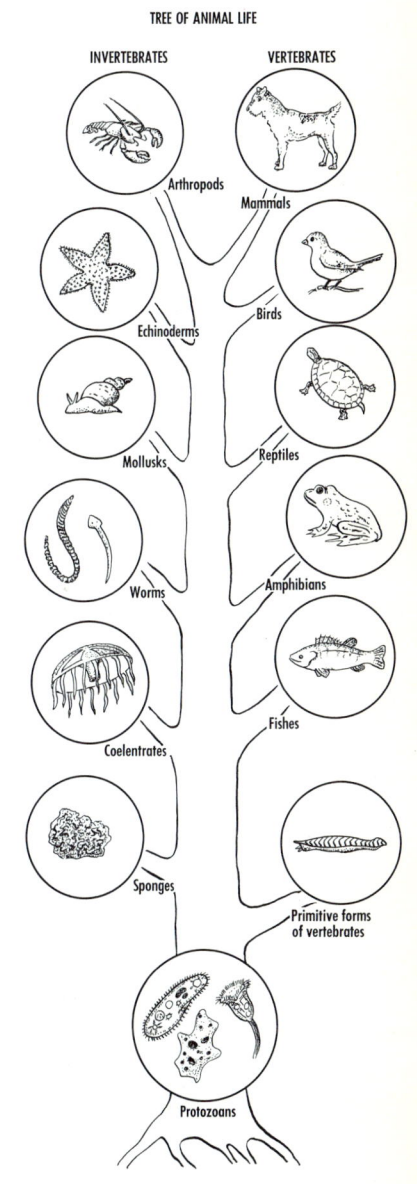

9-17 What characteristic divides the two groups?

9-18 Finger coral (top left), sea urchin (top right), jellyfish (lower left), sea anemone (bottom right). Which animal is not a coelenterate?

and corals. The only one of economic importance is the coral group which builds the coral reefs. It is very possible, though, that you have seen some of the others. Several examples of coelenterates are shown in Fig. 9-18.

Hydra, the animal we observed in a previous chapter, is a freshwater member of this phylum. Let us study this organism in detail in the next activity.

OBSERVE

Place a hydra in a small glass dish. Use a hand lens to see how it moves. Observe its tentacles. Place the dish with the hydra under the low power of a microscope or microprojector. Compare the animal with the diagram. Identify as many parts as you can. If you can get some water fleas, place a few in the dish. What happens to them when they touch the hydra's tentacles? How does a hydra capture its food?

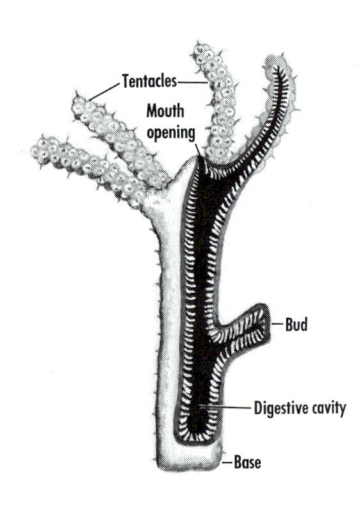

As you have observed, the body of the hydra consists of an outer layer and an inner layer which enclose a hollow, saclike cavity. Food is swept into the mouth and enters this saclike cavity. Digestive juices change the food into soluble material which is absorbed by the cells lining the cavity. These cells, in turn, pass some of the food material to the outside cells. At the same time, the cells in the outer layer are taking in oxygen and passing it to the inner cells. Thus, the hydra's cells show a greater division of labor than those cells making up a sponge.

The *tentacles* (TEN-tuh-kuhlz) are armlike structures which appear at the top of the hydra. These tentacles contain threadlike stingers which kill any small organisms they touch. How does this process help the animal?

In observing the hydra carefully, you probably realized that one way it moved was by tumbling over on its tentacles, similar to the way you would do a cartwheel.

Worms have a tubelike body. There are three main groups of worms: *flatworms*, *roundworms*, and *segmented worms*. Many of these are parasites and are found only inside a variety of other organisms. Yet, most worms are quite harmless to man.

The freshwater flatworms, *planarians*, illustrated in Fig. 9-19, are scavengers that eat dead organic matter. In fact, you can often collect planarians by tying a piece of raw meat on a string and leaving it in a pond for a few hours. You can study planarians in the following activity.

9-19 How do planarians move from place to place?

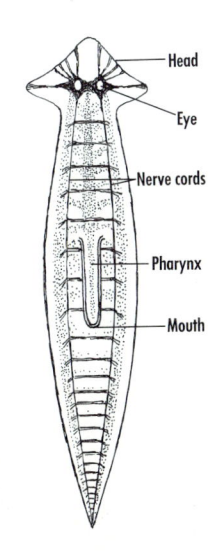

DESCRIBE

Place a few planarians in a shallow dish. Give them a few minutes to adjust to their environment. Take a little piece of raw liver and place it in the shallow dish on the side away from the worms. Observe how the worms approach the food. Notice the location of the mouth. Can you also describe how the worms move from place to place?

Planarians have a simple nervous system connected to a "brain." For our purposes, this is the first group to show this amount of specialization. They also have two light-sensitive *eyespots.* Can you set up an experiment to find out whether planarians move toward or away from light? Be sure to use a control.

Segmented worms have a highly organized body. You probably do not think of the lowly *earthworm,* a segmented worm, as a very highly organized animal. Compared to the vertebrates, it certainly is not. Compared to some of the other invertebrates we have studied thus far, the earthworm and other segmented worms, however, are quite complex. For example, the earthworm has a circulatory system with blood, a digestive system, and a well-developed nervous system so that it can even be taught to do a few very simple things. Let us examine the inner structure of an earthworm to see if we can find some of these special organs. The following activity will help you to understand the system in an earthworm.

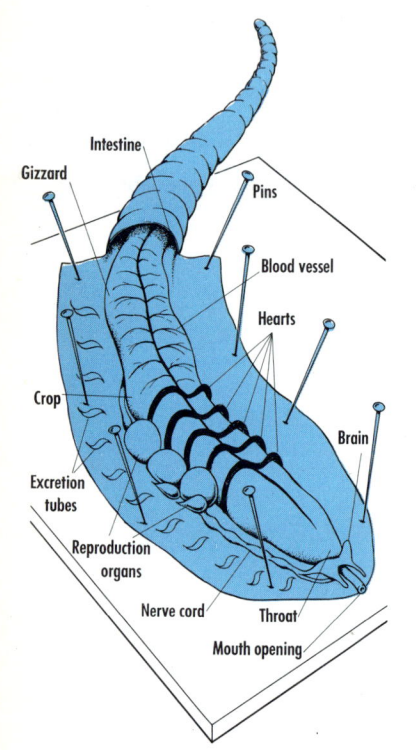

OBSERVE

Place a preserved earthworm in a dissecting tray or on a block of soft wood. Push a pin through each end to hold it in place. With a single-edge razor blade, cut just deep enough into the skin of the upper, light-colored area to open up the body cavity. Using scissors, cut along the top of the body and pin open the walls as shown. Keep the inside of the worm moist by dropping some water on it from time to time. Compare the body organs you see with the earthworm diagram. Identify as many parts as you can. What systems are you able to locate?

Earthworms are important because they loosen the soil as they move through it. By ingesting soil along with any dead organic matter in it, they keep burrowing deeper and deeper, allowing air and water to penetrate. Earthworms also serve as a source of food for birds as well as other organisms.

Sandworms and *leeches* also belong to this group. Sandworms live along the shoreline of oceans, burrowing in and out of the sand. Other segmented worms may be found in lakes or on the bottoms of ponds. Leeches, commonly called "bloodsuckers," sometimes firmly attach themselves to the skin of swimmers.

Mollusks make up an important phylum of invertebrates. Snails, clams, and oysters belong to the phylum called *mollusks* (MAHL-uhsks). Many members of this group are used as food by man. The drawing in Fig. 9-20 shows the parts of a clam from external and internal views.

The clam is a complex animal with body systems that are highly developed. The clam's diet includes protozoans, small algae, and tiny animals which are carried in through tubes called *siphons* (SYE-fuhnz). Water enters the lower siphon and passes out through the upper siphon. Oxygen is carried in with the incoming water and food, and the *gills* absorb the oxygen from the water. Now, examine a clam by carefully dissecting one in the activity below.

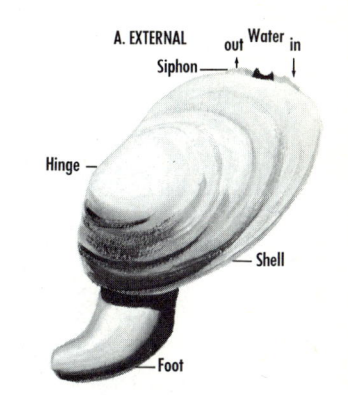

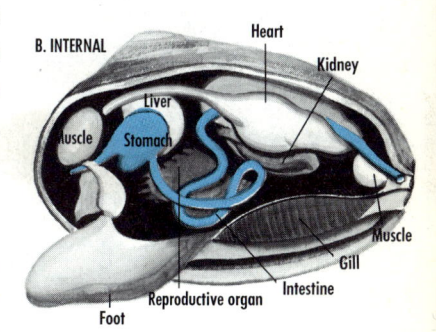

COMPARE

Carefully open a clam by inserting a knife on each side of the hinge and cutting the muscle. Now compare the inside structure with the drawing (Fig. 9-20) of the internal view. Locate the *foot, mantle, siphons, gills,* and the muscles that open and close the shell of the clam. What is the purpose of each of the structures observed? Dissect the clam very carefully and see which organs you can identify.

An examination of the shell will help you to see the *mantle*, the layer that lines the inside of the shell. The mantle lining is covered with microscopic *cilia* which help push the water through the siphons.

9-20 What body systems can you identify in a clam?

You can observe snails in your class aquarium. Several should always be kept because they keep the glass clear by cleaning algae from the sides with their tongues. Perform the activity on snails next.

EXAMINE

Collect some snails and observe them carefully. How do they move? If you place them on a clear piece of plastic, and tilt the plastic, which way do they move? Try to feed them. How do they ingest food?

260 / UNIT 3 MAN EXPLORES LIVING ORGANISMS

The squid and octopus are not covered with a shell like the clam, snail, or oyster, but they are also mollusks (Fig. 9-21). Their outside covering is the mantle. However, internally they do resemble the others. This is a good example of how internal structure serves as the basis for classification.

Animals in the starfish group live in the sea. The members of the *echinoderm* (ih-KYE-no-*duhrm*) phylum have an internal skeleton formed of hard plates, usually topped by spines (Fig. 9-22). Starfish usually have five *arms* attached to a *central disk*. The lower surface of the disk contains the mouth which opens into a *stomach*. The stomach can be pushed out to digest food outside of the body. The arms contain *digestive glands* and *tube feet* with which the starfish moves. It also uses its tube feet to open the shells of mollusks on which it feeds. Other echinoderms include sand dollars, sea urchins, and sea cucumbers.

The arthropods make up the largest animal phylum. There are nearly one million species belonging to the *arthropod* (AHR-thro-*pahd*) group. The largest number, nearly 700,000 species, are insects. In addition to having jointed legs (the main characteristic of this phylum), all animals in the arthropod group have segmented bodies which are covered by an outside skeleton. This skeleton serves as protection and prevents extreme water loss.

As you might imagine, the outside skeleton has the disadvantage of being heavy and thus slowing down movement. To permit growth, the animal must shed its outer skeleton several times during its lifetime in a process called *molting*. Arthropods increase in size between the shedding of the old skeleton and the hardening of the new skeleton.

9-21 How does a squid resemble a clam or oyster?

9-22 A starfish is shown here attacking a clam. How does the starfish open the shell of a clam?

Arthropods are divided into several main classes. Crabs, lobsters, shrimp, and crayfish belong to the class called *crustaceans* (kruhs-TAY-shuhnz). The crayfish, found in fresh water, can be studied in the activity.

OBSERVE

Obtain a crayfish from a pond, or use a preserved one. Study the external or outside structure and notice the special structures used to carry out various activities. Identify the parts that are labeled in the drawing. How is a crayfish adapted to carry on food-getting and movement? How is it protected from its enemies?

Notice that the crayfish has many *appendages* (uh-PEN-dih-juhz), jointed structures attached to the body. The small appendages attached to the abdomen are used for swimming and sending water through the gills, enabling the crayfish to obtain oxygen. The large appendage in back, called a *flipper*, helps the crayfish to swim backwards rapidly. The *antennae* (an-TEN-ee) [sing. *antenna* (an-TEN-uh)] are the "feelers" attached to the front part of the body. The mouth parts are adapted for holding, cutting, and grinding food. Its front *claws* are used for defense and getting food.

Internally, the crayfish has a nervous system, a complete digestive system, and a well-developed circulatory system including a "heart." It has an open circulatory system with no blood vessels to carry blood throughout the body. Instead, blood is carried from its "heart" by arteries into open spaces around the organs. Eventually, the blood returns to the "heart" area. Its nervous system is more highly developed than that of the worms, and it has a developed sense of touch and smell.

Many of the crustaceans, like lobsters, shrimp, and crabs, are important food sources for man. Their major importance, however, is that the smaller members of this group serve as a food source for other sea life.

Spiders belong to another class of arthropods, the *arachnid* (uh-RAK-nihd) class. Arachnids have eight legs and a joined

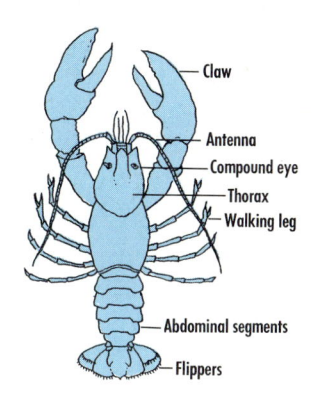

9-23 Examples of arthropods. Left to right: sow bug, body louse, tick.

head and thorax region. Although many people consider them harmful, spiders are really very valuable to man. They aid us by eating many harmful insects which they capture in the webs they spin. Other organisms in this class are sow bugs, ticks, and body lice (Fig. 9-23).

The largest group of arthropods is the insect class. *Insects* have three distinct body regions, the head, thorax, and abdomen. They have six legs, one pair of antennae, usually two pairs of wings, and breathe by means of tubes called *tracheae* (TRAY-kee-ee). Figure 9-24 shows some common insects. We can observe some of them in their natural habitats by doing the next activity.

9-24 What common features can you identify in these common insects shown below?

OBSERVE

An interesting habitat for many kinds of insects can be made by locating a plant on which insects are feeding. Carefully dig up the plant, placing the plant in a small flower pot. The flower pot is then placed in a deep pan, and soil is filled in around the plant. A glass lantern chimney is placed over the plant and covered with a piece of wire screening, as shown in the diagram. The insects on the plant should then be able to carry on their normal activities. Observe how the insects feed, how they move about in their environment, and how they carry out their other functions.

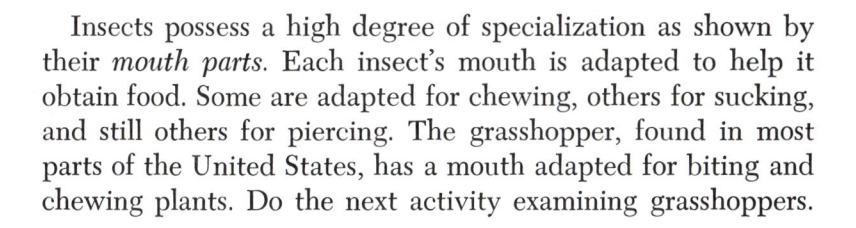

Insects possess a high degree of specialization as shown by their *mouth parts*. Each insect's mouth is adapted to help it obtain food. Some are adapted for chewing, others for sucking, and still others for piercing. The grasshopper, found in most parts of the United States, has a mouth adapted for biting and chewing plants. Do the next activity examining grasshoppers.

EXAMINE

Study the external structure of a living or preserved grasshopper. Are all its legs the same size? What adaptations do you notice? How many wings does the grasshopper have? Now carefully pull off the legs and wings of the grasshopper. Lay the parts out on a sheet of white paper and examine them. Use a hand lens to look at the small openings along the side of the abdomen. The body movements of the insect cause air to enter the tracheae through these openings. How is a grasshopper adapted to carry on its life processes?

Most insects must pass through several stages before becoming adults. Between the egg stage and the adult stage, most insects go through a series of changes in body form known as *metamorphosis* (met-uh-MAWR-fuh-sihs). When the insect passes through four distinct stages known as the *egg, larva, pupa* (PYOO-puh), and *adult,* the process is called *complete* metamorphosis. No stage resembles another. This type of metamorphosis is seen in the fly, moth, mosquito, and butterfly (Fig. 9-25) on the next page.

9-25 Complete metamorphosis in the Cecropia moth. Give the main stages.

The main stages in *incomplete* metamorphosis are *egg*, *nymph* (NIHMF) and *adult*. However, each nymph undergoes several changes, becoming more and more like the adult, as in the case of the grasshopper (Fig. 9-26). You can learn about the way insects are grouped in the next activity.

9-26 Incomplete metamorphosis in the grasshopper. Give the main stages.

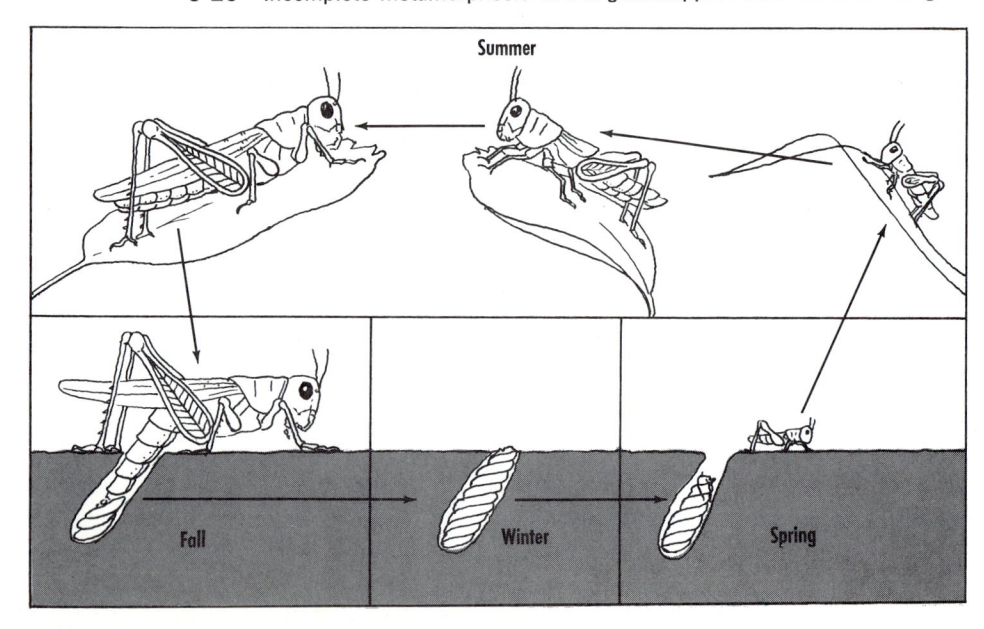

IDENTIFY

Collect many different types of insects. You can preserve them by placing them in ordinary rubbing alcohol. Find the name of each insect that you have collected. By inspecting its mouth parts, see if you can predict what it eats. Look carefully at the legs and wings and describe them.

Make a chart that is similar to the one shown below so that those insects resembling each other are in the same group. **Do not write in your book.**

Main Characteristics of Insects			
Name of insect	Mouth parts	Type of wings	Kind of legs

Insects are both helpful and harmful to man. Insects are most helpful in their ability to pollinate flowers. Bees also supply us with honey and wax. The silkworm provides us with silk. Some, like the praying mantis and ladybug, kill other harmful insects.

Insects are harmful in many ways because they attack not only grain, fruit, vegetables, and trees, but man himself. Many species carry disease. Termites destroy buildings by eating wood. Moths and beetles destroy our clothing, and ants invade houses and destroy food. Other insects bite, sting, and generally annoy us.

Scientists study the life cycles of insects to control them. Plant scientists want to learn where different species of insects lay their eggs, where they get their food, and during which stage they appear to be weakest. In this way, scientists can use various *chemicals* or *natural enemies* to control many of the insect pests.

Millions of dollars are being spent to find new methods of controlling insects. The main difficulty lies in their small size, their ability to reproduce very quickly, and their ability to adapt to variations of temperature, moisture, and even poisonous insecticides.

REVIEW

1. Name the two main kinds of metazoans and give an example of an animal in each group.
2. List the three main groups of worms and give an example of an animal in each group.
3. Name the three main classes of arthropods and give an example of an animal in each group.
4. List five body structures which distinguish the insects from other arthropods.
5. What are the two main forms of metamorphosis found among insects?

E/What Are the Characteristics of Vertebrate Animals?

All vertebrates have a similar body structure. As we learned before, a vertebrate is an animal with a backbone. There are five major classes of vertebrates: *fishes, amphibians, reptiles, birds,* and *mammals.* Each group and its members have different characteristics. However, each vertebrate has a highly organized body, an internal skeleton, a head region which contains a mouth, eyes, and nostrils, two body cavities (one for the brain and the other for the rest of the body organs), and usually two pairs of limbs.

Vertebrate organs are organized into *systems* which are groups of organs working together to perform one general function.

The classes of vertebrates show interesting adaptations. It is believed that they first lived in the sea and later moved to the land. Many adaptations were necessary to make this step. Some of the cold-blooded vertebrates became warm-blooded. The various body systems changed in structure as the animals became adapted to live in different habitats.

Fishes are found in a water habitat. The most primitive fishes have skeletons made up of *cartilage* (KAHR-tih-lihj) instead of bone. Sharks and rays have skeletons made up of cartilage and, therefore, are quite primitive. Sharks, however,

9-27 Why are sharks considered primitive fishes?

are like true fish in many ways. Their jaws are lined with razor-sharp scales which act as teeth. Water enters through the mouth and surrounds the gills. This helps them to get the oxygen dissolved in the water. This group includes the largest living fish, the whale shark, which reaches a length of 50 feet. Sharks eat other fish and are usually at the top of the food pyramid as consumers. (See Fig. 9–27.)

Rays are flattened creatures often living partly buried in the sand. They have a whiplike tail that stings and wounds animals which come near it. They feed on mussels and clams. (See Fig. 9-28.)

The bony fishes are the only true fishes. The true fishes include a wide variety of fresh and saltwater species. They range in size from the small guppies you may have in your home aquarium to the large sturgeon which can weigh as much as 2000 pounds.

They all have a bony skeleton and gills for taking in oxygen. Most bony fishes have a protective outer layer of *scales* not found in the more primitive fishes. They have no eyelids, and their eyes are usually large in order to admit the greatest amount of light. Fish have many types of *fins* which are used for swimming and balance.

Most bony fish also have a *swim bladder* which helps them to stay at a particular depth. If they want to rise, they increase

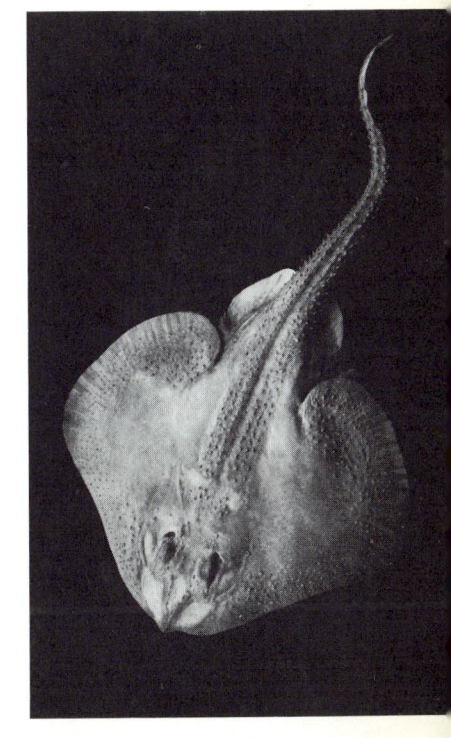

9-28 How does the body shape and tail of the sting ray adapt it for survival in the ocean?

the amount of air in the bladder. If they want to go deeper, they decrease the amount of air.

All fish have a two-chambered *heart*. Their brains are rather small, though they appear highly developed when compared with the invertebrates. Compared with the rest of the vertebrates, however, their nervous systems are poorly developed. You can become familiar with how fishes behave by doing the next activity.

EXAMINE

Obtain a fish from a local market or observe a fish in an aquarium to study its external features. Particularly notice its streamlined body, its mouth, eyes, and fin structure. The *lateral line,* which can be seen on either side of the body, contains nerve endings which help the fish to feel vibrations. Look carefully at the gills and notice that an opening goes to the mouth. Compare the fish with the illustration below and locate the external structures shown. Describe how the fish swims, how it obtains food, and how it obtains oxygen from the water.

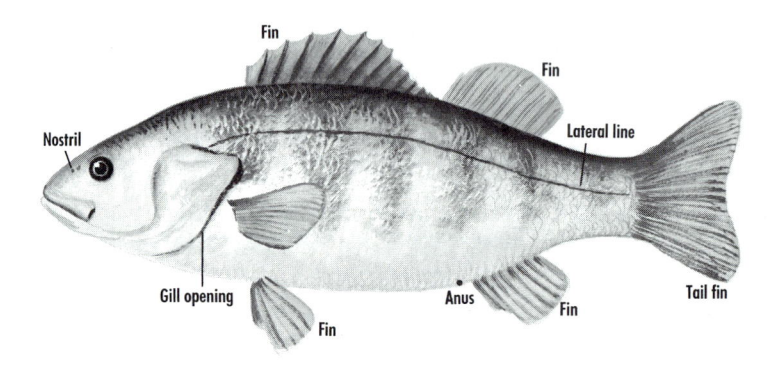

Fish are very important to man as a food source. In addition, there are many products, like cod liver oil and fertilizers, which are obtained from fish.

Amphibians are adapted for life in water and on land. Frogs and toads belong to the class of vertebrates called *amphibians* (am-FIH-bee-uhnz). The name amphibian means "having two lives." These animals spend part of their life in water and part

9-29 Why is a frog classified as an amphibian?

of it on land. The body is covered by a thin, moist skin which does not have any scales or other covering. The moist skin prevents them from drying up when they are out of the water. Feet, if present, are webbed, and the toes do not have claws. The young amphibian, usually a plant eater, has a two-chambered heart. The adult, usually a meat eater, has a three-chambered heart.

Amphibians are nearly always found near the shores of ponds, swamps, or low meadows. Since they lay their eggs in water, they usually do not go far from the water.

Of all amphibians, toads are the best adapted for living on land. They are the most valuable. They eat harmful insects and should not be destroyed. Once they reach the adult stage, they usually leave the water and spend the rest of their lives on land.

Frogs are the most common member of this group. They, too, eat harmful insects. In addition, frog's legs are considered a food delicacy by many people. Frogs are often used for classroom study. They are easy to handle, and their organs can be seen easily.

Certain characteristics of the frog are worth noting. Externally, the frog's legs are different in size. The weaker front legs are used for support, while the stronger hind legs are adapted for swimming and leaping. The frog's mouth is large, containing a sticky tongue which can be flipped out to catch insects, as shown in Fig. 9-29.

Internally, the frog has a fully-formed digestive system, a well-developed circulatory system, and a three-chambered heart. Do you recall how many chambers the fish heart has? The frog's brain is also more highly developed than that of the fish.

The frog can carry on respiration in several ways. When it is quiet and needs little oxygen, it absorbs oxygen through its moist skin. The blood vessels are near the surface and the exchange of oxygen and carbon dioxide take place there. When the frog is active, it can take in oxygen through its mouth by enlarging the mouth cavity. The exchange of oxygen and carbon dioxide then takes place in the mouth lining. Finally, when the frog needs a great deal of oxygen, it uses its lungs. You can study amphibians by doing the activity that follows.

DESCRIBE

If a live frog is available, put it in a jar containing two or three inches of water. Describe how it uses its legs in swimming. Note the movement of the floor of the mouth. What is the purpose of this? Examine the eyes and see if you can observe a membrane which moves up over the surface of the eye. Gently touch the top of the eye with a pencil. What happens? What help would this be to the frog? Drop some live beetles or other insects in the water. How does a frog capture its food? Compare the animal with the illustration and locate the structures shown.

The life history of the frog takes us from the water to the land. Figure 9-30 shows stages of development in frogs. The young tadpole has a long tail and breathes through gills. Gradually, the tail is absorbed into the body, and hind legs appear. When the front legs come through, the tadpole frog begins to breathe in the air. Its tail gets smaller as its front and hind legs become larger. The change from egg to adult takes about three months. If you obtain frogs' eggs and place them in an aquarium tank with water, you can watch them develop.

Frogs survive the winter by means of *hibernation* (hye-buhr-NAY-shuhn), a period of inactivity. As the temperature becomes colder, the body rate slows down greatly, allowing

9-30 Trace the steps in the life cycle of a frog.

the frog to survive with less food and oxygen. It now uses the food stored in its fat cells.

If the pond becomes very warm during the summer, the frog sometimes buries itself in the moist, cool mud or lies quietly on the bottom of the pond. This quiet stage during the summer is called *estivation* (*es*-tih-vay-shuhn).

Reptiles represent the first true land vertebrates. The *reptile* class has many more extinct species than living members. Reptiles are higher in the development of the vertebrates than either fishes or amphibians. They are cold-blooded animals. Reptiles have a dry skin, usually covered with scales, which hold moisture. Their feet, if present, have claws. Claws protect them and help them to move about more easily. Reptiles do not go through a tadpole stage. They have lungs for breathing from the time they are first hatched. Almost all reptiles reproduce by laying eggs.

9-31 Poisonous snakes: cottonmouth moccasin (top left), copperhead (top right), coral snake (lower left), rattlesnake (lower right).

Snakes, turtles, lizards, crocodiles, and alligators are common members of this group. The snakes are the most widespread and numerous. Snakes are found mostly in tropical regions where they are more active in warm temperatures.

In the United States, the most common poisonous snakes are rattlesnakes, moccasins, copperheads, and coral snakes, which are shown in Fig. 9-31. However, most species are nonpoisonous and essential in maintaining a balance in nature. Some common nonpoisonous snakes of North America are shown in Fig. 9-32.

All snakes are legless and have dry skin covered with scales. The large mouth with two rows of teeth is well adapted for getting food. The jaw bones are not attached directly to each

9-32 Non poisonous snakes: milk snake (top left), garter snake (top right), coral king snake (bottom left), hog-nosed snake (bottom right).

other so that the opening in the back of the mouth can be as large as the opening in the front. If the snake can take an animal into its mouth, the snake can swallow it alive without much difficulty.

Snakes can obtain food in several ways. Some, as mentioned, can swallow their prey alive and whole. Others, called *constrictors,* crush their food first and then swallow it. Poisonous snakes first inject poison into the animal and swallow the victim after it dies.

Snakes contain fully-developed body systems. In fact, the digestive juices of their stomachs are much stronger than ours since they must digest whole animals. Snakes can go without food for a long time.

274 / UNIT 3   MAN EXPLORES LIVING ORGANISMS

The turtle is a common reptile in the United States. Different species are found on land, in fresh water, and in the ocean. The sea turtles are the largest and some may weigh over 1000 pounds. The largest land turtle, weighing about 300 pounds, is the giant *Galapagos* (guh-LAP-uh-guhs) *tortoise*, seen in Fig. 9-33.

Although crocodiles and alligators belong to the same group, they differ from one another in small ways. The crocodile has a greenish-gray color; the alligator is a brownish-black. The crocodile is better adapted for living in water because it has a slender body and a more pointed head (Fig. 9-34). Alligator skins are used in making shoes, handbags, and luggage.

Birds are a very successful class of vertebrates. Birds are found in almost every type of habitat. They have several advantages over the other classes studied. Birds are warm-blooded animals with a four-chambered heart, lay eggs which are protected by a shell and incubated in a nest. Their bodies are covered with feathers that are water repellent. Their bones are light and contain small air-filled spaces. This gives them firm support as well as a light skeleton. In all birds, the front legs are developed as wings which are used for flying. You can study birds in the next activity.

COMPARE

Carefully observe a bird to see its flight adaptations. Divide a sheet of paper in half. Head one column Bird, and the other Airplane. Compare the similarities between the two. Refer to biology books, encyclopedias and other sources for information. How does your list compare with those of your classmates?

The mouth of the bird contains a bony, toothless beak which is modified for eating different types of food. Figure 9-35 illustrates a few of the types of beaks. Can you guess what a bird might eat by looking at its beak?

The feet of birds are also modified to help in obtaining food. Refer to Fig. 9-35 to help you guess what activity each type of foot is adapted to perform.

9-33 The Galapagos tortoise. Why is its slowness not a hindrance to its survival?

9-34 How are crocodiles adapted for living in water?

Since birds use a great deal of energy while flying, their body systems must be highly efficient. The feathers act as insulation against heat loss. Since birds have no sweat glands, most of the heat is lost through the lungs. Their breathing rate is rapid compared with ours. The amount of air the lungs can hold is increased by a system of *air sacs*. Air sacs extend from the lungs into the chest area and the abdomen. This increases the efficiency of respiration and helps control the bird's body temperature.

Since birds do not have teeth to chew food, the food is swallowed whole and stored in the *crop*, located above the *stomach*. The food passes into the upper portion of the stomach where digestive juices begin to break it down. Then, it passes into the *gizzard* where it is ground with the aid of pebbles which the bird swallows with its food. The digestive juices continue to act upon the food.

The four-chambered heart is highly efficient. It separates the oxygen-poor blood from the oxygen-rich blood returning from the lungs. The excretory system of the bird, like that of desert animals, is highly efficient. Birds lose very little moisture. You can learn about the feathers of birds in the next activity.

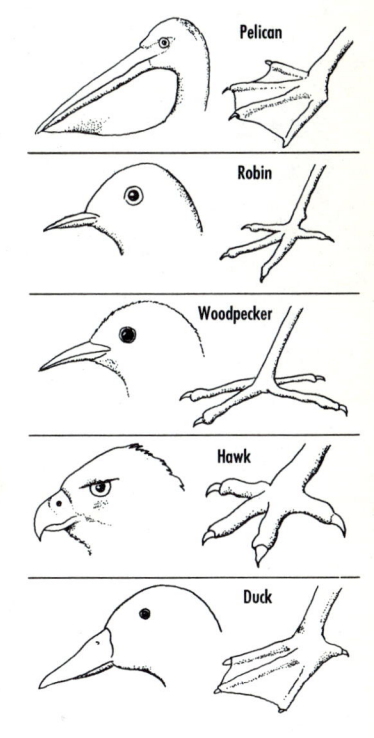

9-35 How are the beaks and feet of birds adapted for food-getting?

OBSERVE

Examine a bird's quill feather and compare it with the drawing. Locate the *vane, rachis* (RAY-kuhs), and *quill*. Cut a small piece from the vane of the feather and mount it in a drop of water on a slide. Cover with a cover glass and examine the feather with the low power of a microscope or microprojector. Make a drawing to show how the fine parts of the vane are held together with tiny hooks.

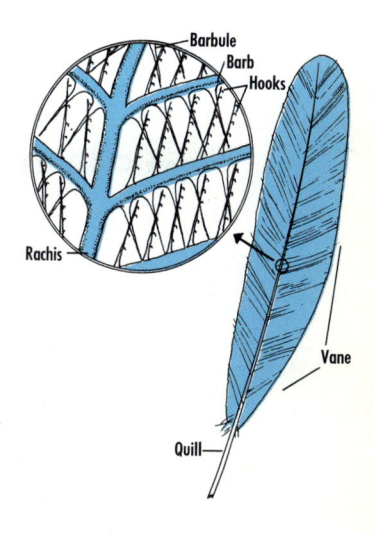

Many birds exhibit a seasonal movement from one region to another. This is called *migration* (mye-GRAY-shuhn). Birds follow the same route every year and usually arrive at a certain point about the same time every year. Although some people think birds migrate due to cold weather, migration is probably due to the need for food.

9-36 Why are these porpoises classified as mammals?

9-37 To what group of mammals does the kangaroo belong?

Birds are important to man. Not only do we eat their eggs and meat, but wild birds are used by man for clothing and ornaments. In addition, birds keep many insects and rodents in check and destroy weed seeds.

Mammals are the most highly-developed form of animal life. The habitats of *mammals* may be found all over the earth's surface. Some, like the whale and the porpoise (Fig. 9-36), live in the sea. Bats fly through the air. However, because of certain common structural features, they are all classified as mammals. Some common characteristics are as follows:

1. The young are born alive with the young developed and nourished internally. (Exceptions: duckbill platypus and spiny anteater.)
2. The young are nourished by milk-secreting glands of the female.
3. Hair or fur is present as an outer body covering.
4. The brain is highly developed.
5. They have a four-chambered heart.
6. They are warm-blooded, air-breathing vertebrates.

The *pouched mammals,* mainly found in Australia and New Zealand, make up one order. At one time, these animals were present all over the earth, but because they could not compete with other mammals, their number gradually decreased. The kangaroo, wombat, Tasmanian wolf, and common opossum are members of this order. The female of these carries her young in an abdominal pouch (Fig. 9-37).

Shrews and moles belong to the *insect-eating mammals.* Shrews, the smallest of the mammal group, are almost blind. So, too, are the moles, which live underground in tunnels. Their great value lies in their constant eating of many harmful insects to man.

Bats are *flying mammals.* They have poor eyesight and usually fly at night. Even in total darkness, they can find their way about by responding to the echo of their own sounds. By sending out high-pitched sounds which bounce off objects and return to them, they can find where the objects are and change their movements accordingly. Although most bats are carnivores and eat many harmful insects, one species, at least, feeds on blood.

Rodents are *gnawing mammals*. Rats, mice, squirrels, chipmunks, beavers and rabbits are examples of this order. Most of these fairly small animals are hard to control and cause man much trouble. They multiply rapidly, eat our crops, carry disease, and generally are great pests.

All rodents have large teeth adapted for cutting and grinding plants. This helps them to eat various food like rough grasses and nuts.

The *carnivores* are a large order of flesh-eating mammals. This order includes all dogs, cats, weasels, raccoons, and foxes (Fig. 9-38). Among its water-living members are seals and walruses. Carnivores are usually third in the food pyramid, eating animals which live on plants or other smaller animals.

Sea-dwelling mammals include the whale, porpoise, and dolphin. Though they spend all their life in the sea, these animals breathe with lungs. Even the blue whale, the largest mammal on earth, must come to the surface to get air.

The *hoofed mammals* are our main source of food. Pigs, cows, sheep, deer, and horses are examples of this group (Fig. 9-39). Many have been domesticated by man since prehistoric times.

The *primates* are the highest order of mammals. Among its members are man, monkeys, apes, and lemurs. Most primates are tree dwellers. Man, because of his ability to walk erect, can live on the ground. With his superior brain, he has been able to develop a language and pass on his knowledge from one generation to the next. His hands, with a thumb and fingers, have helped him to build objects and carry on activities not possible in the lower animals.

9-38 Why are the fox (top) and lion (below) called carnivores?

REVIEW

1. List five general characteristics of all vertebrates.
2. What are the two main classes of fish?
3. Why are frogs and toads called amphibians?
4. Define the terms: (*a*) hibernation, (*b*) estivation, (*c*) migration.
5. What are six characteristics that are common to all mammals?

9-39 Why are sheep valuable?

THINKING WITH SCIENCE

A. *On a separate sheet of paper, write the numbers 1 to 15. Match the letter of the correct term in Column A with its definition in Column B.* **Do not write in your book.**

Column A

a. algae

b. amphibians

c. arthropods

d. bacteria

e. coelenterates

f. colony

g. conifers

h. fermentation

i. fission

j. fungi

k. mammals

l. mollusks

m. penicillin

n. reptiles

o. worms

Column B

1. Green plants that produce seeds in cones instead of in flowers

2. Microscopic single-celled plants lacking chlorophyll

3. Group of plants which includes mushrooms

4. Filaments or sheets of single-celled plants having chlorophyll

5. Animals having a saclike body and tentacles

6. A large group of bacterial cells in a culture

7. Tubelike animals having the first true body systems

8. Soft-bodied animals that usually are enclosed in a shell.

9. A method of reproduction by a cell dividing in half

10. The most highly developed group of vertebrates

11. The first group of vertebrates adapted to live entirely on land

CHAPTER 9 THE DIVERSITY OF LIFE / 279

12. Breakdown of sugar which produces carbon dioxide and alcohol

13. Jointed-leg animals making up the most numerous animal phylum

14. A substance produced by mold growth which stops bacterial growth

15. Vertebrate animals that spend part of their lives in water and part of their lives on land

B. *Write the answers to the following in your notebook. Be sure to use complete sentences and correct spelling and grammar.*

1. How do molds and mushrooms digest food?

2. Describe the process of conjugation in Spirogyra.

3. Why do metazoans have a division of labor in their specialized cells?

4. Describe how a hydra ingests a water flea.

5. How does a clam obtain its food and oxygen?

6. Explain the difference between complete and incomplete metamorphosis in insects.

7. How is a fish adapted for life in a water environment?

8. Describe the life cycle of the frog.

9. Explain how a bird is adapted for flight.

10. Compare the differences in the structure of the heart of a fish, frog, bird, and mammal.

RESEARCH IN SCIENCE

1. Plan an experiment to find out in what parts of a plant starch is stored.

2. Make a display of cloth samples made from different plant fibers.

3. Make a chart summarizing the characteristics of each of the animal groups studied in this chapter.

4. Report on how a common insect pest was introduced into this country and how it is being controlled.

5. Make a chart comparing the skeletal structures of the main classes of vertebrates.

6. Design an experiment to find out the speed with which snails move, using inches per hour (iph) as the unit.

chapter 10
The Continuity of Life

A/How Are Characteristics in Living Organisms Transmitted?

No two organisms are exactly alike. Have you ever seen a litter of puppies or kittens? Did you notice how you were able to tell each puppy or kitten from the others? Look around your classroom and observe how each of your classmates differs from each other and any other individual. A difference in living organisms is called *variability* (*vay*-ree-uh-BIHL-ih-tee). Since no two living organisms are exactly alike, how are the differences produced?

In this chapter, we shall study *heredity* (huh-RED-ih-tee), the science of the way characteristics pass from parents to offspring. Heredity involves a careful study to explain the similarities and differences among organisms. We shall explore such questions as: Which characteristics are closely related to inheritance? How is it possible for two brown-eyed parents to have a blue-eyed child? How are identical twins formed? How can we create new plant and animal varieties?

When we studied asexual reproduction in a previous chapter, you learned that each organism produced another organism like itself. The microscopic ameba, you learned, splits by fission and produces two similar amebas. Although the ameba consists of a single cell, and our bodies are made up of billions of cells, man's cells, too, go through a somewhat similar process of cell division.

The cell is the unit of heredity. In our previous study, you learned that the cell is the unit of structure and function in all living things. Do you remember the small nucleus in the cells you saw under the microscope? In the male and female reproductive cells, the nucleus carries the material which determines the characteristics that are inherited. When cells in the reproductive organs are prepared and stained for study, tiny bits of colored structures, called *chromosomes* (KRO-muh-somz) can be seen in the nuclei. The number of chromosomes seen in the cells of different species, both plants and animals, varies. It is the same, however, in every normal plant or animal of a particular species. For example: all fruitflies have 8 chromosomes, all corn plants have 20, all bullfrogs have 26, and all humans have 46 chromosomes in each cell. Figure 10-1 shows the appearance of stained chromosomes in the fruitfly and in human cells.

10-1 Chromosomes in the salivary glands of the fruitfly (left) and in the liver cells of the human body (right). In what ways are their chromosomes similar?

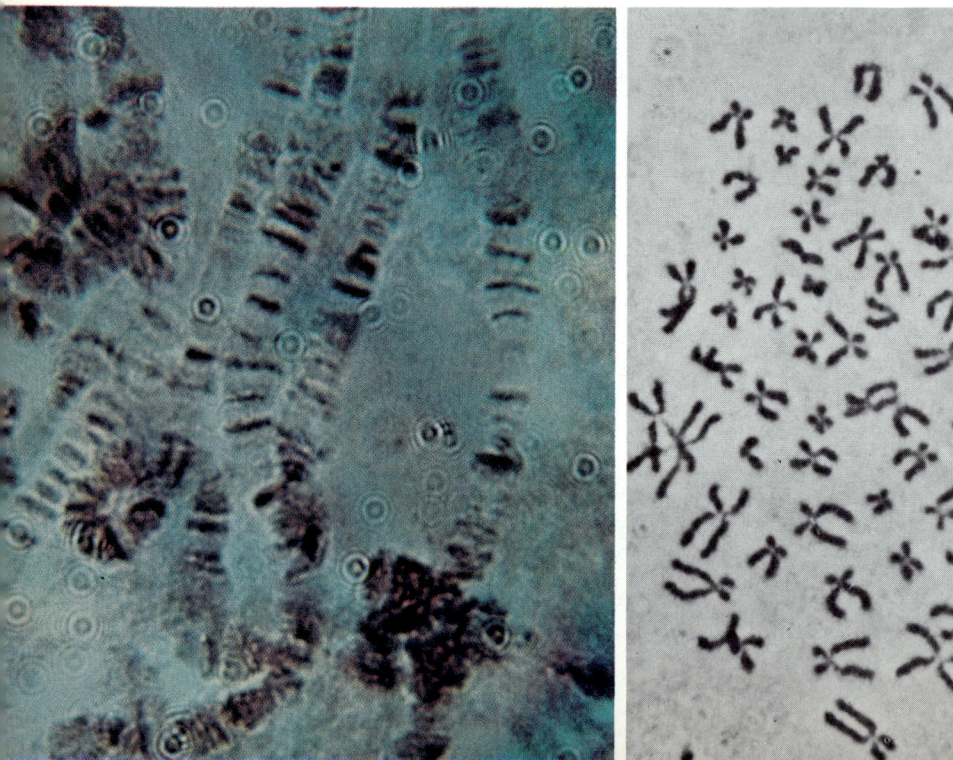

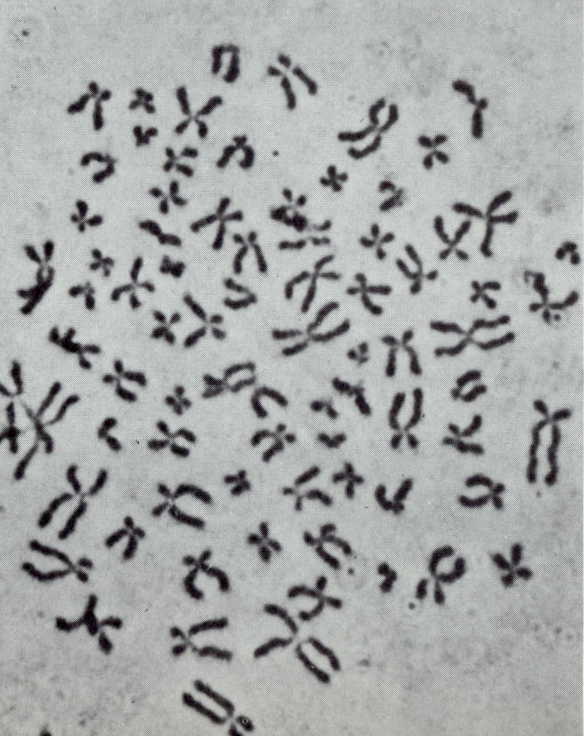

The shape of the chromosomes also varies from species to species. If you could look at a chromosome under an electron microscope, it would appear like a string of beads. Each "bead" or "band," on the chromosome is called a *gene* (JEEN). It is the unit of heredity. They are the carriers of hereditary characteristics. Scientists have learned that the chromosomes are composed of a complex chemical substance called *DNA*.

Cells divide to form new cells. Scientists have studied the process of chromosome division, called *mitosis* (mye-TO-sihs), to determine exactly how it occurs. Before you study this process, look at Fig. 10-2, which illustrates a series of mitotic stages in a cell. In order to separate each stage, scientists have named each of them. These stages are actually a *continuous process*.

The *prophase* (PRO-fayz) *stage* begins with the shortening of the chromosomes and their splitting lengthwise. During the *metaphase* (MET-uh-*fayz*) *stage*, the chromosomes group together in the center of the cell. In the *anaphase* (AN-uh-*fayz*) *stage*, the chromosomes separate; one from each pair moves to the opposite end of the cell. In the *telophase* (TEE-luh-*fayz*) *stage*, the cell membrane begins to form in the middle of the original cell. The chromosomes at either end begin to form a *chromatin* (KRO-muh-*tihn*) *thread*, and the nucleus begins to form a *nuclear membrane* around itself. Before and after these four stages, we have an *interphase* (IHN-tuhr-*fayz*) *stage*, when there is a period of cell growth. Figure 10-3 shows four stages in the division of an animal cell. The following activity will help show you some of the stages of mitosis.

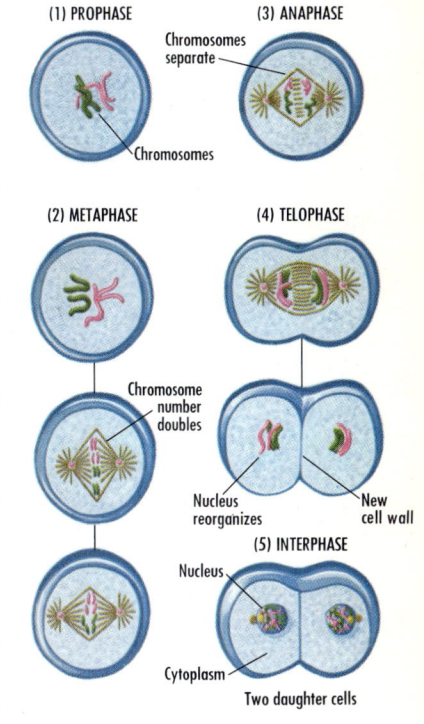

10-2 How can you distinguish between the main stages of mitosis in cell division?

10-3 Identify the stages of mitosis in an animal cell.

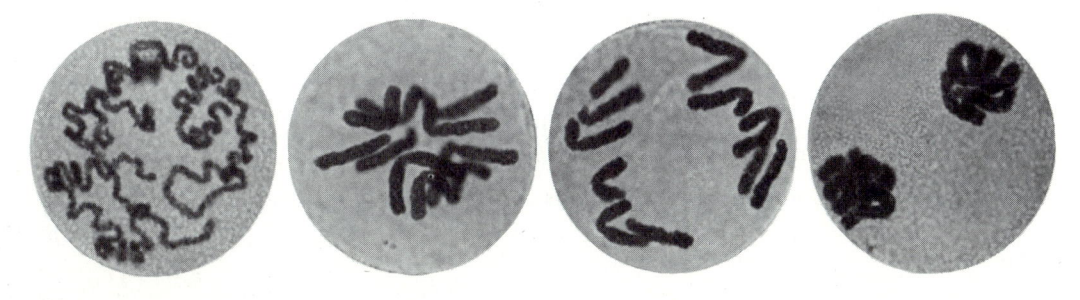

EXPLORE

Examine an onion root tip with the microprojector or microscope or the photographic slides with a slide projector. Find the different stages of cell division shown on the slide. Compare them with Fig. 10-2. Try to trace the phases of a dividing cell.

How many different stages of cell division were you able to find? What happens to the chromosomes? What is the size of each new cell compared to the original cell? How many chromosomes does each new cell have?

Each parent furnishes one cell to form the offspring. Each parent organism is able to produce special cells in the reproductive organs. The male produces *sperm cells* and the female produces *egg cells*. These cells are very small and can be seen only with a microscope. During fertilization, a sperm cell joins with an egg cell to form a new cell called a *zygote* (ZYE-gote). Although the young animal develops inside the mother's body, can you understand how each parent provides chromosomes that carry the characteristics for the offspring? The sperm cell has a nucleus which has chromosomes in it. The egg cell also has a nucleus containing chromosomes. When the two cells join in fertilization, the chromosomes from both parents are combined in the zygote. The offspring, therefore, inherits characteristics equally from both parents.

Sexual reproduction involves the union of a sperm cell with an egg cell. If each cell contained the original number of chromosomes, then the number would be doubled when they combined. For example, man has 23 pairs of chromosomes in each cell. Therefore, the combination of a sperm cell and an egg cell would give us 46 pairs, or 92 chromosomes in the fertilized egg. The 46 pairs of chromosomes do *not* represent the same organism, which in this example, is man. Something must occur to keep the number of chromosomes from increasing every time a sperm cell and an egg cell combine.

Division of sex cells is different from division of other cells. The process of cell division in which the number of chromosomes in the nucleus of sex cells is reduced is called *meiosis* (my-o-sihs). During meiosis, the chromosome number in the

CHAPTER 10 THE CONTINUITY OF LIFE / 285

sex cells is halved. Let us study this process in an organism containing two *pairs* of chromosomes as shown in Fig. 10-4.

The sex cells, with the normal number of chromosomes, first go through the process of mitosis. These cells then divide into two cells, each containing the normal number. You will notice in Fig. 10-4 that in the female the two cells are of different sizes. These smaller cells serve to remove two chromosomes from the developing egg cell.

Each of these cells divides again, but this time by the process of meiosis. Thus, four cells are produced, each having one-half the number of chromosomes found in the original sex

10-4 What happens to the chromosomes in both sex cells during meiosis?

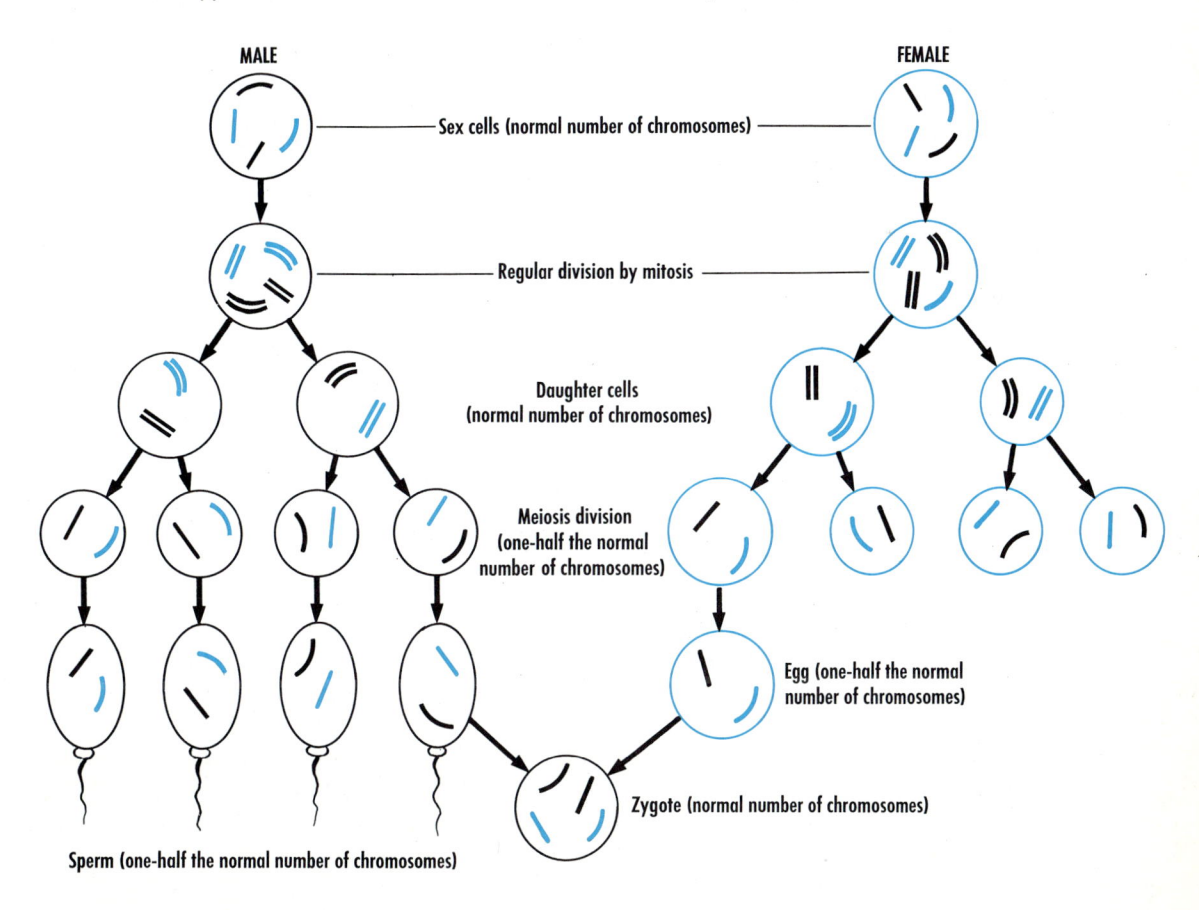

cells. In the female, one large and three small cells form, while in the male, all four cells are the same size. The large female cell develops into the egg cell, while four male cells grow tiny "tails" and develop into the sperm cells. The egg cell, like each sperm cell, now has half the number of chromosomes found in the original sex cells.

Immediately after fertilization, the zygote begins to divide by mitosis. This is a rather rapid process. Since the chromosomes split lengthwise, each cell which is produced in these divisions contains the exact number and makeup of the original zygote. The study of the developing zygote is known as *embryology* (*em*-bree-AHL-o-jee).

Identical twins result from the same zygote which divides and separates after fertilization into two identical cells. Each cell undergoes cell division, and instead of one embryo, two develop. Each embryo has the same hereditary characteristics because it develops from the same zygote. You can learn more about inheritance in the next activity.

COMPARE

Make a list of several characteristics, such as coloring, type of fur, or shape of ears in a family of puppies or kittens. Compare these with the characteristics of their parents. Do all the young have the same characteristics?

Each reproductive cell furnishes genes for the offspring. When the chromosomes from both parents are combined in the zygote, *they arrange themselves in pairs.* That is, there are two chromosomes of *each type* in the zygote. Where does each chromosome in a pair come from? The sperm cell provides one chromosome and the egg cell provides the other chromosome to form each pair. For example, we learned that man has 23 pairs of chromosomes, but two chromosomes do not always appear as a pair. These last two are called *X* and *Y* chromosomes. They determine the sex of the offspring. If two *X* chromosomes are present in the zygote, the young develops into a female. If one *X* and one *Y* chromosome are present in the zygote, the young develops into a male.

To produce a characteristic in the offspring, the genes in a pair of chromosomes have to match up. For example, let us suppose we could see the gene in a certain chromosome carried by the sperm cell that determines hair color. This gene must match up with a gene for hair color in that *same kind* of chromosome in the egg cell. When the two genes are matched up, the hair color of the offspring is determined. Therefore, an inherited characteristic is produced in the offspring when two genes are matched up in a pair of chromosomes.

Inherited characteristics are passed on according to biological laws. *Gregor Mendel (1822–1884),* an Austrian monk, is credited with the discovery of the basic *laws of heredity.* He performed many experiments with plants, mainly garden peas, in which he studied different characteristics. These observations included the size of the plant, the characteristics of the seeds, and the shape and color of the unripe pods. In each case, there were definite differences in the plants.

Having observed these differences, the next step was to control the fertilization. You recall that pollination is the transfer of pollen from the stamen to the pistil of the same species. You also know that the pollen grains contain the sperm cells. Mendel could control, to some extent, the zygote characteristics if he could transfer the type of pollen that reached the pistil. This is known as *artificial pollination* (*pah*-lih-NAY-shuhn). By doing this himself, Mendel was able to control the type of seed which the pollinated plant produced. Let us follow him as he worked with plant size.

Mendel selected two pea plants, one tall and one short. When they matured, he transferred the pollen from one to the other by hand. When the seeds were mature, he planted them. What do you think he found as the seeds grew into new plants? All the plants were *tall*. He repeated this process, crossing a tall and a short plant, until he was certain that a tall plant would always grow from the seeds. This led him to arrive at the *law of dominance.* He stated that the characteristic which appears in the first generation is the *dominant* (DAHM-ih-nuhnt) characteristic. The one which does not appear is the *recessive* (ree-SES-ihv) characteristic. Thus, in pea plants, tallness is dominant over shortness. You can determine inherited characteristics by doing the activity below.

COMPARE

To show an inherited characteristic in different students in the class, use a specially-coated paper called P.T.C. Have each pupil in the class taste a piece of this paper. How many report a bitter taste? How many report no taste? What is the ratio of those who can taste the paper to those who cannot? Would you say that this is a dominant or a recessive characteristic? Can you explain your answer?

CHARACTERISTICS OF PEA PLANTS

Dominant	Recessive
Round seeds	Wrinkled seeds
Yellow seeds	Green seeds
Colored seed coats	White seed coats
Inflated pod	Constricted pod
Green pod	Yellow pod
Axial flowers (on sides)	Terminal flowers (on ends)
Tall stem	Short stem

Mendel then proceeded to take the seeds of a large number of the original crosses of tall and short plants. He planted them to see what would result in the next generation. After careful study and counting the number of tall and short plants, he found that there were about three times as many tall plants as short ones. He also found that when he self-pollinated short plants, they always produced short plants. However, when he self-pollinated tall plants, one third always produced tall plants and the other two-thirds produced tall and short plants.

As a result of these experiments, Mendel discovered the *law of segregation.* This states that if an organism with both a gene for tallness and a gene for shortness is crossed with another organism containing the same genes, one-fourth of the offspring will have two genes for tallness, one-half will have a tall and a short gene (but will appear tall because tallness is dominant), and one-fourth will have two genes for shortness and thus be short. Figure 10-5 explains this in a simple diagram. An organism with both a dominant and recessive gene for the same characteristic is called a *hybrid* (HYE-brihd). Thus, one-half of the tall plants are hybrids. (The characteristic is tallness.) You can learn more about the law of segregation in the activity that follows.

PREDICT

To show how the law of segregation works, take two coins. Shake them in your hand and let the two coins drop on your desk. Repeat this at least 100 times. Record the number of times two "heads" come up, two "tails" come up, and one "head" and one "tail" show. What ratio are you approaching? Combine the results of all your classmates. Does the result come closer to your predicted ratio?

Scientists use a shorthand method to show crosses in organisms. Some convenient symbols can be used to show the above crosses. You know that chromosomes occur in pairs because the zygote inherits one chromosome from each parent. Since height for pea plants appears to be controlled by one set of genes on one particular pair of chromosomes, we can show height by using capital and small letters of the alphabet. Let us use T to indicate the gene for tallness, and t to indicate the gene for shortness. Thus, a pure tall plant would be TT and a short plant would be tt. During meiosis, the normal chromosome number is reduced to one-half the number. Thus, both the egg cell and the sperm cell each have only one gene. Naturally, if the parent cell is pure for this characteristic, all the sperms and eggs would contain the same genes. For example, when the genes in the male cell are written as TT, then every sperm cell must have a T gene. The same is true of the egg cell when it comes from a pure-strain parent.

In Mendel's work, he crossed a pure tall pea plant (TT) with a pure short pea plant (tt), and the results of the crossing are shown on the next page.

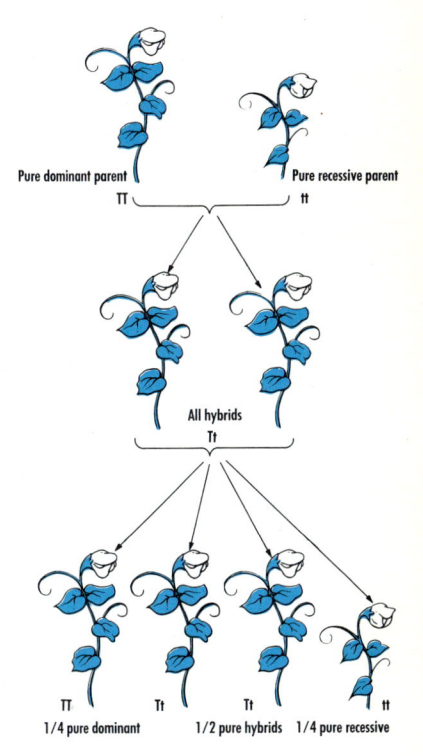

10-5 Trace the inheritance of height in the pea plant.

Cross between *TT* plant and *tt* plant

female → male ↓	t	t
T	Tt	Tt
T	Tt	Tt

100% hybrid tall

Since tallness is dominant over shortness, all of the offspring will be tall. Now let us see what happens when we cross two hybrids.

Cross between *Tt* and *Tt* plant

female → male ↓	T	t
T	TT	Tt
t	Tt	tt

25% pure tall
50% hybrid tall
25% pure short

In the previous case, is it possible to get a short pea plant which is not pure? You can become more familiar with hybrid crosses in the next activity.

COMPARE

A. Consider a cross between a hybrid tall pea plant and a short pea plant. Diagram the possibilities of the cross and summarize your results carefully.

B. Consider a cross between a hybrid tall and a pure tall pea plant. Diagram the possibilities of this cross and then summarize your results again.

Guinea pigs are valuable laboratory animals to show dominant and recessive characteristics. Figure 10-6 shows what might happen if a pure black guinea pig were crossed with a pure white one. Which characteristic is dominant? Diagram the results which appear in the drawing according to the method shown before.

The previous illustrations allow you to predict the possibility of certain crosses. However, one cannot predict what each individual offspring will be like. Let us illustrate this in the human being. We know that females have two *XX* chromosomes, and males have two chromosomes that are combined as *XY*. Therefore, the male sperm cell contains either an *X* chromosome or a *Y* chromosome. Each female egg must contain an *X* chromosome. We can diagram to find the possibility of a male or female offspring.

female → male ↓	X	X
X	XX	XX
Y	XY	XY

XX-50% female
XY-50% male

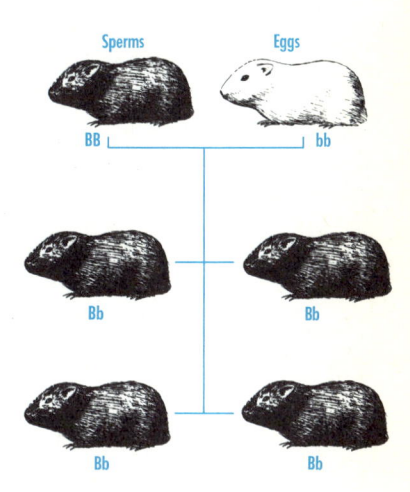

10-6 Why are all the offspring hybrids?

As the results in the diagram show, there is a 50-50 chance of giving birth to either sex. Yet we know that in many families there may be several girls born and no boys, or several boys born and no girls. We must realize that the ratios we are predicting for crosses only hold true when we deal with very large numbers.

Certain characteristics may not be dominant over others. Genes are not always dominant or recessive. In some species, one characteristic may *blend* with another to produce a third characteristic. This blending of two genes is called *incomplete dominance*. The four-o'clock is an example of a plant whose gene for flower color shows incomplete dominance. Figure 10-7 (page 292) is a diagram of the crossing of a pure red (*RR*) four-o'clock with a pure white (*WW*). The result is a hybrid which is pink (*RW*). Again, when two hybrids are crossed, they produce one-fourth red, one-fourth white, and one-half pink. Zinnias and short-horn cattle are other examples of plants and animals that also show this type of pattern or blending for color.

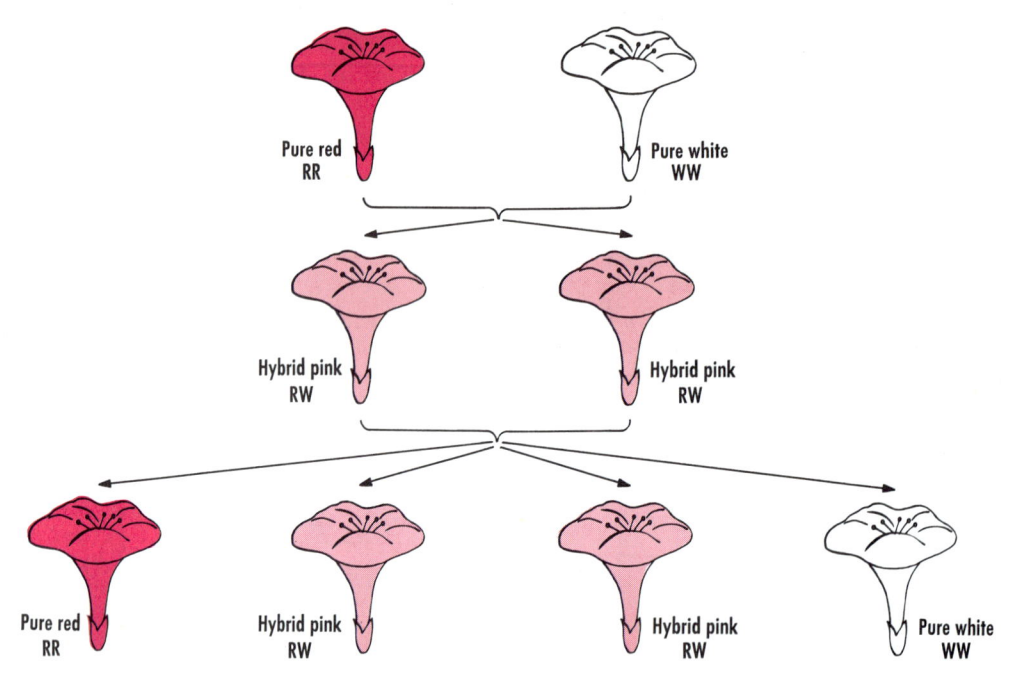

10-7 Pure red and white four-o'clocks produce pink hybrids. Explain.

Can more than one characteristic be crossed at one time? We have been dealing only with crossing one characteristic at a time. Mendel observed that if two characteristics were being studied, such as size of the plant and texture of the seeds, each was independent of the other. He found that pea plants could be tall and produce smooth peas or could be tall and produce wrinkled peas. Each characteristic was inherited separately. This is called the *law of independent assortment*. A species which is hybrid for two characteristics is called a *dihybrid*. Figure 10-8 shows what happens when two characteristics in guinea pigs are crossed. Can you tell how many of each type of animal are produced as a result of this?

The science of heredity is based on averages. The numbers of plants or animals produced in each new generation are not exact numbers that appear in every case. Sometimes all the offspring from one cross have dominant characteristics, and sometimes they all have recessive characteristics. We cannot

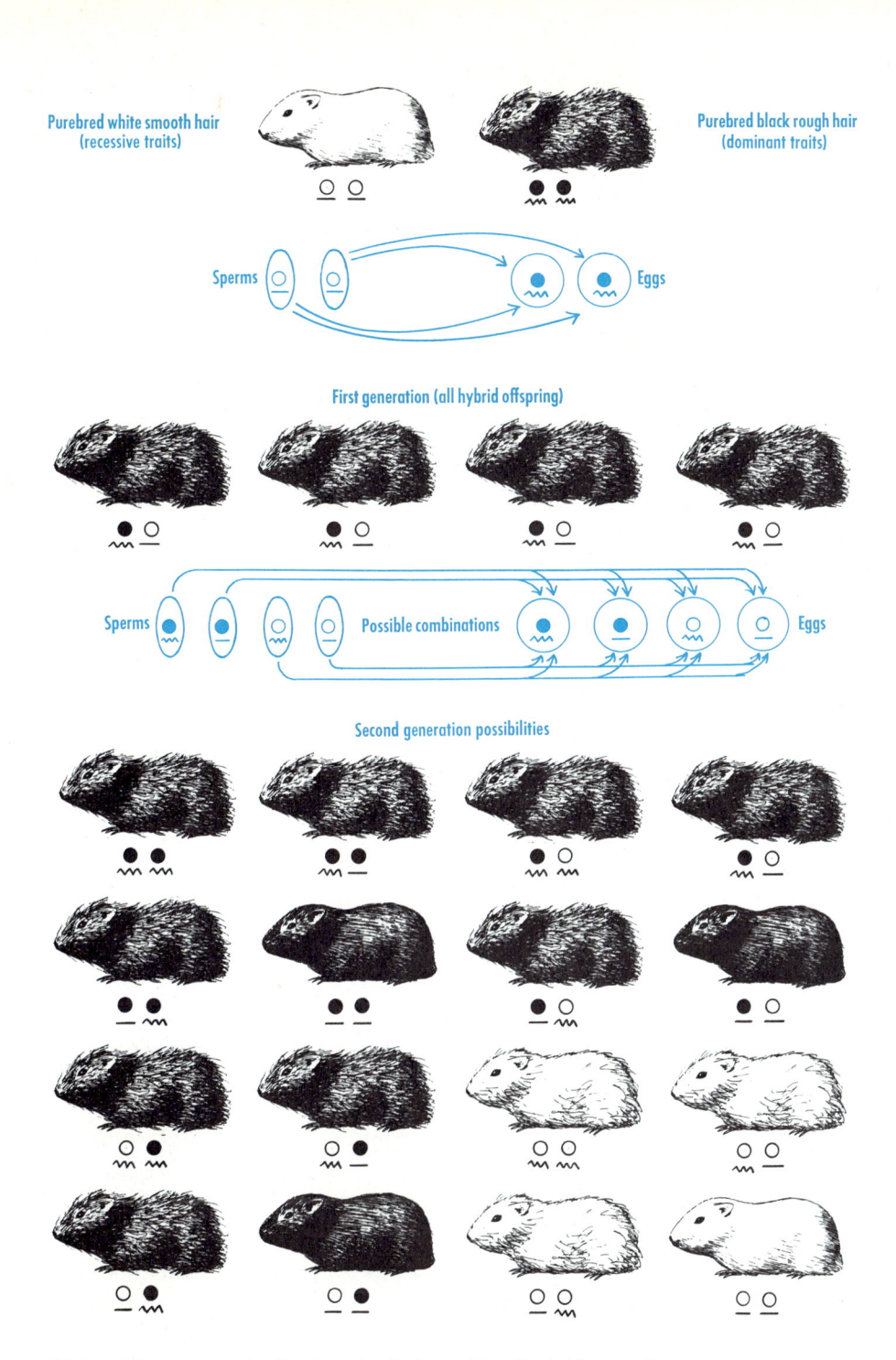

10-8 What types of offspring result from this dihybrid cross?

be sure of the results when only two plants or two animals are crossed. But when many crosses are made, the total number of offspring can be studied and the number showing each combination of characteristics will be found. This is the procedure which scientists use to study heredity in different organisms. It often requires many years of careful and patient observation to find the number of plants or animals having a particular characteristic.

REVIEW

1. What part of the cell contains the materials which transmit inherited characteristics?
2. How many pairs of chromosomes are found in: (*a*) human cells, (*b*) fruitfly cells, (*c*) corn cells?
3. Name the five main stages seen in mitotic division in a typical cell.
4. What is meant by a: (*a*) hybrid, (*b*) dihybrid?
5. Name the three main laws of heredity discovered by Mendel.

B/Why Have Living Organisms Changed?

Living things have changed through the ages. Scientists have known for a long time that there has been a slow change in the kinds of organisms that have lived on earth. For example, *fossils*, the remains of ancient plants and animals, have been found in rock layers. Fossil records are not very complete, but enough fossils have been found, at the present time, to show us that plants and animals which lived millions of years ago are no longer present.

In certain cases, rather complete fossil remains have helped scientists trace the slow development of a particular kind of animal now living. *Eohippus* (ee-o-HIHP-uhs) (Fig. 10-9) is a primitive ancestor of the modern horse. It is believed to have lived on the earth about 45 million years ago. It was a horse-

10-9 Compare characteristics of Eohippus and a modern horse.

like animal about the size of a small dog and had toes on its feet of instead of hooves.

Geographical distribution shows that changes have taken place. In our study of biomes in a previous chapter, we learned that certain plants and animals are adapted for life in specific habitats. That is, organisms usually found in forest areas are not normally found in desert areas. Forms of life, however, found in these similar biomes might be quite different as you will see by doing the activity.

CHECK

Look in biology textbooks and geography books to find out what kinds of typical plants and animals are found in each of the major continents. Show this distribution on an outline map of the world. Make a list of some typical plants and animals growing in each of these regions.

Scientists have long wondered why alligators and magnolia trees, for example, are found in North America and in Africa, but not in South America or in Asia. Why are members of the camel family found in Africa, Asia, and South America, but not in North America? Fossils studied by scientists clearly show that primitive forms of camel-like animals did live, however, in North America in great numbers millions of years ago. What caused these changes in camel distribution?

A study of plant and animal distribution shows that the spreading of organisms from one part of the earth to another is controlled partly by *corridors* and *barriers*. The study of changes in the earth's surface, discussed in detail in another unit, tells us that a wide land bridge, or corridor, once connected North America and Asia in the Bering Strait region. The camel's ancestors probably spread from Asia to North America, Africa, and South America over similar land corridors.

As the earth's surface features changed, some of the land corridors were broken and barriers, in the form of open seas or high mountains, were formed. Each camel-like animal in these different land areas began to change and adapt along different lines. Today, then, we find that the Asian camel has two humps, while the African camel has one. Both of these members of the camel family are adapted for life on the desert, but the llama of South America, a relative, is adapted for life in mountainous regions.

What happened to the camel population in North America? Changes in climatic conditions were brought about by huge ice sheets and snow that covered these land masses for thousands of years. These conditions probably caused these animals to become extinct. Fossils show that camels were animals that were common in North America many millions of years ago.

How does geographic distribution of organisms show us that living things change over long time periods? Scientists explain this as follows:
1. Each kind of organism had a place of origin on earth.
2. As the population increased, the organism spread by means of corridors to other land areas.
3. Changes in the earth's surface brought about barriers which separated the organisms.
4. In time, climatic changes and natural enemies caused organisms in certain areas to become extinct.
5. The organisms left in scattered areas continued to change and became adapted for life in different environments.

Some people do not think that scientists are correct in their conclusions. These people say that a Book of Scriptures states that God made the world and all its living creatures in six days. This means that the world and its creatures were created almost immediately. However, religious scholars who study the Scriptures point out that religious writings are not meant to be scientific. These scholars say that the term "day" in Scripture may not mean our 24 hour day but a "period" of time. This could mean that creation by God of the world with all of its many forms of life came into existence in six "periods" of time. These six "periods" do not, then, correspond to our six days. In this sense, science does not seem to disagree with Holy Scripture.

There are two theories that explain the changes in species. In 1859 *Charles Darwin (1809–1882)*, the great English scientist, set forth his theory explaining plant and animal variations in his book, *The Origin of Species by Natural Selection.* Briefly, this theory states that there is an overproduction of members of a particular species. This leads to a struggle for survival. Food becomes scarce. Also, there is a shortage of those conditions necessary for life. Since variations always occur among the offspring of living organisms, those offspring that inherit desirable variations, survive. These organisms might be taller, stronger, need less food, or adjust better to environmental conditions. Weaker organisms die out. Since more of the stronger organisms mature, they reproduce to pass their desirable characteristics on to their offspring. New forms result which are better adapted, leading to new species. Darwin's theory is called *natural selection.*

Scientists realized the importance of Darwin's contribution, but his theory did not account for all changes and had to be modified. *Hugo De Vries (1848-1935),* a Dutch botanist, further added to our knowledge in this field. While he was working with a flower called the evening primrose, he observed new varieties suddenly appearing among the offspring. When he crossed a new variety, it passed on its characteristics to its offspring.

Genes may undergo sudden changes. In most cases, the genes in the cells of a particular organism remain the same throughout many generations. Sometimes, however, a gene undergoes a sudden change which produces a characteristic that is totally different from any other characteristic that is found normally in that species. This changed characteristic may then be passed onto the offspring, and it is called a *mutation* (myoo-TAY-shuhn). Scientists believe that a true mutation is caused by a sudden change in the chemical make-up of a gene, not by the loss of one.

Many mutations are harmful to the organism. For example, you will remember what we learned about protective coloration. Suppose a protectively-colored animal produces a mutation in which hair or feather color is missing. The animal, called an *albino* (al-BYE-no), cannot escape its enemies readily by blending into the background in which it is found normally (Fig. 10-10). Thus, the chances for survival in this condition

10-10 Why is an albino fawn a harmful mutation?

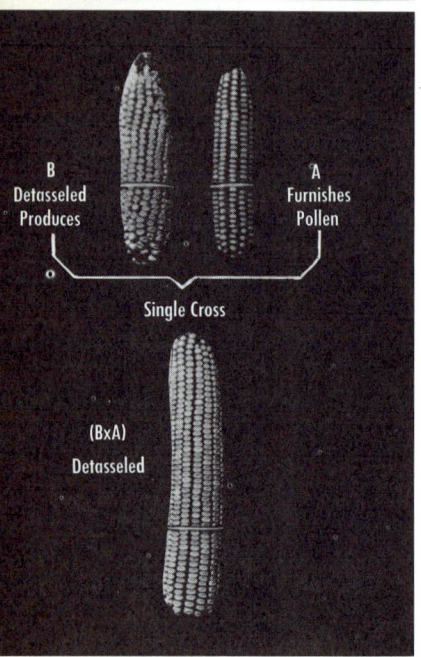

are not very good. This may be why many albino animals are not found in their natural habitats.

On some occasions, the mutation may help the species to become better adapted to environmental conditions. De Vries also concluded that there is a struggle for life and that the new species survive, reproduce, and gradually replace the old. De Vries added the *mutation theory* to Darwin's original theory of natural selection.

Scientists have learned how to improve plants and animals. Plant breeders have several purposes in producing improved plant varieties. Such characteristics as large and early ripening fruit, abundant seeds, large leafy vegetables, and large root crops are desirable. In addition to increasing the yield of the crop, disease resistance is very important as the following example shows.

Let us assume that a farmer loses a wheat crop because of a plant disease. As he looks over his fields of diseased plants, he may find several plants that have survived the disease. If he saves the seeds from these plants and plants them the next year, he would probably find that a few more plants had this desirable characteristic. By this method of *selection*, the farmer will finally develop a disease-resistant wheat strain for that disease.

Another goal of plant breeding is to develop plants that can grow in areas where they could not grow before. A good example is the extension of wheat growth into the Great Plains region. Ordinary wheat varieties are not adapted to withstand the climatic conditions of this region. Wheat, however, developed from plants resistant to these conditions has grown successfully.

Mass selection is used to obtain desirable varieties. This plant breeding method consists of the careful selection of parent plants from a great many individuals. In order to do this, the farmer may look over a large crop of plants and select only those that show favorable breeding characteristics. He then cross-pollinates or self-pollinates these plants and saves the seeds. The following year, the seeds are planted and only the parent plants showing the desirable characteristics are

10-11 Corn tassel is covered with a bag to collect pollen (top); pollen is blown onto the silk of another plant (center); and an example of corn produced (bottom).

saved for seed. By following this method over a period of time, a characteristic that is considered good can be bred into a plant variety.

Another method used by the plant breeder is called *hybridization* (*hye*-brih-dih-ZAY-shuhn). This process requires the use of two different plant varieties to obtain the new plant. The varieties, however, must be closely related or the seeds will not produce new plants. The best example of what has been done by hybridization is the production of hybrid corn. Figure 10-11 shows the cross-pollination process. In this method, plant A is crossed with plant B. Plant M, having desirable characteristics, is crossed with plant N. Now, as the new plants mature, pollination is controlled by man. The pollen from the new plant taken from A and B is crossed with the plant resulting from M and N. These seeds are planted, and a new plant is obtained from the four original ones, A, B, M, and N.

Once the desired plant is obtained, plant breeders begin to inbreed plants. During inbreeding, which is the opposite of hybridization, the plant is always self-pollinated. Thus, through the processes of selection and inbreeding, plant improvements can be made.

Mass selection has also been a method of producing desirable animal breeds. Modern breeds of poultry, for example, have been produced for different purposes. Leghorns are bred for their ability to lay eggs. Plymouth Rocks are used for both egg-laying and meat, and other varieties are bred mainly for fine meat production.

The modern turkey is quite different from the slender bird first found in a wild state by early settlers. Modern turkeys spend their lives eating a scientifically prepared diet building up large muscles that are more suitable for eating by man.

Using similar selective-breeding methods, domestic cattle have been developed by scientists for special characteristics. Beef cattle have low, broad bodies which yield large quantities of steaks and roasts. Other cattle breeds are bred to produce large quantities of milk. Some breeds of cattle are developed both for their tender meat and their milk-producing qualities (Fig. 10-12).

10-12 How has selective breeding improved the Guernsey (top) and the Hereford (bottom)?

REVIEW

1. What are two examples which show that living organisms have changed through the ages?
2. What is meant by a: (*a*) corridor, (*b*) barrier?
3. Name the two main theories which explain how changes occur in species.
4. What are three purposes of plant breeding?
5. List some animals that have been improved by selective breeding.

C/How Are Living Organisms Conserved?

Life on earth is in a critical balance. In any biotic community, all the species are part of a balanced, cooperative organization. Interfering with this balance, even if it means the changing of only one small part, may begin a chain reaction. This could, in time, destroy much of this community. Although man has been responsible for much of the damage to our natural resources, he is not the only one to blame. Destructive insects, plant diseases, certain types of gnawing animals, and natural occurrences like storms, floods, and fires caused by lightning account for much of our losses. You can learn more about the balance of nature in the next activity.

CHART

Make an illustrated chart showing all the possible ways in which man interferes with the balance of nature. Arrange the chart in two columns to show first the ways unbalanced conditions have been brought about and, then, the ways in which these conditions can be corrected.

Three major reasons are responsible for the destruction of our forest lands. For the first reason, we must look at our early settlers and lumbermen. When the first settlers arrived, nearly one-half of the country was covered by forests. However, these

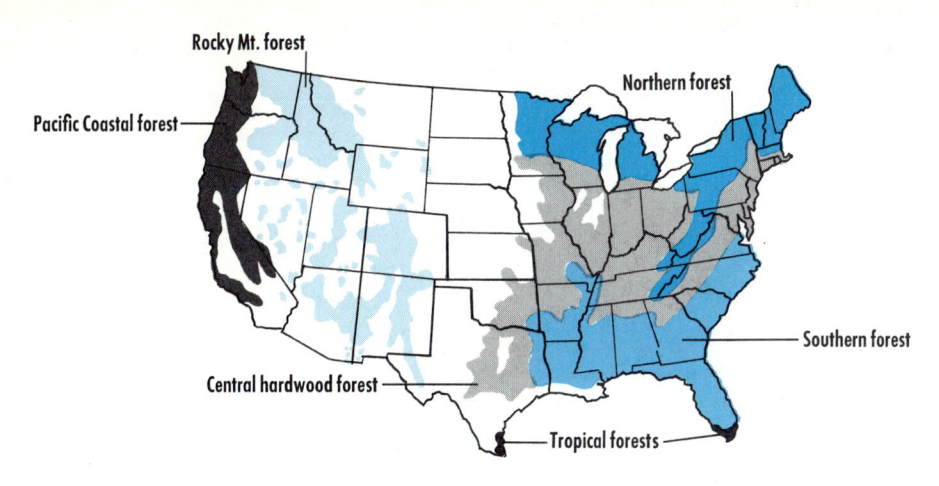

10-13 Locate the main forest regions of the United States.

forests proved very useful in supplying wood for homes, fuel, furniture, boats, and tools. Trees had to be cut to clear land for homes and farms.

New England became the center of the lumber region during the early 1800's. As settlers later moved westward, the lumber industry followed, leaving large treeless areas behind. The peak of this industry occurred in 1907, when over 50,000 sawmills were in operation. Since then, demands have lessened, owing mainly to the increase of products using materials other than wood. Lumber is still a major industry today in the southern and far western regions of the United States. This can be seen from the location of forests in Fig. 10-13.

The second major reason for the decline of our forest areas is forest fires. Of the total of about 500 million acres of timber forests, 83 million acres or about 17 percent are non-productive because of forest fires. These forest fires destroy valuable timber and precious topsoil. Sad, most of all, is the fact that most forest fires are caused by man's carelessness.

Insects and disease are a third major reason. Nearly one-quarter of the one million species of insects depend upon forest trees for their food. It is estimated that in the western forests alone, five to six billion board feet of timber are destroyed every year by bark beetles. Also responsible for much damage are the gypsy moth, which eats leaves, the spruce budworm, white pine weevil, and locust borer. In addition, timber that has been killed by insects dries up, and then becomes a serious fire hazard.

302 / UNIT 3 MAN EXPLORES LIVING ORGANISMS

Diseases caused by various fungi have drastically reduced the numbers of certain kinds of trees. *Chestnut blight* has destroyed almost all the chestnut trees in the eastern United States. *Dutch elm disease* is a serious fungus infection which has threatened the existence of the American elm. *White pine blister rust* has wiped out whole forests of valuable pine trees, and *oak wilt* is threatening our fine oak forests. The activity will help you to understand how diseases of trees upset the the balance of nature.

CHECK

Do some library research or write to your State Forestry Department to find out how a tree disease has affected forest land in your area. Write a report on your findings and tell what steps are being taken to control the disease.

Our wildlife has also declined. In the early settlement of this country, great buffalo herds roamed the plains. Antelopes, elks, and moose were found in great numbers in the northeast. Today, game laws are needed to protect their numbers. Passenger pigeons, California condors, and other birds once filled the sky. Today, some of these birds are already extinct and others are nearly gone. A few forms of wildlife near extinction are shown in Fig. 10-14.

10-14 Left to right: condor, bald eagle, and whooping crane. Why are they almost extinct?

Every animal and plant needs certain conditions in order to live. Clearing forests and farms for cities, however, has caused the destruction of much wildlife in these areas.

Overhunting, overfishing, and *overtrapping* are also responsible for the decline of many species. The antelope, elk, bighorn sheep, and large numbers of waterfowl and birds were shot by hunters. Common fur animals like the marten, beaver, muskrat, and certain foxes are scarce today because of overtrapping. The numbers of shad, sturgeon, and Pacific halibut have been seriously reduced by overfishing.

The drainage of swamplands has killed many species. Swamp areas where mosquitoes breed had to be drained as cities were built. In addition, shallow lakes and marshland were drained for farming purposes. The geese, ducks, and other birds which inhabited this community either died or migrated. Since so many of their natural habitats throughout the country were destroyed, these animals found it difficult to survive and today many are almost extinct.

Oil pollution in our waters has also been responsible for wildlife destruction. Coastal waters polluted in this way are very harmful because the oil dumped by ships is carried in by the tide and gathers along the shore. Oil leaking from wells drilled offshore result in the death of many thousands of fish and birds.

Man is now protecting forests and wildlife. Because man has not always used his natural resources wisely, many organisms needed for our environment have been lost. In recent years, man has tried to correct many of his wasteful practices. We will discuss some of the steps that are presently being taken to protect both our forests and wildlife.

First and foremost is a program of *education.* Everyone must realize the need for protecting our remaining timber and national forests. Therefore, carrying out a wise program of replanting our forests and protecting them from being destroyed is a long and difficult process.

Simple rules must be obeyed by all who enter forest areas. Road signs, like those in Fig. 10-15, are one way to help you remember. Never throw lighted cigarettes or matches on the ground or along the road. When camping, be sure that camp-

10-15 How are signs helpful in preventing forest fires?

304 / UNIT 3 MAN EXPLORES LIVING ORGANISMS

fires are well protected and completely out when no longer needed. As you travel in forest areas, be on the alert for any fires. If you see a fire, or see anyone starting a fire in an unprotected area, notify the authorities at once. Following these few, simple rules can save millions of acres of your precious forest lands yearly, and a saving of many millions of dollars as well.

Our federal and state governments spend large amounts of money to help prevent large forest fires. Over 90 percent of our forests are under organized protection. Fire towers are manned by forest rangers who watch for fires during the summer. Many more need to be built. Forest rangers, reporting on a fire from two different towers, make it easier to locate the origin. Better forest roads help men and fire-fighting equipment to get through faster.

In addition to fire protection, our tree-cutting methods are changing. Both private and public forests are now being cut scientifically. *Selective cutting* has been introduced, the results of which can be seen in Fig. 10-16. Using this method of cutting, only trees of a certain age and size are cut down. Young trees are given an opportunity to mature. *Block cutting*, also shown in Fig. 10-16, is another helpful practice. A logger can cut trees in a certain area, but he must leave clumps of trees in the area. Trees in and around these blocks soon spread their seed into the cut region. If grazing is controlled, the seedlings soon begin to grow. Eventually, this area that was cut down is covered with trees once again. The activity helps to illustrate how cattle can destroy trees.

COMPARE

Examine the ground under trees in a park or wooded area where cattle are allowed to graze. Compare the growth of shrubs and young trees in that area with a protected area where growth is not disturbed. Make a map of the two areas showing the distribution of shrubs and young trees. Which area will maintain itself over a long period of time? Which area must start from a grass-weed stage? Explain the reasons for your answers.

10-16 Block cutting (top), selective cutting (below). What is the difference?

Reforestation is under the control of the United States Forest Service. It is estimated that we have about 135 million acres of land without satisfactory tree cover. About half of this area will be able to restock itself. The other half will have to be restocked by man. The Forest Service is planning to reforest about 25 million acres during the next 20 years. Although this is a costly procedure, it will eventually provide recreational areas, timber areas, and wildlife preservation.

There is a movement underway in the United States to provide more private lands for recreational use. The growing population has taken more and more land for homes, roads, and industrial areas, leaving less for recreational purposes. Increase in leisure time, however, resulting from a shorter work week increases the need for additional recreational areas.

Scientists are working hard trying to discover methods to control the insect pests which destroy much of our forests. Although sprays and poisons are helpful, spraying whole forest regions is costly. Also, this practice does harm to useful plants and animals as well as the insects.

Scientists are trying to import the natural enemies of these pests into the area. The ladybird beetle, brought in from Australia, is quite successful in controlling the citrus-fruit scale. Certain wasp species also are very effective in destroying many insect pests.

By using all these methods to protect our forests, many thousands of acres now bare will once again be covered by seedlings. Eventually, another forest crop will grow.

The protection of our wildlife is very important. Federal and state laws have been passed to protect these animals, which are considered government property. By issuing hunting and fishing licenses, states are able to limit the length of the hunting and fishing season. The number and size of the "kill" is also controlled.

Establishing national and state *game preserves* has given protection to many animals in the United States (Fig. 10-7, p. 306). No hunting or fishing is allowed in these areas. Similar preserves are also maintained for wild birds where man may enjoy the wildlife but cannot destroy it.

10-17 How do game preserves help protect wildlife?

Many states, finding that certain animals are becoming scarce, import them from other areas and breed them. Deer were brought back to Vermont in 1900 after becoming extinct. Today, Vermont law restricts the deer hunting season. Because of overtrapping, beavers almost disappeared in Pennsylvania. In 1920, beavers were brought from Canada and reintroduced to the state. By 1950, they were so numerous that trapping under regulation was again allowed.

Game management practices are restoring the balance of living organisms. Although wildlife in the southern United States has little difficulty finding food in winter, animals in our northern regions have this problem. You can help by placing food outside on top of the snow. Building and maintaining winter feeding stations is both interesting and helpful. As a result of practices such as these, the trumpeter swan is no longer considered in danger of extinction. The whooping crane, too, is making a comeback. Whether they continue to increase, however, is not known.

In our western states, game birds have been introduced and protected. The ring-necked pheasant, shown in Fig. 10-18,

CHAPTER 10 THE CONTINUITY OF LIFE / 307

was brought into the western grasslands area in about 1900. The few original pairs multiplied so much that in South Dakota alone, game officials have estimated over 50 million pheasants have been killed by hunters. Yet, the number of pheasants is still increasing.

The United States Bureau of Biological Survey is a leading federal agency concerned with wildlife protection. This agency places bands on the legs of various birds to study their migration patterns, their numbers and diseases. Fur animal conservation is also studied. All of this information is available to the public.

The Audubon Society is a well-known national group which has long been concerned with bird protection. This society carries on a wide educational program to aid in wildlife preservation. The activity below will make you aware of animals that have become extinct.

10-18 Why is the ring-neck pheasant a good game bird?

CHECK

Look in natural history books in the library to find out about bird and animal species that are extinct or nearly extinct in North America. Study the game laws in your state to see how rare species are being protected. Make a report on what other conservation agencies are doing to save various bird and animal species that are in danger of becoming extinct.

Hatchery programs produce large numbers of game fish. Many adult fish are kept in special spawning pools in fish hatcheries. In the small area of these special kinds of tanks, many more eggs can be fertilized than in lakes or streams. As the young develop, they are carefully cared for in the hatcheries. In summer, they are taken by special trucks and placed in rivers and streams. You can understand why water pollution is such a serious problem. Why try to restock polluted streams and rivers?

There are also laws which protect fish from being removed at a destructive rate. Although each state controls its own waters, certain rules seem to be general, such as special seasons for fishing and a minimum-size requirement. A fish, caught under a certain size or length, must be put back. The number of a particular species a fisherman can catch is also limited.

Pond drainage as well as the use of large nets is prohibited in fishing areas. These laws are necessary if we are to continue to have enough fish for food, recreation, and other purposes. Removing large fish from a pond or lake, however, is actually a necessity. Can you explain why?

Everyone should be concerned with the proper use and management of living organisms. This is important to insure an ample wildlife supply for the future and to preserve the beauty and balance of nature. Further knowledge of animal life cycles, the food cycles existing in different habitats, and other interrelationships can help in establishing a long-term program of conservation.

REVIEW

1. Name three factors which contribute largely to forest destruction.
2. Give some examples of nearly extinct birds or mammals.
3. Name two methods of tree-cutting which help greatly to save our forests.
4. List three agencies in the United States concerned with forest and wildlife protection.
5. Of what value in conservation are: (*a*) game preserves, (*b*) fish hatcheries?

THINKING WITH SCIENCE

A. *On a separate sheet of paper write the numbers 1 to 15. Some of the following statements are true and some are false. Rewrite the statement, changing the terms in italics if necessary, to make them all true.* **Do not write in your book.**

1. The union of a sperm cell and an egg cell to form a new individual is called *hybridization.*

2. The process of cell division in which the number of chromosomes is reduced by half is known as *mitosis.*

3. Areas in which animals are protected from man's destructive activities are called *game preserves.*

4. A new characteristic in an organism which results from a change in genes produces a *mutation.*

5. The characteristic which appears in the first generation of a hybrid cross is called *recessive.*

6. The theory developed by Darwin to explain changes in species is called the *mutation theory.*

7. The method of breeding by selecting desirable offspring from a great number of parents is called *mass selection.*

8. Cutting trees for lumber by selecting only trees of a certain size and age is known as *block cutting.*

9. The blending of two characteristics to produce a third characteristic in the offspring is known as *incomplete dominance.*

10. In mitosis, the chromosomes are formed in the *anaphase stage.*

11. The science of the way characteristics pass from parents to offspring is called *heredity*.

12. Two factors which have reduced our forest lands are destructive lumbering and *forest fires*.

13. When a sperm cell unites with an egg cell, a new cell, the *zygote,* is formed.

14. *Identical twins* result from the same zygote which divides after fertilization.

15. Mendel found that in pea plants, shortness is a *dominant* characteristic.

B. *Write the answers to the following in your notebook. Be sure to use complete sentences and correct spelling and grammar.*

1. Why is the nucleus of the cell considered to be the part that is most important in heredity?

2. What is the difference between cell division in mitosis and cell division in meiosis?

3. Explain why both parents are equally responsible for the hereditary characteristics in the offspring.

4. Explain how the pairing of chromosomes determines the sex of a developing animal.

5. Briefly explain: (*a*) the law of dominance, (*b*) the law of segregation, (*c*) the law of independent assortment.

6. How do corridors and barriers affect the distribution of plants and animals on earth?

7. Briefly explain what is meant by Darwin's theory of natural selection.

8. How did De Vries' theory of mutation add to our understanding of the theory of natural selection?

9. Give some examples showing how scientific plant breeding has improved the usefulness of certain plants.

10. Give some examples of how scientific animal breeding has improved the usefulness of certain animals.

RESEARCH IN SCIENCE

1. Obtain some fruitflies and devise means of studying them to find out how inherited characteristics are transmitted.

2. Arrange a field trip to a fish hatchery. Report on fish raising and stocking of streams and lakes.

3. Do a research project in the library on how beavers are important in conservation.

4. Make a chart showing how insects and fungus pests destroy trees.

5. Prepare a report on the work of one of the following agencies: (*a*) National Audubon Society; (*b*) National Wildlife Federation; (*c*) United States Forest Service; (*d*) United States Fish and Wildlife Service.

6. Obtain a packet of albino corn seeds and plant them. Find the ratio of green to white corn kernels produced.

READINGS IN SCIENCE

Carson, Rachel, *Silent Spring*. Houghton-Mifflin, 1962. The harmful effects of chemical controls of insects on the balance in nature are discussed. This book is important in the study of ecology.

Clemons, Elizabeth, *Tide Pools and Beaches*. Alfred A. Knopf, 1964. An introduction to one of the most fascinating of ecological environments — the living things that inhabit the tidal pools of both coasts of the United States.

Cosgrove, Margaret, *Bone for Bone*. Dodd, Mead and Company, 1968. Relationships of vertebrate life are described. A comparison of all major groups is presented.

Cosgrove, Margaret, *Strange Worlds Under a Microscope*. Dodd, Mead, 1962. Discusses early history, use, care, and optical principles. Contains a good chapter on microtechniques for beginners.

Crosby, Alexander L., *Junior Science Book of Pond Life*. Garrard, 1964. The author begins with the construction of his own artificial pond. He describes photosynthesis as the basis of the food cycle involving phytoplankton and zooplankton, progressing upward through invertebrates, fish, amphibians, reptiles and mammals.

Davis, Joseph, *Finding Out About Mammals*. Home Library Press, 1963. Surveys the mammalian world in relation to its classification. Sketches and colored photographs illustrate examples of various groups.

Glemser, Bernard, *All About Biology*. Random House, 1964. Develops the theories of the beginning of life. Includes basic biological concepts such as cell theory, germ theory of disease, and photosynthesis.

Lanyon, Wesley C., *Biology of Birds*. Natural History Press, Doubleday, 1963. Discusses evolution, flight, distribution, environment, and development from egg to bird.

Laycock, George, *America's Endangered Wildlife*. W. W. Norton, 1969. Author tells specifically what events have brought eagles, blue whales, prairie chickens and other animals to their present difficulties. The attempts to save them are discussed.

Laycock, George, *Wild Refuge*. Doubleday, 1969. The book presents an introduction to the United States system of wildlife refuges. Author tells how each was established and which species of wildlife is maintained.

Lorus, J., and Milne, Margaret, *The Senses of Animals and Man*. Atheneum, 1962. A treasury of information on the senses of animals and man. Written in a nontechnical style.

Marteka, Vincent, *Bionics*. Lippincott, 1965. A simple treatment of a new field which attempts to equate animal behavior and senses with technological counterparts. Discusses biological clocks, rhythms, navigation, and migration.

Mohr, Eugene V., *The Study of Reptiles and Amphibians Made Simple*. Doubleday, 1963. Outlines each reptile and amphibian group, giving characteristics, feeding, reproduction, ecology, and behavior.

Riskin, Edith, *Watchers, Pursuers, and Masqueraders*. McGraw-Hill, 1964. The books brings out the importance of mimicry and protective coloration for survival.

Schneider, Leo, *You and Your Cells*. Harcourt, Brace and World, 1964. The author's gift for explaining complex scientific ideas and concepts to young readers is again shown in this introduction to cellular and molecular biology.

Selsam, Millicent E., *Plants That Move*. William Morrow, 1962. Discusses the movement of flowers, leaves, and vines. It also tells how seeds are scattered and how some plants capture insects. Some easy-to-do experiments are included.

Teale, Edwin, *The Strange Lives of Familiar Insects*. Dodd, 1962. Tells many interesting things about the lives and habits of common insects. Contains excellent illustrations.

Zeichner, Irving, *How Life Goes On*. Prentice-Hall, 1961. Describes the reproductive processes from simple protozoans to humans in a simple direct manner.

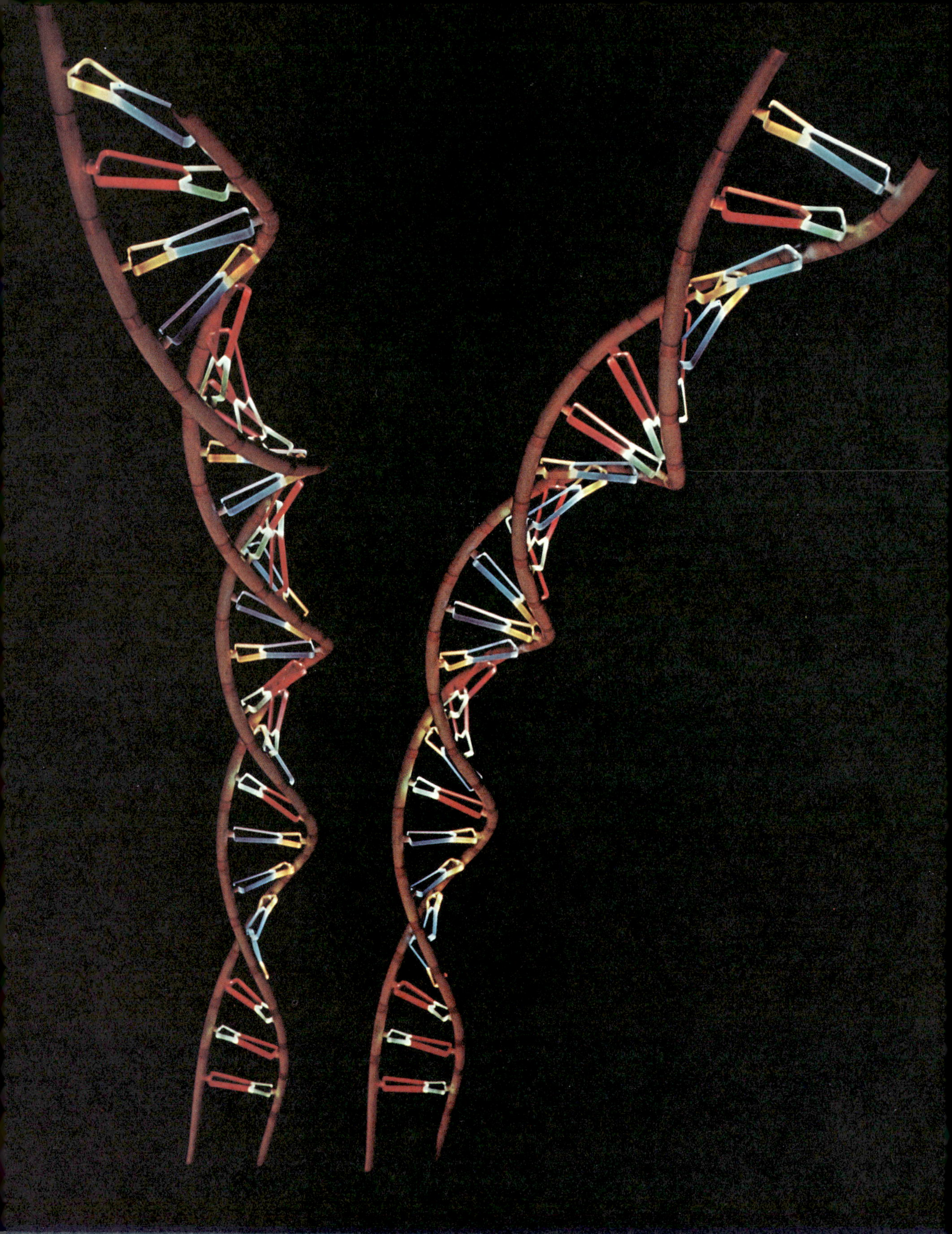

unit 4

Man Explores The Human Body

Although the early Greek philosopher-scientists studied the structure of the body, their writings showed a limited knowledge of the internal functions of the body. Their explanations of how the human body carried on its life processes were often incomplete and misleading. Some theories stated that food was "cooked" in the intestine and in some way was changed into blood. Blood was one of the four "juices" thought to control all activities of the body. When these "juices" were present in correct amounts, the person was thought to be healthy. If the balance among the four "juices" was upset, the person became ill.

When the city of Alexandria became the Greek capital of Egypt, its kings established scientific centers called museums. Here, scientists from every country could come to live and

316 / UNIT 4 MAN EXPLORES THE HUMAN BODY

work free of charge. All branches of science were studied in the museums, and the medical sciences, especially, advanced in many areas. The best known of the Alexandrian anatomists was *Claudius Galen (131–210)*, a famous physician.

Galen performed many experiments. Because dissection of dead people was frowned upon, he used apes instead of human bodies for his experiments. He thought that digested food was carried to the liver where it was changed into blood. Undigested food was changed to waste and removed through the large intestine. He stated that the kidneys removed extra water from the blood. Galen was the leading authority on the structure and function of the body for over a thousand years.

A young Belgian anatomist, *Andreas Vesalius (1514–1564)*, who was teaching in Italy, performed many dissections. As he made more and more observations, he found that Galen was in error on many points. Vesalius's findings became the foundation for the modern science of anatomy.

The early scientists knew that food and blood were somehow related, but the circulation of blood was one of the mysteries they were not able to solve. Their theories supposed that food was made into fresh blood in the liver. Blood was then carried to all parts of the body through the veins. They did not understand the function of the heart. They assumed that it only warmed up the blood! It was not until *William Harvey (1578–1657)*, an English physician, studied the circulation of blood that an answer was found.

The invention of the microscope, in the seventeenth century, provided scientists with a number of theories about the body. *Marcello Malpighi (1628–1694)*, an Italian scientist supplied evidence supporting Harvey's theory of circulation. Through Malpighi's discovery of the capillaries that connect the arteries and veins in the tissues, Harvey's theory was accepted.

With the aid of the microscope, other scientists studied the nervous system. They found that all the different parts of the body were connected to the brain by fine nerves. *Marie Francois Bichat (1771–1802)*, a French scientist, presented the first scientific organization of the body into tissues, organs, and systems. He described 21 different tissues and proved that the tissues were essential materials of the body.

In this unit we will study the human body and the role energy plays in its maintenance.

chapter 11
Energy for the Body

A/Why Does the Body Need Food?

The body acts like a machine. To perform all its activities, the body must have energy just like any machine. You have learned that in order for a machine to work, energy must be used. The source of energy for many machines is the chemical burning of fuel. This is known as a "downhill" reaction for this reason. As the energy-rich fuel burns, it releases energy to produce work, forming energy-poor carbon dioxide and water.

The living cell is also an energy-converting machine. It changes the chemical energy in food into forms with very little loss of heat. The living cell is remarkable, however, in that it can carry on "uphill" reactions as well as the usual "downhill" reactions found in nonliving machines. Thus, the cell not only uses up energy, but it can store energy in the body for future use. You might think of a living cell as a tiny battery that can be charged to store energy, and then discharged to release the energy.

Food is the source of energy in the body. The fuel needed by cells to furnish energy is the food we eat. Again, the living body has an advantage over a nonliving machine. Although a machine can use only one kind of chemical fuel, like coal or gasoline, the body releases energy from almost any kind of food. However, *carbohydrates* (kahr-bo-HYE-drayts) are the chief energy foods used in the body.

One common form of carbohydrate is sugar. If sugar is burned in air, it produces energy in the forms of heat and light. It yields carbon dioxide and water at the same time. The word equation shows this reaction.

$$\text{sugar} + \text{oxygen} \rightarrow \text{carbon dioxide} + \text{water} + \text{energy}$$

Yet, living cells do not light up when this same chemical reaction takes place in them! In fact, a cell performs its functions only within a narrow temperature range. The cell, therefore, uses the energy from food as chemical energy. This energy can be used to do work at low temperatures.

The body uses food for growth and repair. You know that an engine wears out with use. It must be repaired or parts replaced from time to time. The body also wears out, but replacement of a tired muscle or a worn-out gland is not possible. Instead, the body repairs itself by building new cells to replace those that are worn out in certain tissues. For example, if you cut your finger or break a bone, the damaged cells will be repaired or replaced with new cells.

You are still growing, and your body needs millions of new cells every day to make this possible. As you increase in size, your bones become longer. You add muscle tissue, and your organs get larger. The materials and energy needed for this growth come from the food you eat. That is why it is important for you to eat a daily supply of the proper foods.

The body needs food to maintain its temperature. Humans are warmblooded animals, like cats, dogs, and other mammals. This means that our bodies remain at about the same internal temperature, regardless of how cold or warm the air around us may be. Heat for the body is produced as a waste product in the energy-conversion cycle in the cell. Heat serves to keep the average temperature of the body at about 98.6°F. By doing the following activity, you will see that food produces heat energy for the body.

CHAPTER 11 ENERGY FOR THE BODY / 319

OBSERVE

Shell several whole peanuts, or other nuts, and remove the brown skin from them. Place a few of the nuts in an evaporating dish or small aluminum dish and arrange the equipment, as shown. Fill a small beaker half full of water. Stir the water with a thermometer to find its temperature. Ignite the nuts with a Bunsen burner. Put the dish under the beaker to allow the flame from the burning peanuts to warm the water. When the nuts have burned, stir the water with the thermometer. Note the temperature of the water in the beaker.

What happened to the temperature of the water? Where did this heat energy come from? What does this show about the heat energy of nuts? Try different foods to determine the amount of heat energy a weighed amount releases when burned.

The energy in food is measured in Calories. A *Calorie* (KAL-o-ree) is the amount of heat needed to raise the temperature of one liter of water one degree Celsius. The Calorie used to measure the energy in food is equal to *1000 small calories*. It is written with a capital letter to distinguish it from the small calorie which is used in physical science.

It takes a large number of Calories to supply the daily needs for the body. A daily requirement of about 2500 to 3500 Calories is probably about average. However, the number of Calories needed depends mainly on your age, body weight, and on physical activities (Fig. 11-1). The following table shows some average daily Calorie requirements.

Average Daily Calorie Needs	
	Calories
Child from 10 to 12 years	2000
Girl from 12 to 14 years	2200
Boy from 12 to 14 years	2600
Girl from 15 to 18 years	2600
Boy from 15 to 18 years	3000
Man doing moderate work	3200
Man doing hard work	3500–4500
Man doing heavy labor	4500–5000

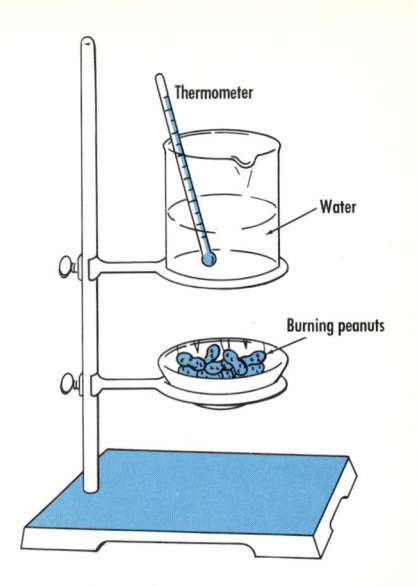

11-1 Would there be a difference in the number of Calories required by the students (above and below) in their activities? Why?

The figures in the table are all averages and will differ for each person. They include all the energy needed for both the involuntary and voluntary activities of the body. (The involuntary activities are those necessary to keep the body alive, like the beating of the heart, breathing, and food digestion). The number of Calories a day needed to furnish the energy just to stay alive is called your *basal metabolism* (me-TAB-o-*lizm*). The basal metabolic rate in a healthy person is about one Calorie an hour for every kilogram of body weight (1 kg = 2.2 lbs.). The formula for figuring out your average daily basal metabolism is:

$$\boldsymbol{M = Kg \times 24}$$

In the formula, *M* is the total number of Calories needed, *Kg* is the body weight in kilograms, and *24* is the number of hours per day.

SAMPLE PROBLEMS

A. Find the average daily basal metabolism for a boy weighing 132 pounds.
Solution:
1. 1 Kg = 2.2 lbs.
2. $\frac{132}{2.2}$ = 60 Kg
3. M = Kg × 24
4. M = 60 × 24
5. M = 1440 Calories

B. Find the average daily basal metabolism for a girl weighing 120 pounds.
Solution:
1. 1 Kg = 2.2 lbs.
2. $\frac{120}{2.2}$ = 55 Kg
3. M = Kg × 24
4. M = 55 × 24
5. M = 1320 Calories

Your voluntary activities include all those that are not a part of your basal metabolism. Walking, sitting, doing your homework, and playing are all voluntary activities, and they require energy. To find out about how many Calories you use each day for voluntary activities, subtract your basal metabolic rate from the average daily Calorie needs for your particular age group shown on page 319.

Estimate the amount of food you eat and the number of Calories in the food. In this way, you can get some idea of whether you are receiving your daily Calorie requirements. If you eat less food than your body requires for your activities, you may be tired and run down. If you eat more food than you need, the excess food is changed into fat and stored in the body. Thus, if you continue to overeat, your weight will increase. You may not know exactly how many Calories there are in the various foods you eat, so we have listed a few common food examples.

Foods Producing About 100 Calories a Portion	
Apple	1 large size
Bread, white	1½ slices
Butter or margarine	1 tablespoon
Cake, no icing	1 small piece
Cheese	1 ounce
Egg, boiled	1 large size
Ice cream, chocolate	2 scoops
Jam or jelly	2 tablespoons
Meat, lean	1½ ounces
Milk, nonfat	1 cup
Peaches, canned	2 halves in syrup
Peas, cooked	1 cup
Potato, boiled	1 large size
Soda, cola type	1 cup
Soup, tomato	1 cup

How many Calories do you take into your body each day? The following activity will help you find out.

RECORD

Make up a chart listing the different foods that you eat for a week. Figure out the total number of Calories in the food. Determine the average number of Calories that you take into your body each day. Use a Calorie booklet that includes a wide variety of foods. Be sure to include all food that is eaten between meals and for your meals. Compare the chart that you have prepared with those charts made by other students in your class. How do the charts compare? Explain why differences appear in the charts prepared by your classmates.

Our eating habits have changed. Many Americans are eating more fats and fewer starchy foods than our grandparents consumed. We are also eating greater amounts of meat, fish, poultry, fresh fruits and vegetables, and less grain products, like flour and cereals (Fig. 11-2). The greater amounts of meats, fruits, and vegetables that we are eating have increased the amount of essential body-building substances taken into the body. This is healthful. In order for a substance to be used as food in the body, the food must be digested to supply energy and materials for body cells. These essential substances in foods are called *nutrients* (NYOO-tree-uhnts).

11-2 How have American diets changed in the last fifty years?

REVIEW

1. What is meant by: (*a*) an "uphill" reaction, (*b*) a "downhill" reaction?
2. What is the main energy-producing class of foods used by the body?
3. (*a*) What is the average daily Calorie requirement for most people your age and sex? (*b*) What is the average daily Calorie requirement for a man that is employed in doing heavy labor?
4. (*a*) Give examples of several different portions of food, each of which supplies about 100 Calories for the body. (*b*) Give examples that would supply about 300 Calories for the body.
5. List the three main uses of food in the body.

CHAPTER 11 ENERGY FOR THE BODY / 323

B/What Are the Main Classes of Nutrients?

The body needs five main classes of nutrients. We eat different kinds of foods like fruits, vegetables, meats, and cereals. The nurtrients in foods, however, fall into five main classes: *carbohydrates*, *proteins*, *fats*, *minerals*, and *vitamins*.

It is possible to eat enough food and still be undernourished. This can happen if you do not receive all the nutrients in your diet. When the foods you eat supply the body with the essential minerals and vitamins, as well as the necessary nutrients, you have a *balanced diet*.

How can you be sure you are eating a balanced diet? Scientists have discovered a list of basic groups needed by everyone for maintaining good health. The amount you should eat depends on your age and activities. You should have at least one food daily from each of the four groups shown in Fig. 11-3.

11-3 Identify these food groups that are essential for a balanced diet.

324 / UNIT 4   MAN EXPLORES THE HUMAN BODY

Carbohydrates furnish energy for the body. *Carbohydrates* are composed of carbon, hydrogen, and oxygen. They are the chief source of energy for the body. Sugar and starch are the best known carbohydrates. You can test for starch by doing the activity below.

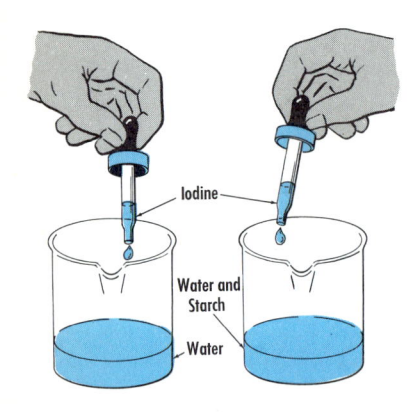

TEST

Put a pinch of starch in a beaker, a pinch of table sugar in a second, and nothing in a third. Fill each beaker one-fourth full of water and stir. Add a few drops of iodine solution to each beaker, as shown. Why was plain water used in one beaker? What color is the water plus iodine? What color is the sugar solution plus iodine? What color is the starch mixture plus iodine? This color is a standard test for starch. Test other foods, like crackers, white bread, raw potato, and others for starch and record your results.

Although sugars are also carbohydrates, the iodine test will not show their presence. In fact, different kinds of sugars require different chemical tests. One group is known as the simple sugars. *Glucose* (GLOO-kose), found in such foods as corn syrup, honey, and in some fruits, is an example of this type. Another group is the complex sugars. *Sucrose* (soo-krose), found in ordinary table sugar and in many fruits and vegetables, is the best example. You can test for a simple sugar in the next activity.

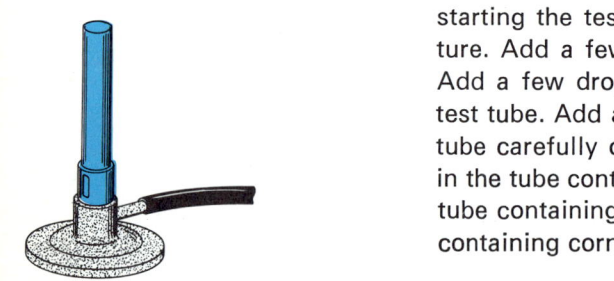

TEST

Mix equal parts of Fehling's solution A and B in a beaker before starting the test. Fill three test tubes one-third full of the mixture. Add a few drops of table-sugar solution to one test tube. Add a few drops of corn syrup or Karo solution to the second test tube. Add a few drops of plain water to the third. Heat each tube carefully over a burner, as shown. What color is produced in the tube containing plain water? What color is produced in the tube containing table-sugar? What color is produced in the tube containing corn syrup solution? This color is used as a standard test for glucose sugar. Test other foods, like fruits, candy, and vegetables, for glucose. Record the results after you have tested each substance.

In a normal diet, more than half the foods you eat contain carbohydrates. All the cereal grains are rich in starches. Most fruits and vegetables contain a great deal of sugar. Candy and other highly sweetened foods are obviously a rich source of sugars.

Carbohydrates are excellent foods because they carry on "downhill" reactions easily. As can be seen in the following equations: When sugars combine with oxygen, they change into carbon dioxide and water, yielding at the same time large amounts of energy.

$$C_6H_{12}O_6 + 6O_2 \rightarrow 6CO_2 + 6H_2O + \text{energy}$$
(glucose) (oxygen) (carbon dioxide) (water)

$$C_{12}H_{22}O_{11} + 12O_2 \rightarrow 12CO_2 + 11H_2O + \text{energy}$$
(sucrose) (oxygen) (carbon dioxide) (water)

Proteins furnish materials to build and repair the body. Proteins are very complex chemical compounds. We know that they are composed of carbon, hydrogen, and oxygen, and in addition, there are large amounts of nitrogen. Usually present in proteins is some sulfur and phosphorus. An approximate formula for *hemoglobin* (HEE-mo-*glo*-bihn,) the protein that is present in red blood cells is $C_{3032}H_{4816}O_{872}Fe_4S_8$. Can you identify the atoms?

Proteins are giant molecules. Scientists are interested in the basic units, *amino* (uh-MEE-no) *acids,* of which these molecules are composed. There are more than 20 different kinds of amino acids in the body. Ten of these are called essential amino acids because they must be in the diet to maintain normal health.

Proteins that are not used directly to form new protoplasm are excreted. Some are changed into sugars and stored as energy food in the body. Foods, like meats, eggs, milk, and beans, supply proteins in the diet. There are different types of proteins in various foods, however. Egg white and meats contain a protein called *albumin* (al-BYOO-mihn), *gelatin* is found in gelatin desserts, and *legumin* (le-GYOO-mihn) is found in peas and beans. All cereals, like wheat, rice, and barley, contain *gluten* (GLOO-tuhn), and milk products furnish us with *casein* (KAY-seen). You can learn how to test for protein in the following activity.

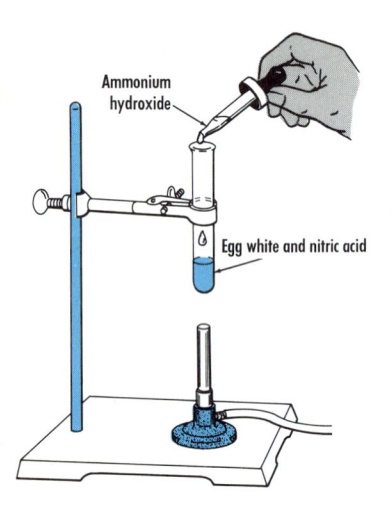

EXPLORE

Put a small piece of egg white in a test tube. Add a few drops of nitric acid, as shown. **(Caution: Nitric acid is dangerous to handle. In case of accident, immediately wash in running water. Sponge off with dilute ammonium hydroxide.)** Heat the test tube in a small flame, but do not let the acid boil. Keep the test tube pointed away from you. Pour off the acid. Carefully add a few drops of ammonium hydroxide to the egg white.

What was the color the egg white when heated in nitric acid? What was its color after the ammonium hydroxide was added? How would you test an unknown substance to find out if it contains proteins?

Fats are another source of heat and energy. Fats are chemically like carbohydrates, except that they contain less oxygen in each molecule. Like starches and sugars, fat molecules contain carbon, hydrogen, and oxygen. Two fats found in butter have the formulas $C_{16}H_{32}O_2$ and $C_{18}H_{36}O_2$.

Fats can be divided into two types, depending on the amount of hydrogen present in the molecule. If you look at the formulas of the fats in butter, you see that they have twice as many hydrogen atoms as carbon in each molecule. These fats are called *saturated*. The other type of fat molecules with less than twice the number of hydrogen atoms as carbon is called *unsaturated*. Some oleomargarines contain unsaturated fat molecules with the formula $C_{18}H_{34}O_2$.

Many of the foods we eat contain a combination of fats which give many foods a particular flavor and taste. All fats and oils can be identified by their greasy property. Test for the presence of fats in the next activity.

COMPARE

Put a drop of oil on a square of white paper. Put a drop of water on another square of paper. Let the papers dry and hold them up to the light. What difference do you see in the two papers? How would you test an unknown substance to find out if it contains fats?

Butter, lard, and oleomargarine are almost pure fats, as are oils obtained from peanuts, corn, olives, and cottonseeds. Most meats, fish, and poultry are good sources of fat in the diet. Fats from these and other products that are not used immediately for energy are stored as fatty tissues in the body.

NUTRIENTS IN COMMON FOODS

Food	Protein	Carbo-hydrates	Fat
Apples	.3%	12.8%	.5%
Bacon	9.6%	—	64.0%
Bananas	1.0%	20.0%	.5%
Beans (baked)	4.8%	19.6%	2.3%
Beans (lima)	5.3%	21.6%	.6%
Beef (lean)	19.1%	—	12.1%
Bread (corn)	6.5%	45.2%	2.0%
Bread (white)	9.6%	57.3%	4.2%
Cabbage	1.2%	5.5%	.3%
Eggs	11.9%	—	9.3%
Fish (salmon)	21.1%	—	11.5%
Ham	15.8%	—	36.9%
Lettuce	.9%	2.9%	.3%
Peas	5.2%	16.7%	.5%
Potatoes	1.9%	20.0%	.1%
Rice	2.3%	23.8%	.1%
Spinach	1.6%	3.2%	.3%
Veal	19.7%	—	7.3%

Minerals furnish important materials for tissues. The body needs many *minerals*, like iron and copper, for normal growth. These and other minerals are found in small amounts in most of the foods you eat. If you do not have enough iron in your diet, for example, hemoglobin in the blood cells cannot be formed properly. A condition generally called *anemia* (uh-NEE-mee-uh) results. In one type of anemia, the blood cells do not carry enough oxygen to the tissues. You may wish to do library research on other types of anemias and report your findings to the class. The table lists the common minerals which are needed by the body.

MINERALS NEEDED IN THE DIET	
Calcium	Building of bones and teeth; clotting of blood; regulation of heart and muscles; health of nervous system.
Chlorides	Formation of hydrochloric acid in stomach; enzyme activities.
Cobalt	Normal appetite and growth; prevention of anemia.
Copper	Tissue respiration; enzyme activities.
Iodine	Regulation of oxidation of food; prevention of goiter.
Iron	Formation of hemoglobin; carrying of oxygen to body tissues.
Magnesium	Regulation of muscle and nerve activities; enzyme activities.
Phosphorus	Formation of bones and teeth; formation of protoplasm.
Potassium	Regulation of heart and muscles; essential for normal growth.
Sodium	Regulation of heart and muscles; prevents excessive loss of water.
Sulfur	Formation of proteins; helps in the formation of protoplasm.
Zinc	Normal growth; tissue respiration.

Most of the minerals in our diet are obtained from fruits and vegetables. However, lean meat, fish, milk, and eggs are also good sources. Liver is a good source of iron, and seafoods are excellent sources of iodine. Milk furnishes most of the calcium needed in the formation of healthy bones and teeth. What part of a food is mineral? The following activity will help you find out.

IDENTIFY

Weigh a small evaporating dish, containing small pieces of bread (or other food) carefully. Record your results. Weigh the empty dish. Heat the dish over a hot flame, as shown. The food will char at first, but keep on heating until there is no more smoke given off. What remains in the dish? What part of the food is this? When the dish cools, weigh it again. Figure out the percentage of mineral in the food as shown below. **Do not write in your book.**
Weight of dish and food before heating _____
Weight of dish and food after heating _____
Weight of mineral in food _____
Percentage of mineral in food _____

Vitamins are needed to regulate body activities. *Vitamins* are complex chemical substances normally present in foods. They are necessary for growth and good health. We have already learned that nutrients are needed for energy and for growth and repair of the body. However, even if you had a sufficient supply of these nutrients, you would not be healthy without the proper vitamins. Vitamins seem to act as regulators in the body. The activity below shows you how to test for vitamin C.

TEST

Boil a teaspoon of starch in a beaker of water. Allow the solution to cool. Put ten drops of it in a test tube. Add a drop of iodine solution and shake the tube, as shown. Note the color.

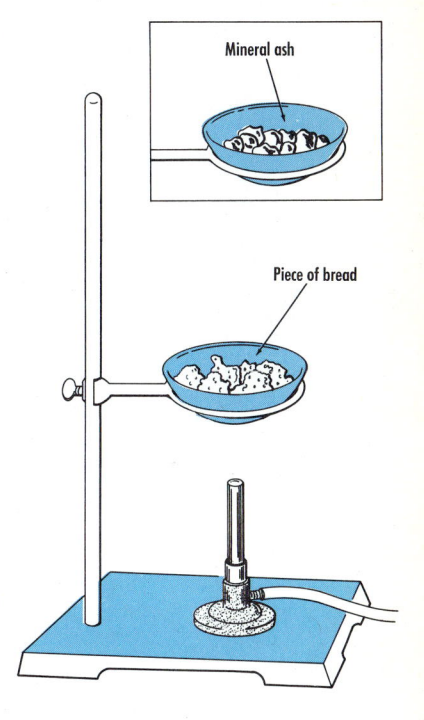

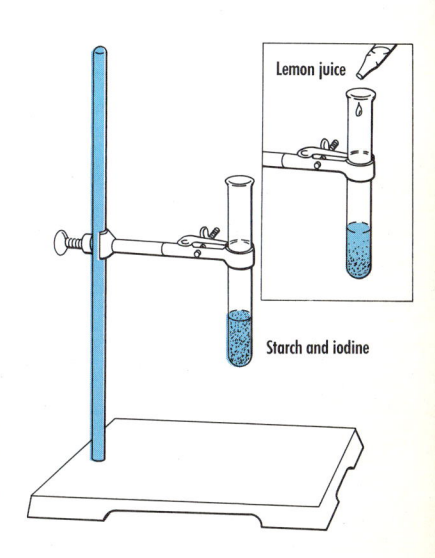

330 / UNIT 4 MAN EXPLORES THE HUMAN BODY

Now, add lemon juice, drop by drop, and observe what happens to the color of the solution. What happened to the color of the starch solution when the iodine was added? What happened to the color as lemon juice was added? This color change is a test for vitamin C. Test other fruit juices to find out if they have vitamin C. Record your results.

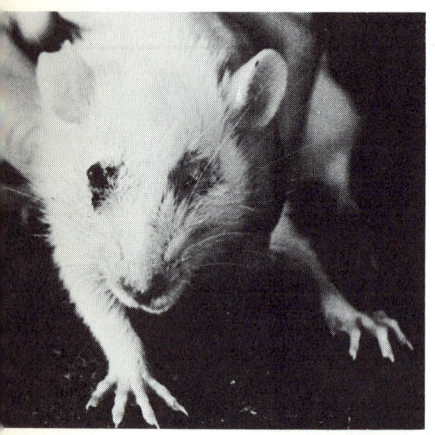

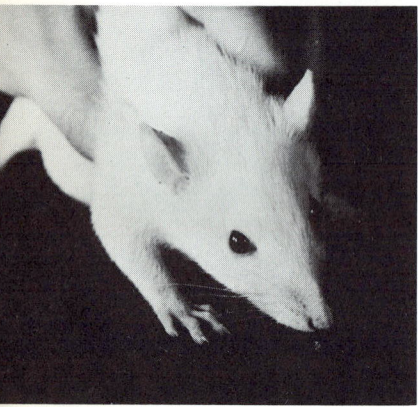

A number of vitamins can now be prepared in concentrated form. These vitamins can be obtained as tablets. However, by eating a balanced diet every day, you can usually obtain all the vitamins which are necessary for normal body health.

Deficiency diseases result from a lack of vitamins. If you do not receive the proper kinds and amounts of vitamins in your diet, the skin, bones, and other parts of the body may develop *deficiency* (dee-FIHSH-uhn-see) *diseases*.

Vitamin A builds up your resistance to colds and helps keep eyes, skin, and hair healthy. Figure 11-4 shows the effects of a lack of this vitamin on an experimental animal. A condition known as *night blindness*, in which a person sees poorly in dim light, may also result from a lack of proper amounts of vitamin A in the diet. This vitamin is obtained from green leafy vegetables and from yellow vegetables. Spinach, carrots, turnips, squash, sweet potatoes, and butter are good sources of vitamin A.

Vitamin B originally was thought to be one vitamin, but scientists now know that it is made up of several parts. Numbers are written after the letter, B_1, B_2, etc., to identify the different vitamins in this group. *Vitamin B_1* prevents a disease known as *beriberi* which is common in many Oriental countries. It also helps keep the nervous and digestive systems functioning properly. Figure 11-5 shows the effects of a lack

11-4 Note the appearance of the eyes in the rat (above and below). How was the condition corrected (below)?

11-5 The rat (left) shows vitamin B_1 deficiency. The same animal (right) after taking sufficient amounts of the vitamin. Name the disease.

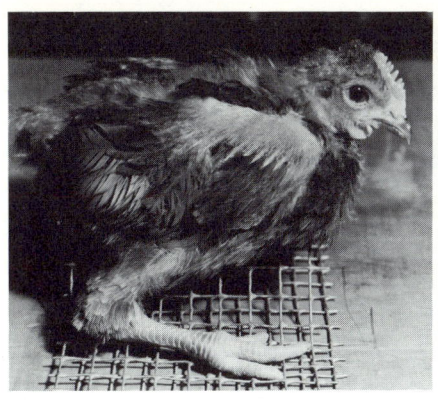

11-6 Describe the appearance of the rooster lacking vitamin B_2 (left) and after taking vitamin B_2 (right).

of this vitamin on an experimental animal. The main sources of vitamin B_1, are yeast, meat, eggs, and whole cereals.

Vitamin B_2 helps keep the skin and the linings of the mouth healthy. A lack of this vitamin results in poor growth in an experimental animal, as shown in Fig. 11-6. Vitamin B_2 is found in meat, green leafy vegetables, milk, peas and beans. Other parts of the vitamin-B group help in keeping normal health.

The effects of a lack of *vitamin C* were known long before scientists learned about vitamins. The lack of this vitamin causes bleeding of the gums and loosening of the teeth. In severe cases, the joints are also affected; the condition is known as *scurvy* (Fig. 11-7). Fortunately, vitamin C is found in practically all fresh fruits and vegetables, but citrus fruits and tomatoes are especially rich sources.

Vitamin D is often called the sunshine vitamin because the body, upon exposure to sunlight, produces a certain amount of this vitamin in the cells of the skin. Normal food sources of vitamin D are limited, but fish liver oils offer a good source. Certain foods, like milk, also show an increase of vitamin D when exposed to ultraviolet light. This vitamin is of particular importance to young animals in helping their bones grow normally. A lack of this vitamin in young animals and children causes a disease known as *rickets*. The effect of a lack of vitamin D on a young child is shown in Fig. 11-8 on page 332.

Vitamin K is found in many green, leafy vegetables like celery. Tomatoes and egg yolks, too, are rich in vitamin K.

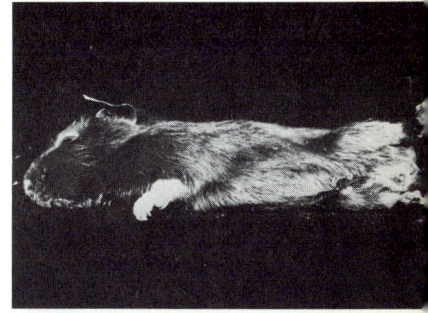

11-7 The guinea pig (above) has scurvy. What vitamin corrected the disease (below)?

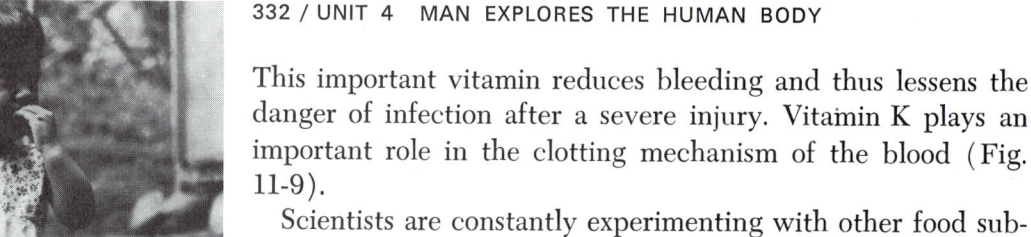

11-8 What vitamin will correct the poor bone structure in this young Indonesian child?

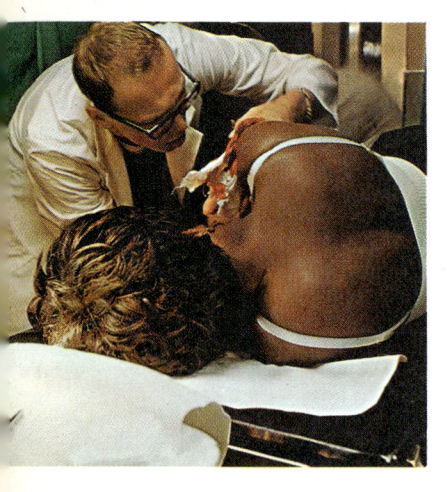

11-9 Why is this woman receiving vitamin K?

This important vitamin reduces bleeding and thus lessens the danger of infection after a severe injury. Vitamin K plays an important role in the clotting mechanism of the blood (Fig. 11-9).

Scientists are constantly experimenting with other food substances that have effects on laboratory animals, like rats and chickens. The lack of some of these substances produces serious deficiency diseases. Therefore, it is possible that there are several other vitamins needed by the body. However, most of these substances are complex chemical compounds, and their effects on the human body are not fully understood. The main vitamins that have been discussed are summarized in the table.

MAIN VITAMINS NEEDED BY THE BODY

Vitamin	Functions in body	Deficiency Disease
A	Proper growth; healthy eyes, skin, and hair.	Night blindness
B_1	Proper growth; healthy heart, nerves, muscles.	Beriberi
B_2	Proper growth; healthy skin, mouth, eyes.	Premature aging, loss of sight.
C	Proper growth; healthy teeth, gums, joints.	Scurvy
D	Proper growth; healthy bones and teeth.	Rickets
K	Normal clotting of blood.	Excessive bleeding

Water is needed for normal functioning of the body. We may not think of water as a nutrient, but without water we could not live. Water is needed in the digestion of food and to excrete waste materials. It is also a part of the secretions of glands. In fact, the body is made up of between 60 and 70 percent water. The amount of water you need daily varies with your bodily activities as well as the temperature and humidity of the air around you. Besides the water you drink, most of the foods you eat supply a considerable amount of water to the body. How can you show the water content of food? What experiment can you make up to answer this question?

REVIEW

1. Name the five main nutrients in foods.
2. What is the main use of carbohydrates in the body?
3. What do proteins furnish for the body?
4. What do minerals furnish for the body?
5. Why is it essential for the body to have vitamins in the daily diet?

C/How Is Our Food Digested?

Digestion is both a physical and chemical process. Most of the food you eat is in a solid form. It must be broken down into a soluble form before it can be absorbed by the blood and carried to the cells of the body. Any substance you swallow that cannot be changed into soluble form cannot be used for food.

The chewing of food in the mouth and the churning action of the stomach and intestines are *physical actions* which help break the food into small pieces. This is necessary so that the action of the *digestive juices* can take place more easily. We can illustrate this in the activity below.

INFER

Fill two test tubes about three-quarters full of water. Put a small amount of meat tenderizer in each. Shake the tubes to dissolve the tenderizer. Add a large piece of hard-boiled egg white in one tube. Add tiny pieces of egg white to the other tube. Leave both tubes in a warm place for several hours. Look at the edges of the pieces of egg white in both tubes. In which tube do the pieces of egg white appear to be dissolving faster? What does this show you of the need to chew food thoroughly?

After swallowing, the food passes through the digestive system by means of a series of contractions of the muscles in walls of the digestive organs. Called *peristalsis* (*per*-ih-STAWL-

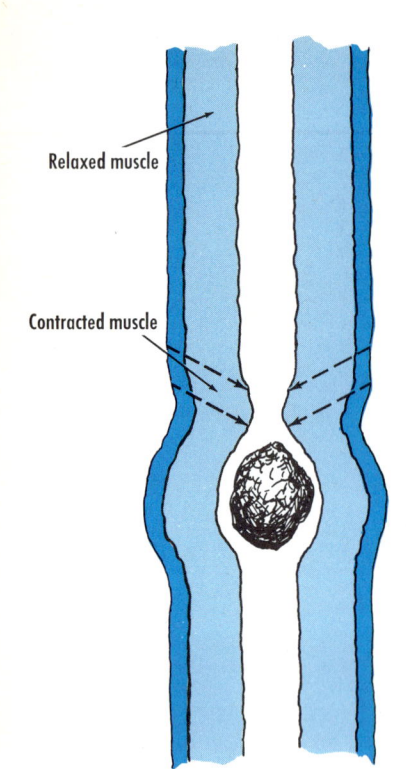

11-10 What is the term for the muscular contractions which force food along the digestive system?

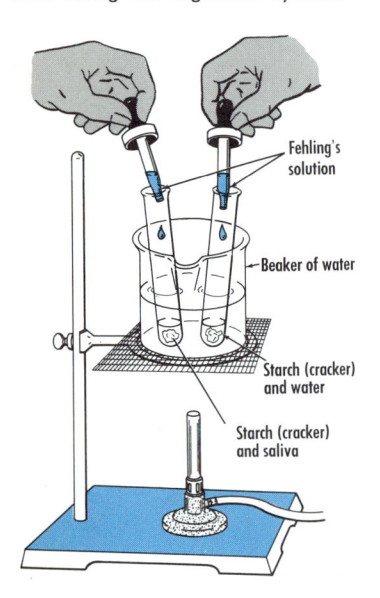

sihs), these are wave-like contractions that move the food along (Fig. 11-10). You can demonstrate how peristalsis works in the next activity.

EXPLAIN

Place a marble in a rubber tube so that the marble can move, but not so wide that the marble rolls through. Press behind the marble to make it move through the tube. What does the marble represent? What does the pressure of your fingers represent? Explain how food passes through the digestive system.

Enzymes act on food in chemical digestion. *Enzymes* (EN-zyems) are chemical compounds, protein in nature, that help substances to combine readily, or help to speed up reactions. Before food can be used in the body, it is essential that *chemical changes* take place. These changes are produced by the enzymes in the digestive juices produced by *digestive glands*. These enzymes act on foods to change them into soluble materials. Only when the food is in a soluble form can it pass through the cell membranes for use in the activities of the cells.

When the food is chewed, *saliva* (sal-EYE-vuh), produced by the *salivary* (SAL-ih-*ver*-ee) *glands* in the mouth, is mixed with it. The saliva moistens the food; it also contains *ptyalin* (TYE-uh-*lihn*), an enzyme, which begins to change starch into sugar. Since digestion begins in the mouth, food should be chewed thoroughly to mix the saliva with it. The action of this enzyme on starch can be shown in the following activity.

EXPLORE

Put a pinch of starch (cracker) in a test tube. Add enough saliva to cover the starch. In another test tube put the same amount of starch. Add about the same amount of water as saliva. Heat the tubes gently for 15 minutes by placing them in a beaker of hot water, as shown. Test the solutions in both tubes for glucose with Fehling's solution.

Did you get a test for glucose in the test tube with starch and water? In the test tube with starch and saliva? What is the action of saliva on starch? Can you devise an experiment to prove that the glucose does not come from the saliva but from the starch?

When chewed food is swallowed, it passes through the *esophagus* (ih-SAHF-uh-guhs) into the *stomach* (Fig. 11-11). The digestive juices produced by the *gastric glands* in the stomach continue the process of digestion. Two enzymes are found in the gastric juice. One begins the digestion of proteins and the other curdles milk to separate the protein in this food. Weak *hydrochloric acid* is also present in the gastric juice which further softens the food and helps in the digestion of proteins. The peristaltic action of the stomach mixes the digestive juices with the food and moves the partly digested food into the *small intestine*. The action of gastric juice on proteins is shown in the next activity.

INVESTIGATE

Make a solution of artificial gastric juice by mixing one gram of pepsin in 50 milliliters of water. Add a few drops of dilute hydrochloric acid to the solution so that it gives a weak acid reaction with litmus paper. Place a piece of egg white in a test tube and fill the tube one-third full of water. Place a second piece of egg white in another test tube and add an equal amount of artificial gastric juice. Leave the test tubes overnight in a warm place. Examine the edges of egg white in both tubes.

Did the egg white dissolve in the water? in the artificial gastric juice? What is the action of gastric juice on proteins?

The final digestion of food takes place in the small intestine. The *liver* and *pancreas* (PAN-kree-uhs) are two digestive glands which are connected by tiny tubes to the small intestine. Both glands, and the *intestinal glands* in the walls of the small intestine, produce digestive juices which act on food in the small intestine (Fig. 11-12). The liver produces the compound *bile* which is first stored in the *gall bladder*. Bile flows from there into the intestine through the *bile duct*. Bile is a brownish fluid necessary for the breakdown of fat molecules. Although it does not contain enzymes, it increases the action of certain

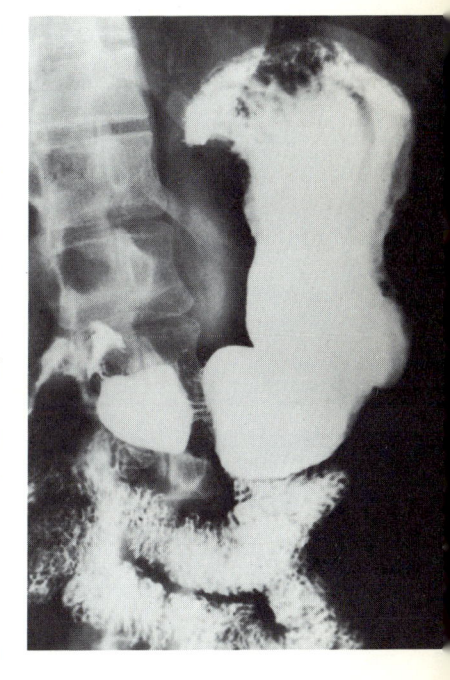

11-11 An X-ray photograph of a normal stomach. What is its function in the digestion of food?

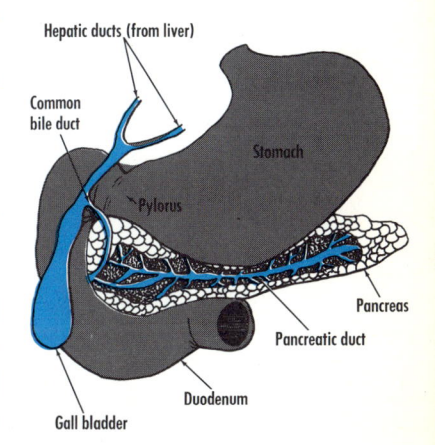

11-12 What role do the pancreas, liver and intestinal glands play in the digestion of food?

enzymes in the digestion of fats. Bile also helps to neutralize stomach acid so that digestion can continue in the intestine.

The digestive juices produced by the pancreas also pass into the intestine through a duct. These juices contain three enzymes. These enzymes continue the digestive action on proteins started in the stomach, act on starch to change it into sugar, and complete the digestion of fats begun by the bile. You can see that the pancreas is a very important digestive gland. The action of pancreatic juice on fats can be shown in the next activity.

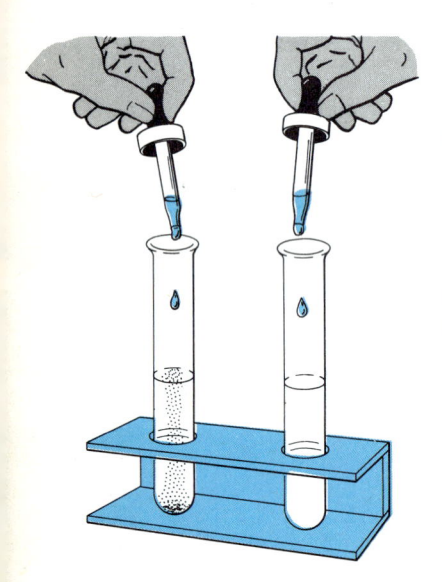

OBSERVE

Make a solution of artificial pancreatic juice. Mix one gram of pancreatin and two grams of baking soda in 50 milliliters of water. Test the solution with red litmus to make sure it has a weak basic reaction. Place a few drops of oil in a test tube and fill it about one-third full of water. Place the same amount of oil in a second test tube and add an equal amount of artificial pancreatic juice as shown. Leave both tubes overnight in a warm place, undisturbed.

Did the oil form a cloudy mixture with the water on shaking the tube? What was the reaction with the artificial pancreatic juice? What is the action of pancreatic juice on fats?

The intestinal glands produce a digestive juice containing four enzymes. One enzyme completes the digestion of proteins begun in the stomach. The others change complex sugars into glucose, a simple sugar. Thus, the combined action of bile, pancreatic juice, and intestinal juice complete the digestion of food in the small intestine.

Digested food is absorbed into the blood stream. When the food is in a soluble form, it can pass through the walls of the small intestine into the bloodstream by means of the process of *diffusion* (dih-FYOO-zhuhn). The inside surface of the small intestine consist of folds, covered with tiny, fingerlike projections, called *villi* (VIHL-eye), as shown in Fig. 11-13. The folds and the villi increase the inside area of the small intestine, allowing the digested food to pass into the blood more readily. In the villi, the soluble food passes into the capillaries. Here, it is carried by the blood to all the body tissues.

Foods that are completely digested no longer resemble foods which are eaten. All carbohydrates, whether eaten in the form of starch or sugars, are broken down into a simple sugar, *glucose*. Proteins go through a number of digestive steps to be finally broken down into simpler forms called *amino acids*. Many different amino acids are needed by the body to build the necessary tissues and cell materials. Fats are broken down into *fatty acids* in the process of digestion. Minerals and vitamins, though necessary, do not go through the same digestive changes that the other main classes of foods do. All the digested nutrients are absorbed into the blood from the small intestine.

Undigested food passes into the large intestine. These materials, plus digestive juices and water, pass into the *large intestine* (Fig. 11-14). Here, most of the liquid is absorbed back into the blood. The solid materials that remain move through the large intestine by peristaltic action to be removed from the body.

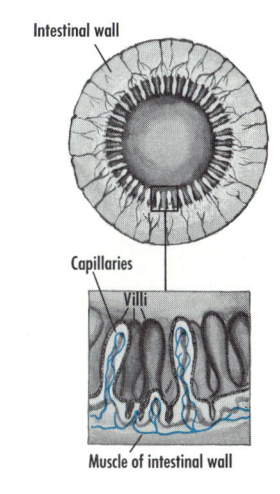

11-13 Describe the inside surface of the small intestine.

SUMMARY OF CHEMICAL DIGESTION

Name of gland	Digestive juice	Enzymes produced	Digestive action
Salivary	Saliva	Ptyalin	Changes starch into sugars
Gastric	Gastric juice	Pepsin	Begins digestion of proteins
		Rennin	Curdles protein in milk
Liver	Bile	None	Begins digestion of fats
Pancreas	Pancreatic juice	Trypsin (TRIHP-sihn)	Continues digestion of proteins
		Amylase (O-mih-*lays*)	Changes starch into sugars
		Lipase (LYE-pays)	Completes digestion of fats
Intestinal	Intestinal juice	Erepsin (uh-REP-sihn)	Completes digestion of proteins
		Maltase, Lactase, Sucrase	Complete digestion of sugars

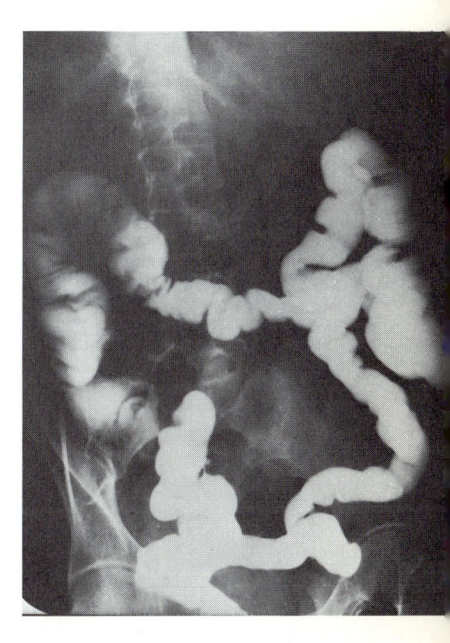

11-14 An X-ray photograph of the large intestine. What is its role?

REVIEW

1. Name two examples of physical actions in the digestion of food.
2. With what nutrient is chemical digestion started in the mouth?
3. Where does the digestion of proteins begin and end?
4. What digestive juice begins the digestion of fat? Which one completes it?
5. What materials are absorbed by the blood from the small intestine?

D/How Is Energy Released from Food?

Living cells release energy from foods. The body is made up of billions of different kinds of cells. With few exceptions, each tiny cell absorbs food and oxygen from the bloodstream. Each cell uses the food to obtain energy for its needs. Until a few years ago, scientists were not certain of how this was done. One theory stated that sugar was combined with oxygen in the cell and "burned."

In the study of the chemical behavior of cells, scientists discovered that two complex chemical compounds were usually present. These are *adenosine diphosphate* (uh-DEN-uh-seen dye-FAHS-fayt), usually written chemically as *ADP,* and *adenosine triphosphate,* or *ATP.* Careful study showed that one of the phosphate groups (PO_4) broke away from the ATP molecule. Scientists learned that the breaking away of the phosphate group released energy for the cell. About 8000 calories (small calories) of energy are produced, and a new compound ADP is formed:

$$ATP \rightarrow ADP + phosphate + energy$$

Glucose and oxygen combine in the cell, releasing energy. *Respiration* (res-pih-RAY-shuhn) is a general name for the process in which energy is released when certain chemicals, like glucose, combine with oxygen. Carbon dioxide and water are given off as waste products. The energy released is used to form ATP. In this manner, the energy cycle in the cell is reactivated (Fig. 11-15). The change from ATP to ADP and back again to ATP can take place almost immediately.

$$ADP + phosphate + energy \rightarrow ATP$$

The cells use ATP, not only for their own energy needs, but also as a source of energy to carry on the many specific jobs for the body. Whether you blink an eyelid or run a 100-yard dash, you need energy for muscular action. This energy is made available in every cell, from the tiny muscles of the eyelid to the large muscles of the body and legs, by ATP.

The exact process in which energy is obtained and converted into *high-energy bonds* in the ATP molecules is not fully understood as yet. It is known, however, that although energy from only *two* ATP molecules is needed to start this whole energy-storing process, *thirty-eight* ATP molecules are produced in the cell as a result. Therefore, new sources of energy are constantly produced in each cell that has a fresh supply of glucose and oxygen available.

Interestingly enough, scientists have found that cells do not store large amounts of ATP. Your body probably contains only two or three ounces of ATP at any given instant. The cells produce fresh ATP as it is needed in the body by changing the ADP into ATP as we have previously described. When you are sitting quietly, as reading a book, you use little muscle energy. The food in a peanut can probably supply all the ATP your body needs in an hour. If you are very active, playing and doing heavy exercise, your cells will actually make in one day, bit by bit, about as much ATP as your body weight!

Enzymes are important in the formation of ATP. In all living cells, the formation of new ATP molecules is brought about by a series of steps. Glucose is changed into energy, released by the action of enzymes in certain parts of the cell. The way an enzyme acts is illustrated in the following activity. (Ash is not an enzyme. It is used to show how one substance acts to help another substance react.)

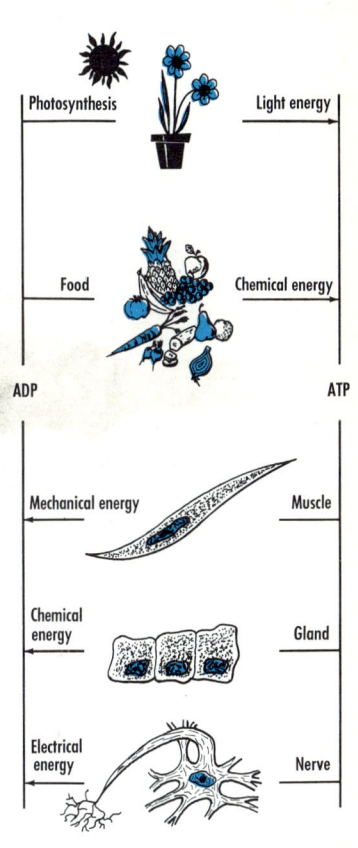

11-15 How is the change from ATP to ADP brought about as illustrated by the diagram.

OBSERVE

Hold a cube of sugar with a pair of tongs or pliers. Try to light it with a match flame. Describe your results. Now, rub a small amount of wood or tobacco ash over another sugar cube and try to light it with a match flame. What results did you get this time? Did the ash itself burn? In what way did the ash seem to help in the chemical reaction of burning the sugar cube?

Enzymes produce changes in the food in cells so that the food combines with oxygen, releasing energy. These changes in the food take place, step by step, with very little release of heat energy. In fact, heat produced by cells is actually a waste product and represents energy which the cell does not use. Until recently, little was known about how enzymes act. However, discoveries in biology and in chemistry are giving some answers about their action in the body. There are hundreds of different enzymes. These help the body to carry on digestion of food, to release energy from the food, and to use the food to build different tissues in the body. It should be remembered that each enzyme acts only on a specific substance and performs its action in a certain order (Fig. 11-16). You can see the action of an enzyme found in fresh pineapples by doing the following activity.

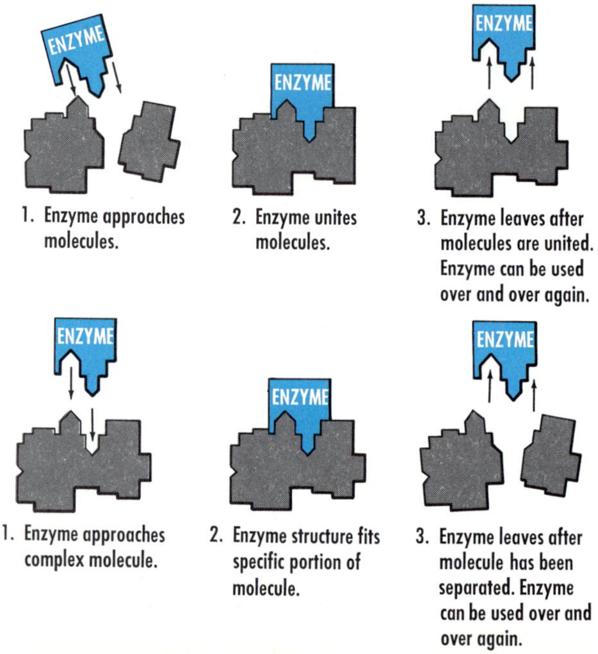

11-16 Why is the structure of a molecule important in the way it reacts with enzymes?

1. Enzyme approaches molecules.
2. Enzyme unites molecules.
3. Enzyme leaves after molecules are united. Enzyme can be used over and over again.

1. Enzyme approaches complex molecule.
2. Enzyme structure fits specific portion of molecule.
3. Enzyme leaves after molecule has been separated. Enzyme can be used over and over again.

ILLUSTRATE

Dissolve a package of gelatin in hot water according to the directions on the package. Pour equal amounts of the mixture into two beakers. Now add one teaspoonful of *fresh* or *frozen* pineapple juice to the hot gelatin in one beaker. Add an equal amount of *canned* or *boiled* juice to the other beaker. (**Do not refrigerate.**) Label the beakers and leave them undisturbed for several hours. What do you observe? Explain the reason for any differences noted.

REVIEW

1. Name two complex compounds containing phosphorus groups found in most living cells.
2. Write the word formula for the change of ATP into ADP.
3. For what purposes is the energy in ATP used by the body?
4. What substance is the starting point for the formation of ATP?
5. Does the action of enzymes in cells release much heat energy? Why is this an advantage to the cells?

THINKING WITH SCIENCE

A. *On a separate sheet of paper, write the numbers 1 to 15. Match the letter of the correct term in* **Section 1** *with its definition in* **Section 2. Do not write in your book.**

Section 1

a. ATP
b. amino acids
c. Calorie
d. carbohydrates
e. carbon dioxide
f. digestion
g. enzymes
h. fatty acids
i. glucose
j. metabolism
k. minerals
l. nutrients
m. peristalsis
n. proteins
o. vitamins

Section 2

1. Nutrients that furnish most of the energy in the body.

2. Chemical substances that speed up chemical changes in the cells.

3. The unit used to measure heat energy in foods.

4. Nutrients that furnish materials for growth and regulation in the body.

5. The contractions of muscles that move food through the digestive system.

6. The more simple chemical units of which all proteins are composed.

7. The essential substances in food needed by the body for proper health.

8. The chemical compound that stores energy in a cell.

9. The process in which food is changed to soluble forms in the body.

CHAPTER 11 ENERGY FOR THE BODY / 343

10. Nutrients needed by the body to prevent deficiency diseases.

11. Substance produced in the complete digestion of carbohydrates.

12. Nutrients that furnish materials for growth and repair of the body.

13. Substances produced in the complete digestion of fats.

14. A chemical substance produced in the respiration cycle.

15. The sum of the activities in the body that are necessary for life.

B. *Write the answers to the following in your notebook. Be sure to use complete sentences and correct spelling and grammar.*

1. In what way is the energy-conversion in a living cell different from that in a machine?

2. Why are enzymes so essential in the necessary activities of cells?

3. Explain how ATP and ADP change from one to another to release energy in the cell.

4. How does the chemical composition of carbohydrates, proteins, and fats compare?

5. Describe chemical tests for: (*a*) glucose, (*b*) proteins, (*c*) starch.

6. Describe the effects on experimental animals of a lack of the following vitamins: (*a*) vitamin A, (*b*) vitamin B_1, (*c*) vitamin C, (*d*) vitamin D.

7. Trace the digestion of a piece of bread (starch and protein) from the mouth until it is completely digested.

8. Trace the digestion of a piece of meat (protein and fat) from the mouth until it is completely digested.

9. Describe the action of the villi in the absorption of digested foods.

10. What happens to the undigested food and other substances in the large intestine?

RESEARCH IN SCIENCE

1. Find out how different nutrients can be separated and identified in a sample of food. Set up a demonstration to show this.

2. Test a number of foods for different nutrients and prepare a chart showing the results of your tests.

3. Using two young rats, set up an experiment under the direction of your teacher to test the effects of a balanced and unbalanced diet. Report your results to the class.

4. Using two young pigeons, set up an experiment under the direction of your teacher to show the effects of a lack of vitamin B_1 in the diet. Report your results to the class.

5. Prepare a report to the class on the effects which light, heat, and oxygen have on various vitamins.

chapter 12
Functions of the Body

A/How Do Cells Function?

Cells are the basic units making up the body. Scientists have determined that *protoplasm* is made up chiefly of water containing dissolved solids like *proteins, sugars, oils*, and *minerals*. Oxygen, carbon, hydrogen, and nitrogen are the main elements in protoplasm. However, no one has yet been able to mix the proper chemical compounds together to produce this living substance. It is found only in living cells, and new protoplasm comes only from other living protoplasm. Let us see how many structures in a cell we can find in the following activity.

EXPLORE

Gently scrape the inside of your cheek with a clean toothpick to remove a film of surface cells. Spread the material into a thin smear on a clean microscope slide. Put a drop of iodine stain on the smear. Let the stain remain for a few minutes. Then, place a cover glass on top of the stained part and examine the cells with a microprojector or microscope.

Make a drawing of the different shapes of the cheek cells you see. Label the nucleus, cell membrane, cytoplasm, and vacuoles.

The cell membrane permits certain substances to pass through it. The cell membrane serves as a protective layer and also outlines the shape of a cell. Because it allows certain substances to pass through it, each cell is able to absorb the materials it needs from the blood and to give off waste products. These materials pass through the cell membrane by *diffusion*.

To understand the process of diffusion, we must know how dissolved substances act. A substance in solution is in the form of molecules. If the molecules are small enough, they are able to pass through tiny pores in a membrane. Larger molecules cannot do so. A membrane that allows the molecules of *all* substances to pass through is called a *permeable* (PER-mee-uh-buhl) *membrane*. One that allows only *certain* molecules through is called a *semipermeable membrane*. Let us observe the action of diffusion through a semipermeable membrane in the next activity.

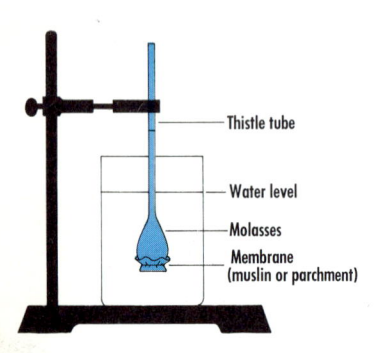

OBSERVE

Fill the bulb of a thistle tube or small funnel with molasses. Tie a piece of parchment of fine muslin securely over the top. Pour some water in a jar and attach the thistle tube to a ring stand with the bulb in the water, as shown. Observe what happens to the level of the liquid in the stem of the tube after several hours. What causes the liquid to rise?

A similar passage of food materials, oxygen, and waste materials take place through membranes in living organisms. Molecules always move from a place where there is a greater amount of the substance toward a place where there is a lesser amount. As food and oxygen are used up in a cell, more substances flow from the blood (where there is a greater amount), through the membrane and into the cell. Similarly, as waste products build up in the cell, they flow from the cell into the blood.

The exchange of water between the cell and its environment is called *osmosis* (ahz-MO-sihs). The next activity will illustrate the process of osmosis.

EXAMINE

Obtain a fresh egg. Carefully chip away the shell on the blunter end, being careful not to puncture the thin membrane underneath. Place the egg in the top of a small beaker or jar filled with water, as shown. The egg should touch the water's surface. Carefully punch a small hole in the upper end of the egg. Push a small glass tube into the liquid part of the egg. Use melted candle wax to seal the tube in the egg. Describe what happens in the tube after several hours.

The egg membrane is a *semipermeable membrane;* the molecules of water can pass through it, but the egg molecules cannot. Hence, an increase of fluids takes place inside the egg as the water molecules pass through the membrane. Water enters the living cell in in this way. Figure 12-1 shows the difference between diffusion and osmosis.

ATP is made in specialized bodies in the cell. Not many years ago, scientists believed that cells were fairly simple in structure. Cells were thought to consist of a nucleus surrounded by cytoplasm, covered with a cell membrane. The functions of the cell also were thought to be simple. In the last few years, electron microscopes and new ways of preparing cells have revealed an entirely new microscopic world. The tiny structures within the cell are specialized parts of a complicated manufacturing process.

The "power plants" in the cells are the tiny, sausage-shaped *mitochondria* (Fig. 12-2). Glucose and oxygen, absorbed from the blood, move to the mitochondria where about 95 percent of the ATP in the body is formed. Most cells contain anywhere from 50 to 5000 mitochondria, depending on their functions in the body. The most active cells, like heart cells, liver cells, and nerve cells, contain the largest number of mitochondria.

The structure of mitochondria is difficult to observe with the ordinary light microscope. Under an electron microscope, the mitochondria are seen as complex bodies. A mitochondrion appears to be made up of two membranes: a smooth one on the outside, and a coiled or folded one on the inside. The change of glucose is believed to take place in these membranes.

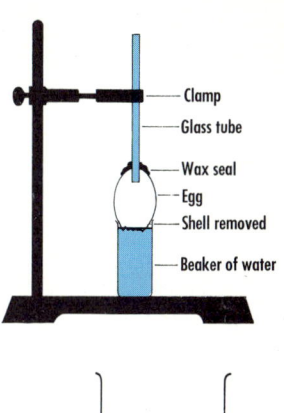

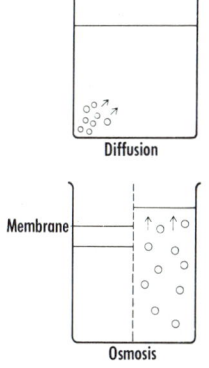

12-1 Explain the difference between diffusion and osmosis.

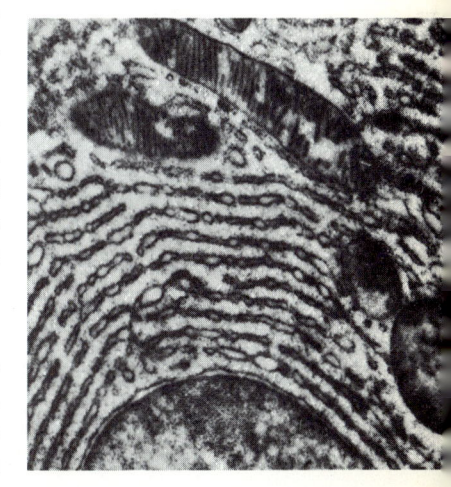

12-2 Locate the mitochondria in the photograph.

Enzymes, found in the mitochondria, act on the glucose molecule. Here is where certain vitamins and minerals must also be present. These substances act as *coenzymes*. In other words, the enzymes do not seem able to cause the necessary reactions to take place unless all of these coenzymes are also present in the cell.

The cell builds proteins from amino acids. A closer look at Fig. 12-2 shows that the tiny channels in the cytoplasm appear to be coated with tiny bubbles or granules. These are so small that they are difficult to see even in an electron microscope photograph. These tiny bodies are called *ribosomes* (RYE-bo-somz) and have a very important function.

You will remember from the previous chapter that the digestion of proteins is completed in the small intestine. The chemical compounds produced in this digestive action are called *amino acids*. The amino acids are absorbed by the blood and carried to the cells.

It is believed that the proteins needed for various uses in the body are put together from these amino acids inside the ribosomes. That is, the ribosomes seem to determine in what order the amino acids are linked together to form proteins for the body. Thus, our cells use amino acids obtained from meats or vegetables to make proteins. Similarly, cattle use amino acids obtained from hay and grain.

Your body is made up of billions of cells, but it does not have all the same cells that it had a few weeks ago. Cells wear out as the body carries on its functions. Part of the food you eat is used to make new cells to replace those that are worn out. In addition, as your body grows in size, new cells are formed to become part of the body. The needs of your body are the needs of the cells because all the life functions take place in the cells.

Similar cells make up a tissue. When you think of a tissue, you probably consider its function in the body. For example, the movement of the body is carried on by *muscle tissue*. Control of the body is carried on by *nerve tissue*. Each cell in the body must carry on all of its life functions in order to stay alive. As part of a tissue, a cell also has a special function to perform.

CHAPTER 12 FUNCTIONS OF THE BODY / 349

The main kinds of tissues in the body, shown in Fig. 12-3, are as follows: *epithelial* (ep-ih-THEE-lee-uhl) *tissue*, making up the skin, the lining of the mouth and nose, and other body cavities; *muscle tissue*, making up the muscles on the outside of the body as well as the internal organs like the heart, stomach, and intestines; *connective* or *supporting tissue*, making up bones and cartilage; and *nerve tissue*, making up the brain, spinal cord and nerves. Blood, a *liquid tissue*, is a special type of connective tissue. You can observe body tissues in the next activity.

OBSERVE

Using prepared slides, study various tissues. Examine their shape and structure. Observe how different tissues are alike and how they are different.

In what way are different tissues of the body alike? In what ways are they different? Why are tissues colored for microscopic study?

Organs function to carry on the life processes in the body. The living body is a complicated form of life in which all the cells making up tissues and organs work together to carry on the body's activities. The table below lists for your reference the main systems of the body. All the systems working together form the living body. Some of the main systems of the body are shown in color plates on pages 350-351.

Systems	Functions
Digestive system	Converts food for use by cells.
Respiratory system	Provides oxygen for tissues, removes CO_2.
Circulatory system	Carries food and oxygen to the body.
Excretory system	Removes waste materials from the body.
Muscular system	Enables the body to move.
Skeletal system	Supports and protects the body.
Glandular system	Provides chemical substances for the body.
Nervous system	Controls the behavior of the body.
Reproductive system	Provides egg cells or sperm cells.
Integumentary system	Provides outer body protection.

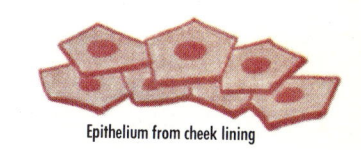

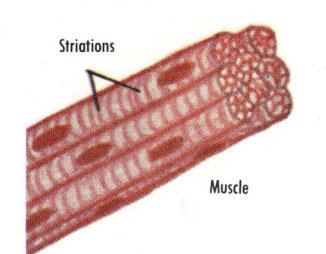

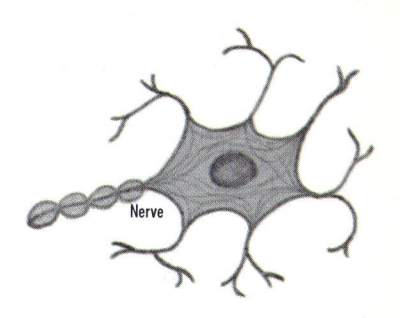

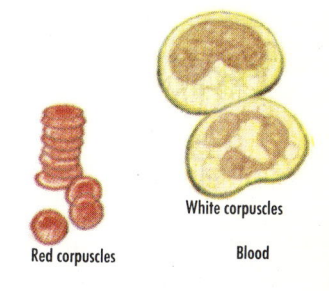

12-3 Describe the various shapes of body cells.

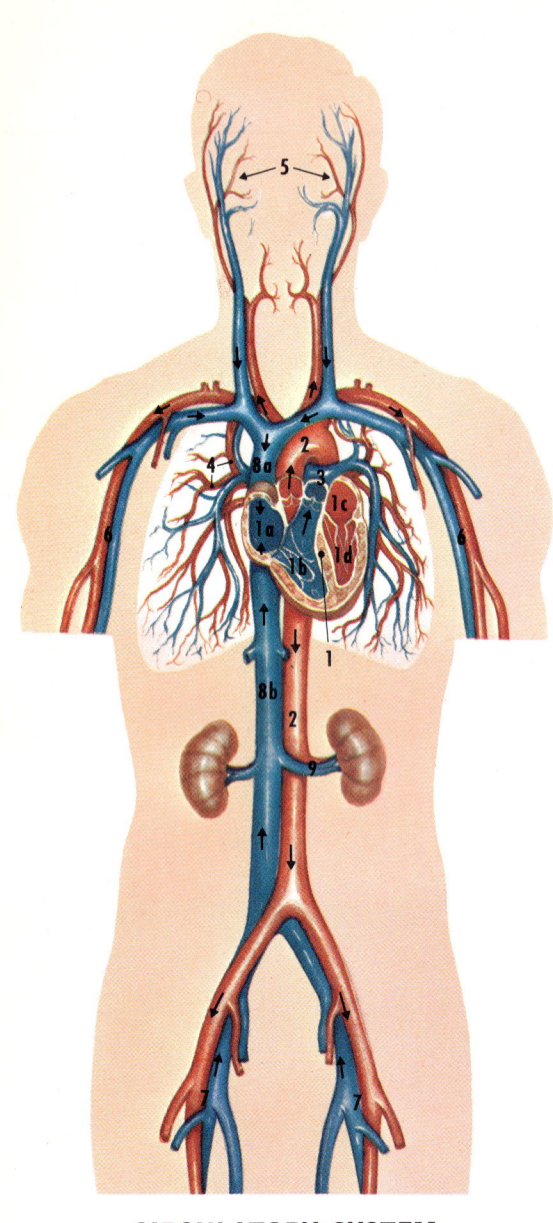

CIRCULATORY SYSTEM

(1) Heart (cut away to show 4 chambers), (1a) Right auricle, (1b) Right ventricle, (1c) Left auricle, (1d) Left ventricle, (2) Great artery from the heart (aorta), (3) Arteries to the lungs (pulmonary arteries), (4) Veins from the lungs (pulmonary veins), (5) Arteries and veins of the head and neck, (6) Arteries and veins of the arms, (7) Arteries and veins of the legs, (8a) Great vein from head and arms (superior vena cava), (8b) Great vein from trunk and legs (inferior vena cava), (9) Arteries and veins of the kidneys. Arrows on vessels show direction of blood flow. Vessels carrying blood with carbon dioxide are shown in blue. Vessels carrying blood with oxygen are shown in red.

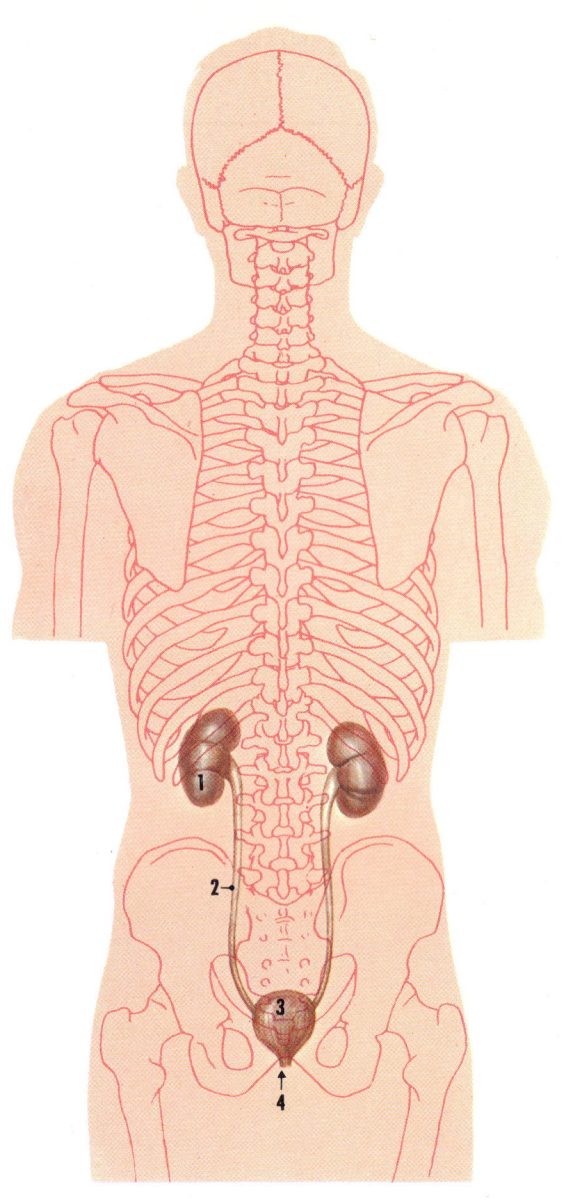

EXCRETORY SYSTEM

(1) Kidney, (2) Ureter, (3) Urinary bladder, (4) Urethra. Also shown is the back view of the skeleton to aid in locating the position of the kidneys.

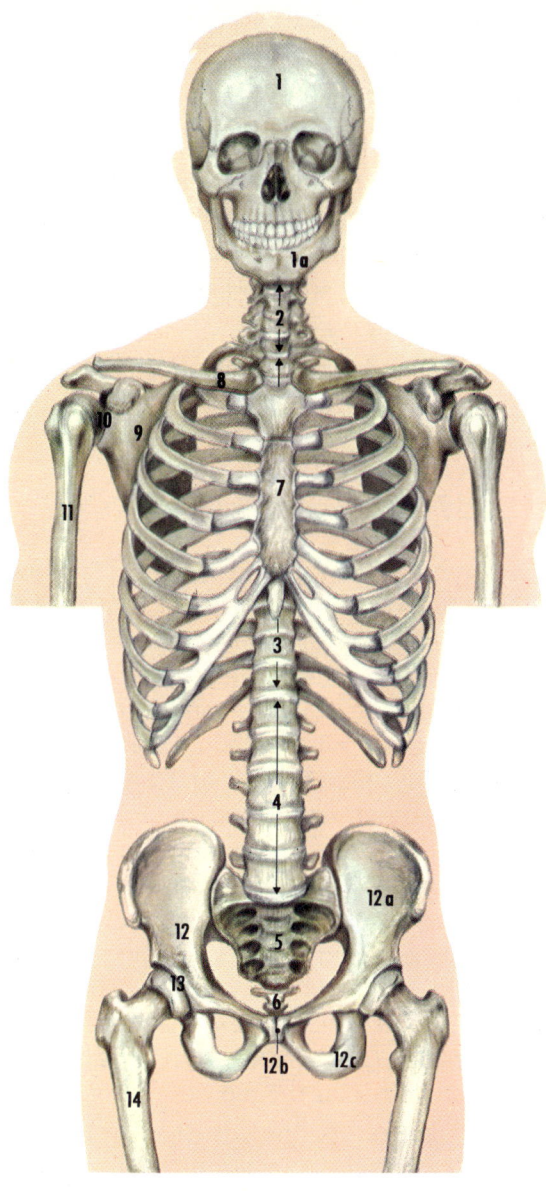

SKELETAL SYSTEM

(1) Skull, (1a) Jawbone (mandible), (2) Cervical vertebrae, (3) Thoracic vertebrae, (4) Lumbar vertebrae, (5) Sacrum, (6) Coccyx, (7) Rib cage (ribs, sternum, and cartilage), (8) Collar bone (clavicle), (9) Shoulder blade (scapula), (10) Shoulder joint, (11) Upper arm bone (humerus), (12) Pelvis, (12a) Ilium, (12b) Pubis, (12c) Ischium, (13) Hip joint, (14) Thigh bone (femur).

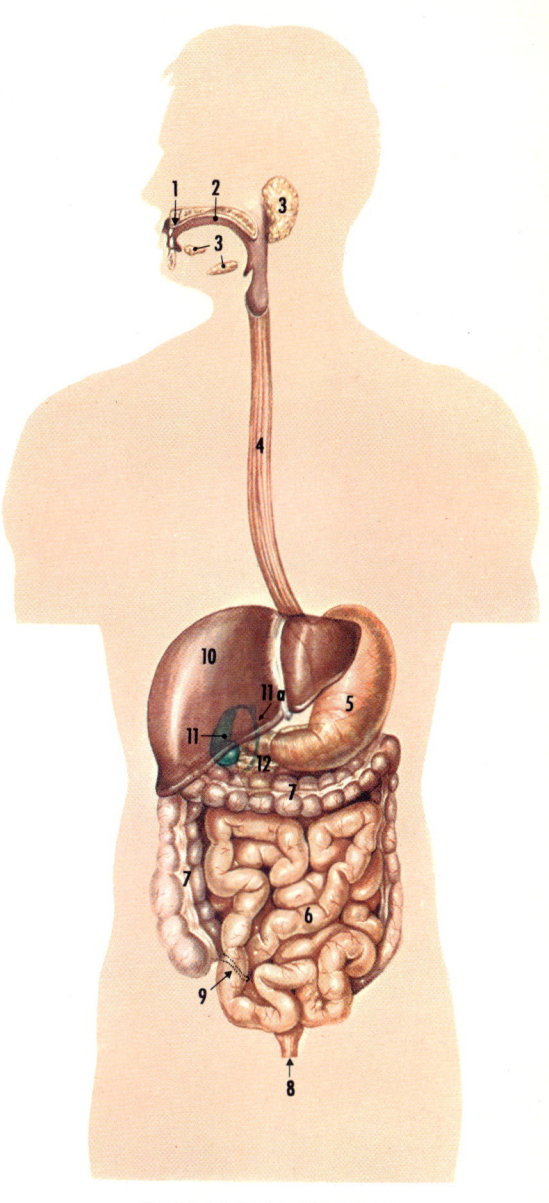

DIGESTIVE SYSTEM

(1) Teeth, (2) Mouth, (3) Salivary glands, (4) Esophagus, (5) Stomach, (6) Small intestine, (7) Large intestine, (8) Rectum, (9) Appendix (shown by dotted line), (10) Liver, (11) Gall bladder (shown as if seen through the liver), (11a) Bile ducts, (12) Pancreas (partly behind the stomach).

REVIEW

1. What is the composition of protoplasm?
2. What are the main parts making up a cheek cell?
3. List the major functions of the cell membrane.
4. Name four main kinds of tissues found in the body.
5. What is the main function of the: (*a*) circulatory system, (*b*) respiratory system, (*c*) excretory system?

B/How Are Nutrients Transported?

Blood is a complex mixture. If a beaker of blood is allowed to stand, we find that it separates into two parts: a liquid called the *plasma* and a solid made up of blood cells or *corpuscles* (KOR-puhs-uhlz). Plasma has a sticky quality, and it is straw-colored. Plasma consists of about 90 percent water and dissolved proteins, digested food, waste products, and minerals.

Certain proteins in the plasma are essential in the clotting of blood. Other proteins play an important role in making our bodies immune to various diseases. The proteins in the plasma give blood its sticky quality. The food found in the blood plasma is in a soluble form due to digestive actions. The waste products are mostly nitrogen-containing substances resulting from breakdown of proteins.

The solid part of the blood is composed of *red blood cells, white blood cells,* and *platelets.* Platelets are irregular-shaped, colorless bodies much smaller than the red cells (Fig. 12-4).

Red blood cells carry oxygen to the tissues. The red blood cells are shaped like hollow discs. They are so small that about 10 million of them spread out would cover about one square inch. An average person has about 25 trillion red blood cells in the body. These cells are red in color because they contain *hemoglobin.*

In most mammals, red blood cells have an interesting characteristic that makes them different from other cells in the

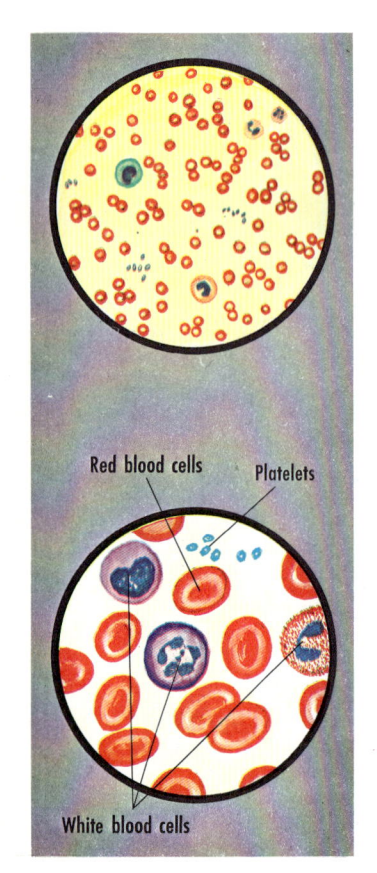

12-4 How are the red and white blood cells different?

body. When red cells are first formed in the red marrow of the bones, they have nuclei like all other cells. However, when they are released into the blood stream, the nuclei have disappeared. In all other vertebrates the red blood cells have a nucleus. Red blood cells last about three months. They must be replaced because they cannot reproduce themselves. The worn out cells are broken up in the spleen and liver. Certain parts, however, are released into the blood to be used in forming new red blood cells.

Hemoglobin is a complex protein containing atoms of iron in its structure. As the red cells move through the blood vessels in the lungs, the iron atoms in hemoglobin combine with oxygen. When the red blood cells reach the tissues, the oxygen is released from the hemoglobin. Oxygen then flows into the cells. In a somewhat similar manner, carbon dioxide is carried by the blood from the tissues to the lungs.

White blood cells help the body fight infections. White blood cells are larger than red blood cells. They can be distinguished by their nuclei that show up well in a stained blood slide. Red blood cells are far more numerous than white blood cells. There are about 1/600 as many white blood cells in the body as red blood cells. In the next activity, stain a drop of your blood on a glass slide to see red and white blood cells.

DISCOVER

Wash the end of your middle finger with alcohol. Hold the tip of the needle in a flame for a few seconds to sterilize it. Stroke the finger to force the blood to the end and prick the skin to get a drop of blood **(Caution: Do not use the same needle on different students.)** Put the blood on a slide and spread it in a thin smear with the edge of another slide, as shown. Allow the blood to dry for a few minutes. Add a film of Wright's stain on the entire slide. Let stand 2-3 minutes. Add carefully, with a dropper, distilled water to the stain to mix the two evenly. Let stand for 3 minutes. Rinse slide off with tap water and let dry. Put a cover glass on the smear and examine the slide with the highest power of the microprojector or microscope.

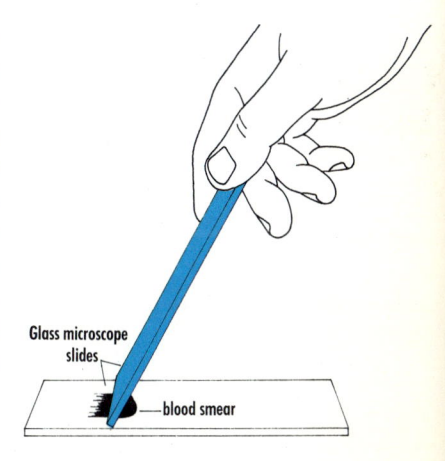

What kind of cells are most abundant on the slide? Can you see any white blood cells? How can you tell the white cells from the red cells?

White blood cells also have an interesting property that makes them different from other cells in the body. They are an important body defense against disease-causing bacteria. They are able to surround and digest foreign materials that enter the body. When an infection occurs in the tissues, the white corpuscles move into the area and destroy bacteria. Perhaps you have noticed the *pus* that forms in some wounds. It is composed of the remains of dead bacteria, white corpuscles, and tissue fluids.

Blood platelets help in the clotting of blood. A small cut in the skin bleeds for a short time and then forms a *clot* that stops the bleeding. This is the normal way. It may seem like a very simple process. There are, however, a number of physical and chemical changes that take place in the blood when a clot forms. When blood platelets flow into a cut, they break up. This releases proteins which react with calcium and other blood substances to produce a protein called *fibrin* (FY-brin). The fibrin forms a fine network of thread, as seen in Fig. 12-5. These threads now trap the blood cells to form a clot. The activity below will show you how clotting takes place.

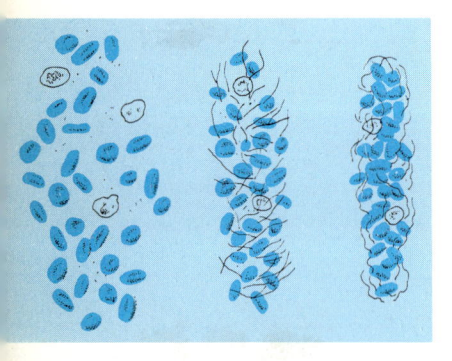

12-5 What is the role of fibrin in the clotting of blood?

EXAMINE

Put a few drops of blood on a clean microscope slide as before. After a few minutes, draw a needle through the drop of blood and lift the needle. Repeat every few seconds until a thread forms between the needle and the drop of blood. Pull some of the threads out to the side of the slide. Examine them with a microscope.

Blood transfusions replace blood lost from the body. In some accidents or diseases large amounts of blood may be lost. The lost blood must be replaced with blood provided by another person by means of a *transfusion* (trans-FEW-zhun). If whole blood is used, the patient receives both the necessary plasma and blood cells. Whole blood can be stored in *blood banks* for use when needed. By using a preservative, whole blood can be stored for as long as 30 days without spoiling (Fig. 12-6).

When whole blood is used for a transfusion, the blood of the person giving the blood and that of the person receiving it must be *typed* before using it. The four human blood types are A, B, AB, and O. Using the wrong type of blood causes the red cells of the patient to clump. This means that the two blood types do not match, and the blood cannot be used in the transfusion. However, if the red cells do not clump together, the blood types match. Investigate the clumping of red blood cells in the activity below.

COMPARE

Place a drop of blood on a microscope slide. Have another student place a drop of his blood alongside the first drop. Mix the two drops together with a clean needle. Add a drop or two of normal saline to the blood. Mix evenly with a needle and spread out over the entire slide. Examine the blood with a microscope or microprojector. Do the red blood cells remain separate or do they form little clumps? Try matching drops of your blood with other members of the class. Keep a record of the pupils whose blood can be mixed with yours. **(Caution: Do not use the same needle on different students.)**

Often a person needs an immediate transfusion to increase the volume of blood in his body. In such cases, plasma, from which the red blood cells have been removed, can be used. If the volume of blood is maintained, new red cells will be formed by the body within a few days. A transfusion of plasma does not require typing. It is the chemical substances present in the red blood cells that react when the blood is not properly matched.

The heart pumps the blood through the body. As you can see, by looking at Fig. 12-7 on page 356, the *heart* is composed of four chambers, two *auricles* (AW-rih-kuhlz) and two *ventricles* (VEN-trih-kuhlz). They are separated from each other

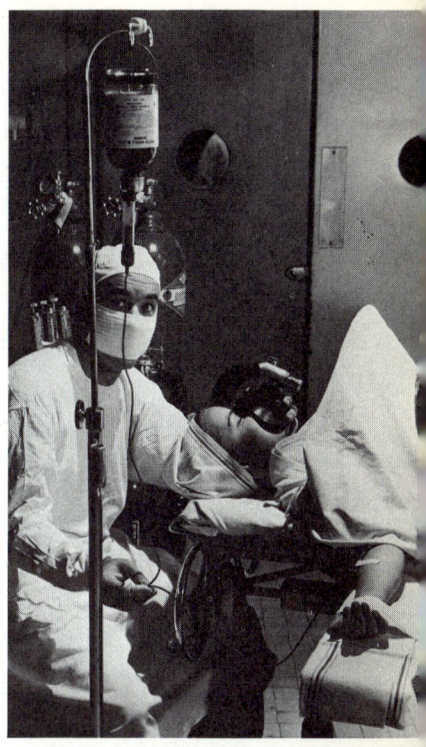

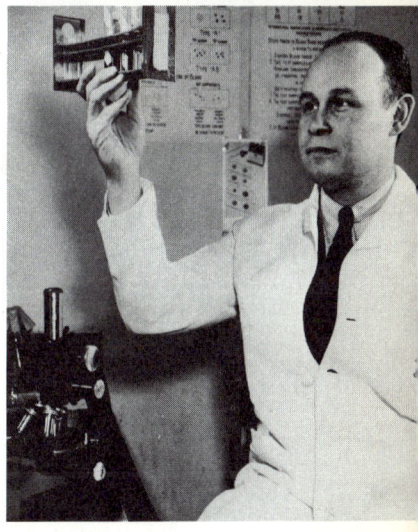

12-6 The blood in the bottle must match the patient's blood. Why? Dr. Charles Drew (below) invented the blood plasma storage bank.

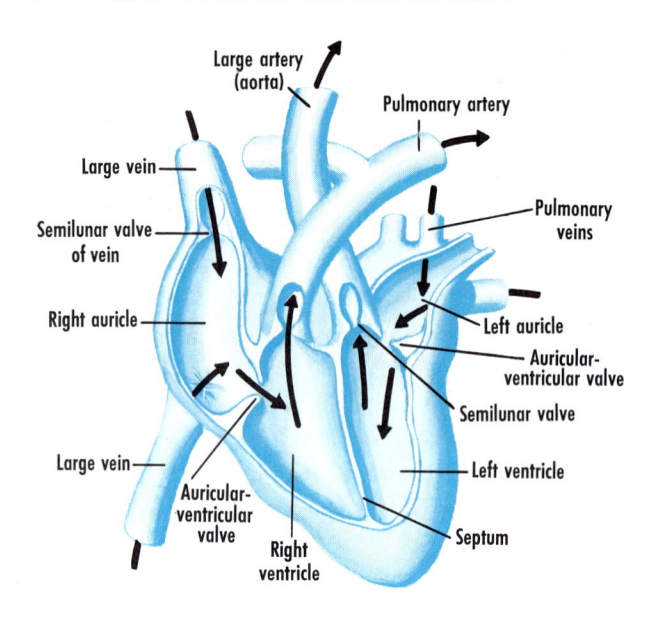

12-7 Can you locate the four chambers of the heart? Name them.

by a muscular partition that divides the heart into the right half and left half. Auricles receive blood; ventricles pump blood. The heart is actually a pump which forces the blood through the body each time it beats. The beat or contraction of the heart can be felt as the *pulse* in some blood vessels. You can count your pulse rate in the next activity.

RECORD

To feel your pulse, place your forefingers against your neck under the jaw, or place the tips of two fingers on the palm side of the wrist near the thumb. Count your pulse rate for one minute. Record the pulse rates for all the students in your class and find the average pulse rate. Now, have a student jump up and down on one foot for 20 seconds. Count his pulse rate after the exercise. It is faster or slower than before?

To trace the flow of blood through the heart, we might start with blood present in the left ventricle. When the heart contracts, about half a pint of blood is forced from the left ventricle into the *aorta* (ay-OHR-tuh). The aorta is a large artery which carries oxygenated blood to all parts of the body. A one-way *valve* between the left ventricle and the aorta keeps the blood from flowing back into the ventricle.

As the blood flows through the aorta, *arteries* branch off to different parts of the body. An artery is a blood vessel carrying blood *away* from the heart. The arteries branch into smaller and smaller tubes and finally become very tiny *capillaries* (KAP-ih-*ler*-eez). Capillaries are important sites of exchange at the cellular level. The capillaries then join to form small *veins* or blood vessels which carry blood *toward* the heart. The tiny veins join to form larger veins. The blood finally returns to the heart through the two *main veins* which open into the right ventricle. The activity below will help you understand how blood cells move through the capillaries.

EXPLORE

Drill a one-inch hole through one end of a thin piece of wood. Wrap a goldfish in wet absorbent cotton but leave the tail uncovered. Tie the goldfish to the piece of wood so that the thin part of the tail is over the hole. Spread the tail apart. Push pins through it on each side so that a clear portion is over the hole, as shown. Examine the tail with the microprojector or microscope. **(Do not keep the fish out of water for more than 15 minutes at a time.)**

Can you see small blood vessels in the tail? What kind of blood cells can you see moving in the capillaries? Are they moving by themselves or are they being carried by the blood?

From the right auricle the blood flows through another one-way valve into the right ventricle. From there it passes through another valve between the right ventricle and the *pulmonary* (PUHL-muhn-er-ee) *artery* and travels to the lungs. The blood then returns through the *pulmonary veins* into the left auricle. From there, the blood flows through the valve (separating the left auricle and ventricle) into the left ventricle. In this chamber of the heart, the whole process of blood circulation starts over.

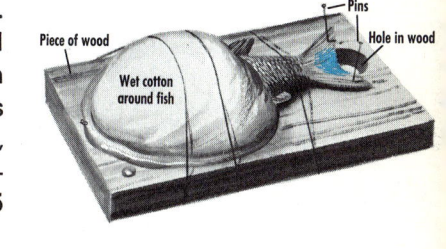

358 / UNIT 4 MAN EXPLORES THE HUMAN BODY

The heart is a cone-shaped organ about the size of your closed fist. It is located under the breastbone in the chest cavity. It usually lies a little to the left of the middle with the tip extending downward and to the left. Since the beat is felt more strongly near the tip, many people have the mistaken idea that the heart is on the left side of the chest. How can you hear the sounds of the heartbeat? Do the next activity and find out.

LISTEN

The sounds of the heartbeat can be heard through a stethoscope. If a stethoscope is not available, a satisfactory substitute can be made by using a thistle tube, a glass Y-tube, and rubber tubing, as shown. When you listen to a normal heart sound, you should hear "lub" and "dup" sounds repeated over and over. The "lub" is due to the contraction of the ventricles and the closing of the valves between the ventricles and auricles. The "dup" sound is caused by the closing of the valves between the ventricles and the pulmonary and aortic arteries. Note the position of the stethoscope in the diagram.

The heart produces pressure in the arteries. To circulate the blood to the arms and legs, as well as to other parts of the body, the heart pumps out the blood under pressure. The pressure is great enough for any artery to lose a great deal of blood if it is cut. To prevent this from happening, if there is an accident, the artery leading to that part of the body should be pressed closed (or pressure applied directly to the wound) to stop the flow of blood. Where an artery is close to the surface of the body, a pulse can be felt. This is called a *pressure point*. The increase in blood pressure that results with each contraction of the heart can be shown in the following activity.

OBSERVE

Attach a U-tube to a ringstand and fill the tube part way with colored water. Attach a rubber tube to one arm of the U-tube

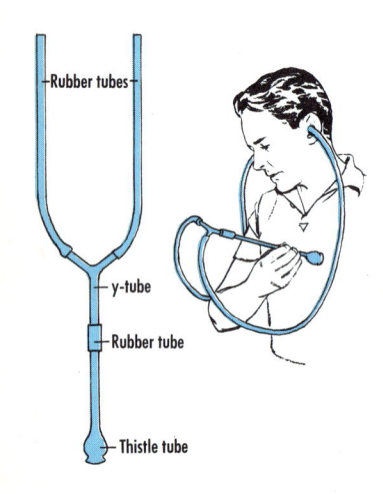

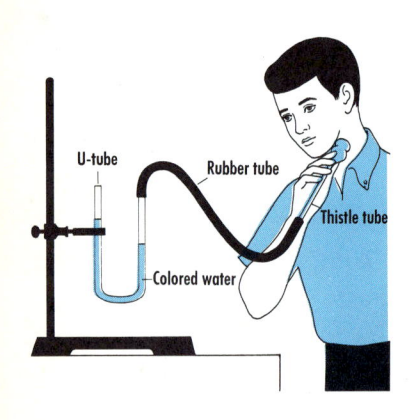

CHAPTER 12 FUNCTIONS OF THE BODY / 359

and a thistle tube or small funnel to the other end of the rubber tube. Hold the open end of the thistle tube tightly against the artery alongside the windpipe in the neck, as shown. **(Hold the bulb, not the stem of the thistle tube. If the tube breaks, it is dangerous.)** What happens to the level of the water in the U-tube as the artery expands with each beat of the heart?

A doctor uses a special instrument called a *sphygmomanometer* (sfig-moh-muh-NOM-uh-ter) to measure the blood pressure in the arteries (Fig. 12-8). An air bag is wrapped around the arm just above the elbow. By increasing the air pressure in the bag, the artery is pressed closed and the blood flow in the vessel stopped. Using a stethoscope, the doctor listens for the first sound of the returning pulse in the artery as the air is released from the bag. This point is known as the *systolic* (sihs-TAHL-ihk) pressure, the greatest pressure of the blood resulting from the contraction of the ventricles. The doctor continues to release the air from the bag until the sound of the pulse is no longer heard. This point is the *diastolic* (dye-uh-STAHL-ihk) pressure, which is the lowest pressure of the blood occurring when the heart is at rest. For this reason, blood pressure is always noted as a fraction, such as 110/70. Find out what blood pressure range is normal for people your age. Does this remain the same throughout life?

Exercise speeds up the action of the heart. The five quarts of blood in your body make a round trip through the heart about once a minute. In mild exercise, this rate of circulation may double, and in heavy exercise it may be four times as fast! Normally, the heart pumps about 10,000 quarts of blood through the body every 24 hours.

How can a living organ keep on beating year after year without rest? Actually, the heart does rest for a fraction of a second between each beat, even though it normally contracts about 70 times a minute. Besides this, the heart has a rich supply of blood to furnish the food and oxygen for its action. Although the heart weighs only about 1/200 of the body's weight, it uses 1/20 of the total blood supply. During a lifetime of 70 years, the heart will beat over two and a half billion times!

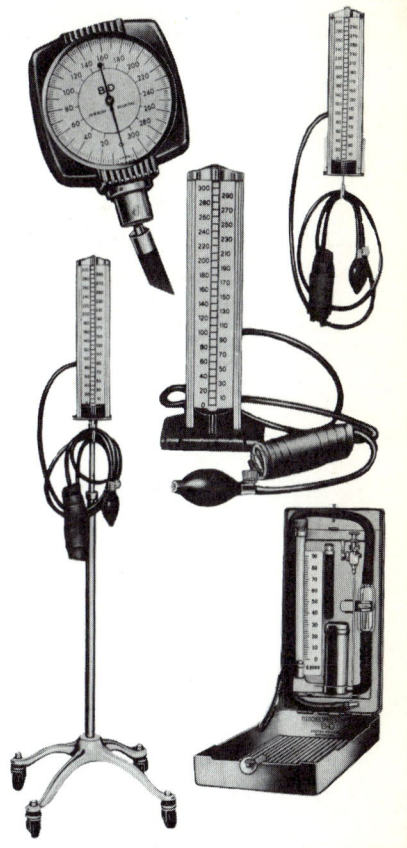

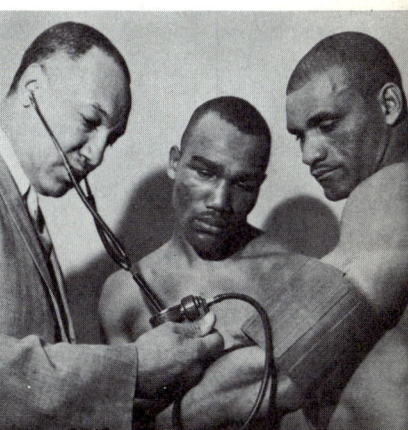

12-8 What instrument does a doctor use to measure blood pressure?

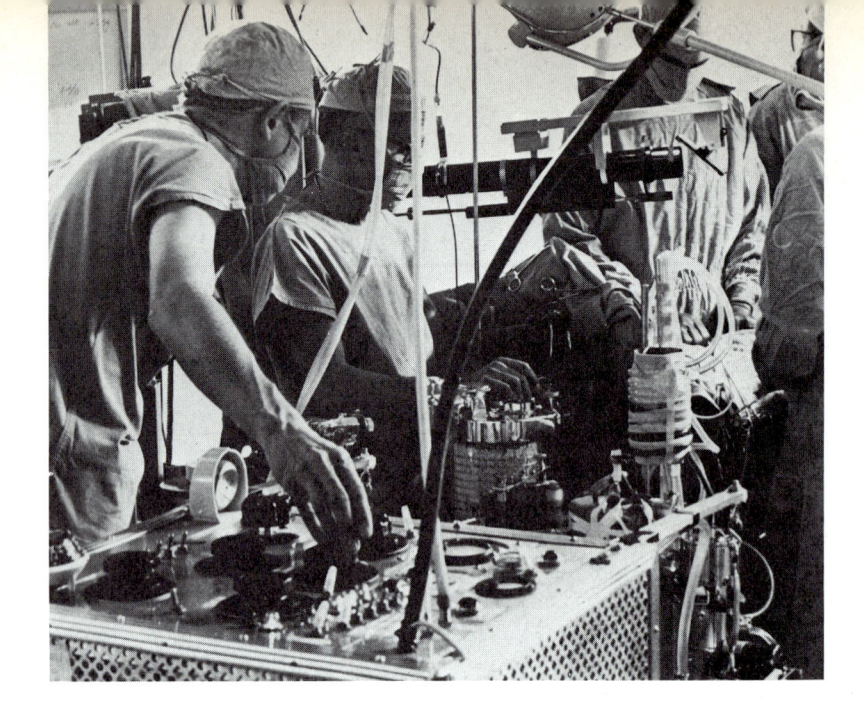

12-9A What is the function of the heart-lung machine?

12-9B Dr. Daniel Hale Williams pioneered in open-heart surgery.

A number of devices designed to keep the blood circulating in the body while the heart action is stopped have been tested. These are of great importance because they allow a surgeon to repair a defect in the heart while the heart is free of blood and not beating. A mechanical heart-lung machine is pictured in Fig. 12-9A.

Breathing is the mechanical part of the process of respiration. When you *inhale,* air flows into the *lungs;* when you *exhale,* air flows out of the lungs. Inhaling and exhaling are two visible steps in breathing. Each time you inhale, air passes through the nose or mouth into the windpipe, or *trachea* (TRAY-kee-uh). Then, air passes into small branches of the trachea called *bronchi* (BRAHN-key), and into tiny *air sacs* in the lungs. The nose is an "air conditioner" where air is warmed, filtered, and moistened by tissue lining the inside. Normally, we inhale and exhale about 16 to 24 times a minute. Exercise speeds up the rate of breathing. The muscles of the chest and the *diaphragm* (DYE-uh-fram), a muscular partition separating the chest from the abdomen, are the main breathing muscles.

The process of breathing takes place because the air pressure in the chest cavity increases and decreases as the muscles relax and contract. When the air pressure in the lungs is less than the surrounding air pressure, the air flows into the lungs, and we inhale. When the air pressure in the lungs is greater than the surrounding air pressure, the air flows out of the lungs, and we exhale. During an average 24-hour day, you breathe over 23,000 times and inhale about 438 cubic feet of air. The volume of air that is present in a breath can be found by doing the next activity.

MEASURE

Fill a large flask or jar about three-quarters full of water. Push two bent glass tubes through a two-hole stopper and place it in a jar, as shown. Be sure the seal is tight. Use a short piece of smooth-edged glass tubing as a mouthpiece. Wash it with rubbing alcohol after each person has used it.

Place a beaker under the overflow tube. Exhale into the jar, emptying the lungs as much as possible. Measure the amount of water in the beaker with a graduated cylinder to determine the amount of air exhaled in cubic centimeters. Determine the amount of air exhaled by each student in the class. Record your results in a table similar to the one shown below. **Do not write in your book.**

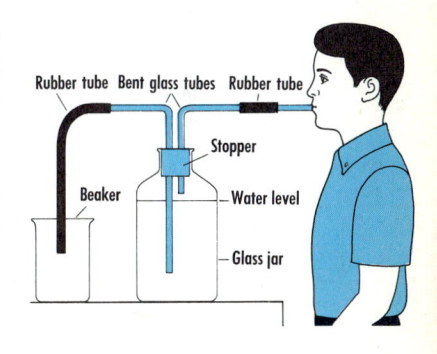

| Amount of Air Exhaled ||
Boys	Girls
1.	1.
2.	2.
3.	3.
Total Number	
Boys _____	Girls _____
Average Lung Capacity	
Boys _____	Girls _____

Can you explain your results?

The blood absorbs oxygen in the lungs. When we inhale, the air reaches the lungs and flows into the air sacs located there. The blood circulates through capillaries around the air

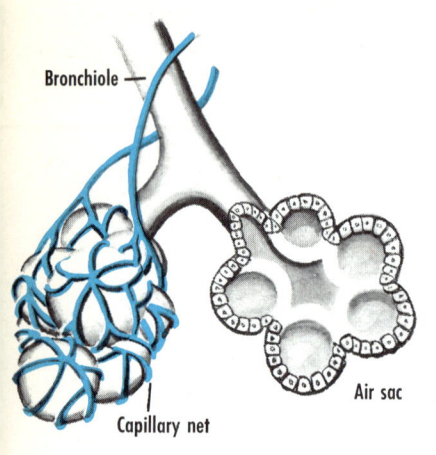

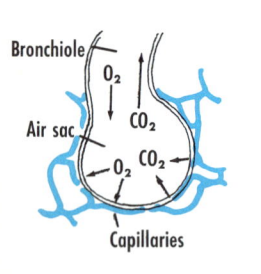

12-10 What is the function of the air sacs in the lungs?

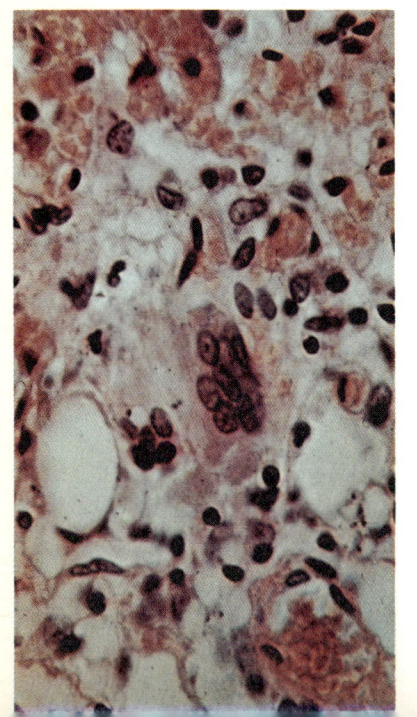

12-11 Note the spongy appearance of lung tissue.

sacs (Fig. 12-10). Blood from the tissues, which is low in oxygen, or *deoxygenated* (dee-AHKS-ih-juhn-*ay*-tuhd), gives off molecules of carbon dioxide which pass into the air sacs. At the same time, molecules of oxygen pass into the capillaries to be combined with the hemoglobin in the red blood cells. The blood becomes bright red in color and is known as *oxygenated* blood.

Therefore, when we exhale, the air contains a higher percentage of carbon dioxide and a lower percentage of oxygen than the air we inhale. You can show that the exhaled air contains a fairly large amount of carbon dioxide by gently blowing your breath through a straw or glass tube into a test tube of clear limewater. The limewater becomes cloudy in a few minutes. The cloudiness of the limewater is a test for the gas, carbon dioxide.

The lungs are complicated organs. If a piece of lung tissue is examined, it looks like a rubber sponge made up of many air sacs (Fig. 12-11). There are millions of air sacs in the lungs. Each air sac is covered by a network of tiny capillaries so small that the red blood cells have to pass through in single file! Every few minutes, the entire blood supply in the body passes through the lungs. If the circulation through the lungs were to stop even for a few minutes, great damage would result in the body tissues due to oxygen starvation. The brain, especially, becomes affected almost immediately.

Air pollution is a serious problem. As you may have learned in the study of the atmosphere, the air we breathe in many parts of the country contains harmful substances known as *pollutants* (puh-LYOO-tuhnts). These are produced mainly by man's conversion of fuels into energy. The fuels used to heat your home, to drive your automobile, and to power many of the industries in your community are among the main sources of *air pollution*.

How does air pollution affect you? Research is being carried on to see if there is a relationship between the increase in *lung cancer* and the increase in air pollution. Since no one yet knows the cause of cancer, air pollution cannot be blamed entirely for this alarming increase in the disease. But medical studies have shown that the lung cancer rate is higher in

12-12 How can smog found over large cities be reduced?

areas where there is a great deal of air pollution. It is also believed that air pollution can shorten the lives of many people that are suffering from *asthma* and other serious respiratory diseases.

Smog is formed in heavily populated areas. Perhaps the most commonly seen result of air pollution is the smog that is produced in the air in many cities. Smog is a mixture of smoke, fog, and heavy particles. Smog occurs when the air is kept from rising by a layer of warmer air over it that acts as a "lid." Air pollutants produced mainly by automobiles and trucks contribute large quantities to this condition. These pollutants are acted upon by sunlight to produce a *photochemical* (*fo*-to-KEM-ih-kuhl) *reaction*. The substances produced in this reaction become the substances irritating to our bodies. The eye irritation that results from smoggy air is only an outward sign. The effects on the respiratory system are not fully known and may be more serious (Fig. 12-12).

Waste products must be removed from the body. The use of food by the body produces waste products which are given off into the blood. The main waste products produced by the activities of the cells are *carbon dioxide, urea* (yoo-REE-ah), and *uric acid.* Carbon dioxide is passed out through the lungs, and urea and uric acid are removed from the blood in the *urine* by the kidneys.

The *kidneys* are bean-shaped organs, located in the lower back of your body. Tubes called the *ureters* (YOO-ree-terz) connect the kidneys with the *urinary bladder* where urine is stored. The urine passes out from the body through another tube, the *urethra* (yoo-REE-thruh).

The process of excretion must take place continually. In a sick person, if urea and uric acid build up in the blood, tissue poisoning occurs. Tissues become filled with these waste products and cannot absorb either food or oxygen. Fever, coma, and death of the patient usually occur if nothing is done to correct the condition.

In severe cases of kidney disease, a mechanical kidney machine is connected to the circulatory system of the body to take over the work of the kidneys. The blood passes through this device, is purified, and then returns to the body. Such a machine cannot be used indefinitely. It can allow the kidneys to rest, however, for a few days so that the disease or injury may be cured. Sometimes, a kidney disease cannot be cured. The machine, by removing the wastes, keeps the patient alive and comfortable.

REVIEW

1. Into what two main parts does the blood separate if allowed to stand?
2. Name the three parts making up the solid blood.
3. What are the two kinds of chambers making up the human heart?
4. The blood absorbs oxygen in what important structures in the lungs?
5. Name the main organs making up the excretory system for urinary wastes.

C/How Is Movement Controlled?

There are two main kinds of muscles in the body. Many of the bodily motions are easily seen. You walk, jump, write, and work with your hands. The muscles that enable you to carry on these activities are known as *voluntary muscles* because you can consciously control their actions. However, there are many actions, like the contractions of the digestive system, movement of breathing muscles, and the beating of the heart, which are not consciously controlled. The muscles that carry on these activities are known as *involuntary muscles*. All muscle tissue is made up of cells that have the ability to contract. We can examine some voluntary muscle cells, using animal tissue, in the following activity.

INVESTIGATE

Boil a small piece of beef until it is very tender. Separate one strand of muscle tissue. Place it on a clean microscope slide. Using two needles, tear the tissue apart into small threads. Remove any larger pieces of muscle. Add a drop of methyl blue stain to the tissue on the slide. Put a cover glass on the stained tissue and examine with a microprojector or microscope. Compare the cells you see with the muscle cells of Fig. 12-3 which is found on page 349.

What is the shape of the muscle cells? How are the cells arranged in the tissue? Can you locate the nucleus in some of the cells? Do you notice any special markings on the cells? Try to describe these particular markings. Can you suggest what their function might be in the muscle cells?

The action of muscles is coordinated. To perform a certain task, the muscles must work together. We refer to this action as *coordination* (ko-*or*-dih-NAY-shuhn). For example, writing may seem to you to be a very simple activity. A small child, however, has great difficulty in writing even a single letter. As a child grows older and practices, his ability to coordinate his muscles improves, and writing becomes an everyday activity. In the same way, you learn to play a musical instru-

366 / UNIT 4 MAN EXPLORES THE HUMAN BODY

ment, to type, or to take part in sports. The nervous system controls all the activities of the body and is responsible for the coordination of body movements.

The arrangement of voluntary muscles is usually in pairs. This allows one muscle to move a part of the body in one direction and the other muscle to move it in the opposite direction. Such a pair can be found in the upper arm.

The *biceps* (BY-seps) muscle is found in front of the upper arm, and the *triceps* (TRYE-seps) muscle is on the back of the upper arm. These two muscles are coordinated to move the forearm. The biceps bends and the triceps extends the forearm. Thus, one muscle relaxes while the other muscle contracts.

Muscles move the bones of the skeleton. The large voluntary muscles are attached to bones by means of *tendons*. These are bands of inelastic connective tissue that form a firm attachment to the bone. Since a muscle moves a bone when it contracts, one end of the muscle has to be fixed; the other end is movable. Determine which ends are fixed, and which are movable in the drawing of the biceps and triceps muscles (Fig. 12-13). As the muscles move the bones, movements in the skeletal system are produced. How do tendons act? Do the following activity to find out.

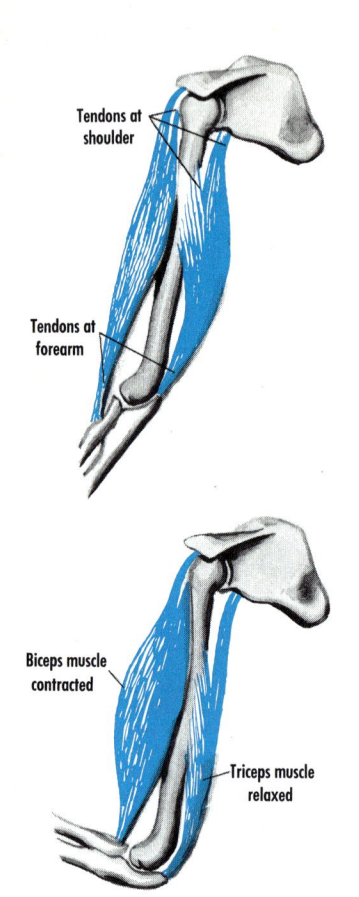

12-13 How do the muscles in the upper arm work together to bend and straighten the arm?

DISCOVER

Obtain a chicken foot from a freshly killed chicken. Trim the skin from the cut end to expose the white, stringlike tendons. Separate these carefully. Use a forceps to pull on each tendon. What happens to the toes in the foot when the tendons are pulled? What furnishes the pull on the tendons in a live chicken?

Nerve impulses produce contractions in muscles. For a long time, scientists were not sure how a nerve impulse or "message" causes a muscle to contract. They knew that nerve impulses cause *electrochemical* (ee-*lek*-tro-KEM-ih-kuhl) reactions in the nerves and muscles. They were not sure, however, how these electrochemical reactions caused muscle cells to contract the way they do.

Today, it is known that a nerve fiber divides into numerous

small branches as it approaches a muscle fiber. Each of these small branches ends in a specially organized structure located on the muscle fiber called the *motor end-plate*. There is a space between the nerve ending and the motor end-plate that is called the *neuromuscular* (*nyoo*-roe-MUHS-kyoo-luhr) *junction*. How does the nerve impulse or "message" actually affect the muscle?

Chemical substances are released by the nerve fiber. As a nerve impulse, traveling along a nerve fiber, arrives at the motor end-plate of a muscle fiber, a substance called *acetylcholine* (uh-*set*-ihl-KO-leen) is produced. This substance transmits the impulse to muscle fibers which begins the process of contraction. After a brief period of contraction, the nerve releases an enzyme called *cholinesterase* (*ko*-lih-NES-tuhr-ays). Cholinesterase is able to neutralize acetylcholine, causing the muscle fibers to relax. The processes of contraction and relaxation take 0.1 second or less.

Since all the fibers of a muscle are covered with nerve endings, the nerve impulse or "message" reaches all of them at the same time. This arrangement causes the entire muscle to contract. However, no voluntary muscle in the body can maintain a steady contraction for a long time. Waste products build up in the cells, and the muscle tires very soon. The muscle, then, no longer responds to the impulses. This failure by the muscle to respond to nerve impulses will last until the waste products are removed by the blood.

There is a serious disease in which acetylcholine is blocked almost completely at the neuromuscular junction. The eye muscles are affected, causing drooping eyelids and double vision. The patient has difficulty in smiling, chewing, speaking, walking and even breathing. To help the patient, a chemical is given which lowers the action of cholinesterase. Lowering the action of cholinesterase allows the level of acetylcholine to rise in the patient. Improvement in the patient is fast and effective.

Patients now take the drug, or similar ones, by mouth in the form of tablets. A person must be careful, however, not to take too many. An overdose causes muscle twitching and stomach cramps due to the presence of too much acetylcholine.

Bones make up the skeleton of the body. The human *skeletal system* consists of 206 *bones*. They vary in size and shape from the curved, immovable bones in your skull to the long, heavy bones in the leg. Bone tissue protects the internal organs and supports the body. The activity below will help you become familiar with the structure of a bone.

EXAMINE

Obtain a large bone from your meat market. Keep it moist and fresh by wrapping it in waxed paper. Examine it and feel the ends of the bone. Saw through the bone at one end. Describe the appearance of the cut end of the bone. Does the bone have the same hardness all over? Where is the hardest part and where is the softest part?

Most of the long bones in the body are designed to carry a great deal of weight. Yet, they are light enough to move easily. For that reason, the enlarged ends of these bones are spongy, and the inside is hollow to make the bone lighter. Since bones are made up of minerals and living cells, we can separate the minerals and living cells by following the directions in the next activity.

EXPERIMENT

Place a chicken leg bone in a tall jar and add dilute hydrochloric acid to cover the bone. **(Caution: Hydrochloric acid is dangerous to handle. In case of accident, sponge off with dilute ammonia and wash in running water.)** Leave the bone in the acid for several days. Remove the bone and wash it in water. Examine its hardness.

Put another piece of bone in an evaporating dish. Heat with a strong flame, as shown, until the bone burns down to an ash. Examine the material left.

Were you able to bend the bone after soaking it in acid? What part of the bone tissue was removed by the acid? What was removed in the burning of the bone? What does this show about the composition of bone?

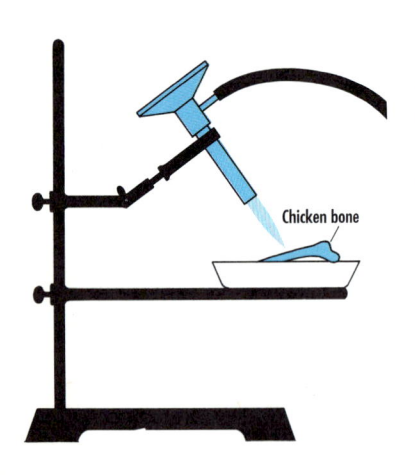

After examining bones, you may wonder how such a solid substance can be alive. Part of the bone is mineral and is nonliving, but the living cells are found in the mineral layers (Fig. 12-14). As the bone cells grow and divide, they deposit minerals in the form of *calcium phosphate* in the spaces between them. This process is most active while a child is growing, but continues at a reduced rate during a person's lifetime.

When a bone is broken, the bone must be set. The two ends are placed together by a physician and kept in a cast to keep the broken ends from moving. After a period of time, the bone cells deposit new layers of minerals in the break, and the bone becomes strong again.

Sometimes a part of a bone is destroyed by disease or accident. Scientists have developed a bone substance to replace the missing part. It is prepared from animal bones by treating them with chemicals to dissolve away the living parts of the bone that would decay. This leaves the mineral part which can be kept, until needed, to replace a part of a bone in the body. The body forms new cells and blood vessels in the piece of bone, and it soon becomes a living part of the body.

Joints in the skeleton permit movement. Your bones can move because they fit together at *joints*. The ends of the bones are covered with thick pads of smooth *cartilage* (KAHR-tih-lihj) so that the bones can move without rubbing. A thin membrane filled with fluid surrounds the entire joint to lubricate it and to allow freedom of movement. The bones are held together at a joint with strong bands of connective tissue, called *ligaments* (LIHG-uh-muhnts). You may have suffered a *sprain* when ligaments were stretched or torn by a sudden twist at a joint. You can learn more about joints from an animal source in the next activity.

12-14 A microscopic view of bone tissue. What important mineral is deposited by bone cells during your life?

INFER

Obtain an animal bone joint from your meat market. Using a pair of scissors and a sharp knife, carefully cut the flesh from the joint. Do you see any ligaments and cartilage? Now, cut the joint apart and look at the ends of the bones. Are they rough or smooth? Why do the bones in joints move easily without friction?

Four main kinds of joints are found in the body: *ball and socket* joints, *hinge* joints, *partially movable* joints, and *immovable* joints (Fig. 12-15). The ball and socket joints in the hip and in the upper arm permit motion in practically all directions. Hinge joints are found in the knee, elbow, and in the fingers. The ribs are joined to the backbone by means of partially movable joints. The bones of the skull form immovable joints. In the activity below, you can observe the motion of bones.

DESCRIBE

Place your arm on a table with the palm up. Feel the bones in the forearm between the wrist and the elbow. Turn the hand so that the palm is down. Describe the motion of the two bones. What kind of joint does this illustrate? Compare the movement of this joint with joints in the elbow and shoulder. What types of joints are these?

In addition to allowing the body to move, the skeletal system supports the body and protects the vital internal organs. Could you stand up straight without a backbone? The bones in your backbone and in your legs support the weight of your body. The bones of the skull protect the brain and the bones of the chest protect the heart and lungs.

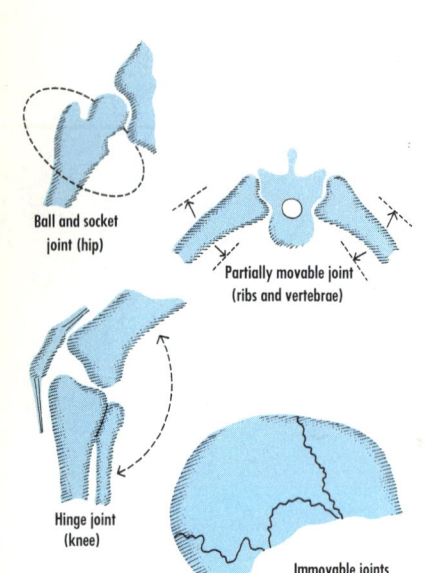

12-15 Which of the four types of joints allows the greatest amount of movement.

REVIEW

1. What are the two main kinds of muscles in the body?
2. Why are voluntary muscles arranged in pairs?
3. Why are the long bones spongy at the ends and hollow in the center?
4. What minerals compose the nonliving part of a bone?
5. Name the four main kinds of joints in the body.

THINKING WITH SCIENCE

A. *On a separate sheet of paper, write the numbers 1 to 15. After the number of each question, write the letter of the term that correctly completes the statement.* **Do not write in your book.**

1. The blood cells which carry oxygen to the tissues are the: (*a*) blood platelets, (*b*) white corpuscles, (*c*) red corpuscles.

2. A tube which carries urinary wastes from a kidney to the bladder is called a: (*a*) urethra, (*b*) urea, (*c*) ureter.

3. A group of similar cells performing a similar function is known as: (*a*) a(n) system, (*b*) organ, (*c*) tissue.

4. A blood vessel that carries blood away from the heart is called: (*a*) a vein, (*b*) an artery, (*c*) a capillary.

5. The strong bands of tissue that hold bones together at joints are: (*a*) tendons, (*b*) ligaments, (*c*) muscles.

6. The small, irregularly shaped bodies in the blood which help in clotting are the: (*a*) blood platelets, (*b*) white corpuscles, (*c*) red corpuscles.

7. The living substance making up cells is called: (*a*) plasma, (*b*) protoplasm, (*c*) vacuole.

8. The tissue making up the skin and other body coverings is called: (*a*) connective, (*b*) epithelial, (*c*) muscle.

9. The system which removes waste materials from the body is the: (*a*) excretory, (*b*) digestive, (*c*) nervous.

10. The breathing muscle that separates the chest from the abdomen is the: (*a*) trachea, (*b*) auricle, (*c*) diaphragm.

11. The oxygen-carrying compound found in red blood cells is called: (*a*) hemoglobin, (*b*) mitochondrion, (*c*) acetylcholine.

12. The type of body joint which permits the greatest amount of motion is the: (*a*) hinge, (*b*) partially movable, (*c*) ball and socket.

13. The chemical substance released by a nerve cell which triggers the contraction of a muscle cell is: (*a*) hemoglobin, (*b*) acetylcholine, (*c*) cytoplasm.

14. The tiny structures in a cell which are the "power plants" in the cell are the: (*a*) vacuoles, (*b*) ribosomes, (*c*) mitochondria.

15. Molecules of some substances which pass through a membrane while others do not illustrate the process of: (*a*) conduction, (*b*) osmosis, (*c*) pollution.

B. *Write the answers to the following in your notebook. Be sure to use complete sentences and correct spelling and grammar.*

1. What are the main parts making up a cell and the functions of each part?

2. Why do certain substances flow into or out of the cell during diffusion?

3. What are the main life functions carried on by most cells?

4. How is hemoglobin adapted to carry oxygen to the tissues?

5. Trace the circulation of the blood through the heart starting in the right auricle.

6. What does systolic and diastolic pressure show?

7. How does the process of breathing take place in the body?

8. What are the main waste products produced by the cells, and how are they removed from the body?

9. Why must the actions of muscles in the body be coordinated?

10. How does a nerve impulse cause a muscle fiber to contract and relax?

RESEARCH IN SCIENCE

1. Using clay, make life-size models of different organs in the body.

2. Make a report on how mechanical heart-lung machines or mechanical kidney machines operate.

3. Make models of joints in the body by whittling them out of wood and fastening the parts together with wire.

4. Find out about the different blood types and how they are recognized. Prepare a report on this information to the class.

5. Prepare a report to the class on recent advances in organ transplants.

chapter 13
Growth of the Body

A/What Changes Occur in the Body?

The body grows in height: You know that you have grown taller since you were a child. But how much taller have you grown each year? If you were three feet in height when you were about three years of age and are now five feet in height, you have grown about 24 inches in ten years. You could say that you grew about two and a half inches a year. Between the ages of three and thirteen years, growth is fairly even each year. In the next three or four years, however, you will grow more rapidly until you have almost reached adult height. You can chart some changes in height in the following activity.

CHART

Make a chart of the heights of ten children six years of age, ten children thirteen years of age, and ten young adults sixteen years of age. Determine the average height of the six-year-olds, the thirteen-year-olds, and the sixteen-year-olds. When does the greatest increase in height take place?

Between the ages of twelve and sixteen, girls usually grow faster than boys. You will find that the girls in your class are probably taller than the boys. Would this also be true in a senior high school class? Probably not, since the boys grow rapidly from about fourteen years of age to seventeen or eighteen. By this time the boys are usually taller than the girls. From about twenty to twenty-five years of age, your body grows more slowly until you reach the size of an adult. Do the activity comparing the average height of boys and girls in the 12-14 age bracket.

COMPARE

Compare the height of boys and girls by selecting at random five girls and five boys of the same age in your class. Measure their heights and make a chart. Is the average height at this age greater for boys or for girls?

You should remember that any figures shown in your charts are average figures. Each person grows at a different rate, some developing earlier and some growing up later. Figure 13-1 on page 376 shows a chart of the growth of four brothers from birth to twenty years of age. Did you notice that although there is a common pattern of growth for all four boys, each one grew at a little different rate?

The body increases in weight. You know that as you grow taller, you also increase in weight. Scientists have measured the growth of thousands of people and have found average figures for weight increases at different ages. As with height, the increase in weight is greatest during the first three years of life. There is a slowing down to a more even gain in weight each year until about the age of ten. The next four or five years show a rapid increase, followed by a slower, more even growth until adult size.

Scientists have also measured the rate of growth of the organs in the body. Scientists find, with few exceptions, that they follow the same general pattern of growth as the rest of the body. The weight of the liver, heart, kidneys, and other organs increases rapidly during the first few years of life. This

DIFFERENCES IN RATE OF GROWTH OF FOUR BROTHERS

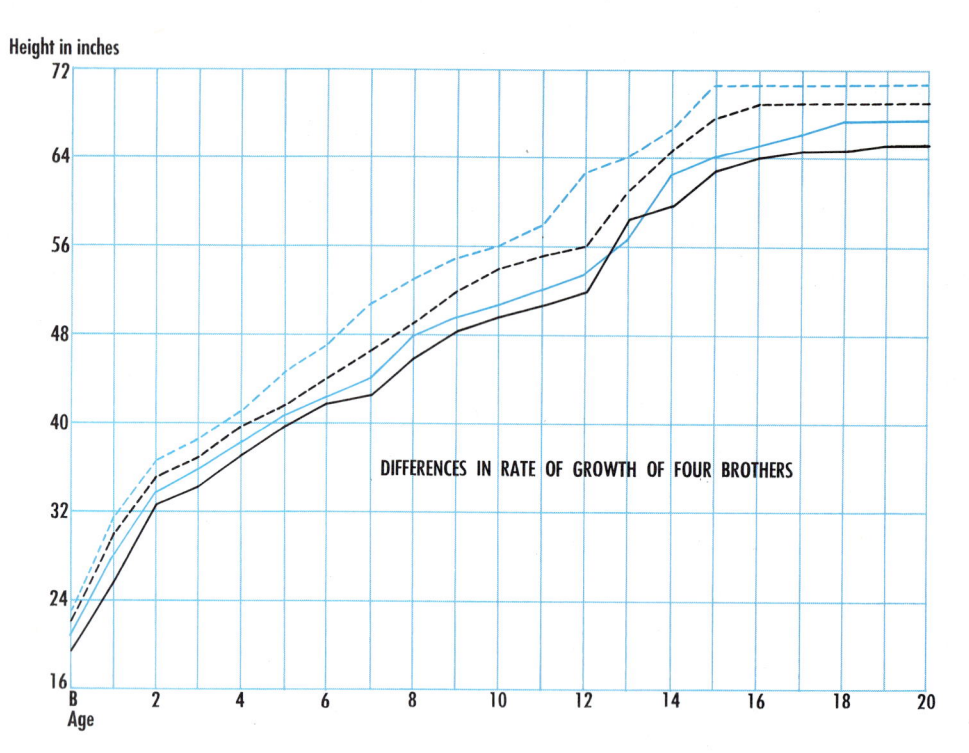

13-1 Compare the growth rate of four brothers on the chart.

is followed by a slower, more even growth in childhood, and then increases rapidly between the ages of ten and fifteen. The brain is one organ that does not follow this pattern. The brain of a four-year-old child weighs about four-fifths of what it will weigh in the adult. The bones of the middle ear are structures that also remain about the same size throughout a person's life.

The body changes rapidly during adolescence. During the period of *adolescence* (*ad*-uh-LES-uhns), many important changes take place in your body. The boys begin to develop hair on the face which may soon become a beard. Their voices become deeper. They develop large muscles, and they become better coordinated. At the same time, the girls develop into young women. These physical changes during adolescence change you from a child into a young adult. You are now ready to take on more mature responsibilities at home and in school (Fig. 13-2).

13-2 Do all young people develop at the same rate? Explain.

The time when adolescence starts is different among boys and girls. Some of you already may have entered this period and others may not be adolescent for another year or two. One thing is sure, however. All young people go through these changes sooner or later. It is often called the "awkward age" because the body is growing so rapidly. This period also may produce different attitudes toward family and friends. Adolescence is a trying period for the adolescent and for everyone around him!

The body develops muscular coordination. Scientists have studied babies and children to find out how coordination of muscle activity develops. For example, careful studies show that you develop the ability to coordinate the muscles of your head first. This is followed by coordination of the muscles of the arms, the legs, and other parts of the body. At birth a baby has practically no control of the voluntary muscles in the body. At about four weeks of age, the baby's eyes follow an object held in front of him. By sixteen weeks of age, most babies lift their heads, and by twenty-eight weeks, babies are able to reach for objects.

378 / UNIT 4 MAN EXPLORES THE HUMAN BODY

Not all babies learn to coordinate various muscles at the same age because each child develops at a different rate. Scientists have discovered, through the study of large numbers of children, that there are certain *patterns* of behavior for children of different ages.

There is a definite pattern in the development of coordination required in muscle control for performing accurate tasks (Fig. 13-3). You probably learned to write by printing letters, rather than writing them as you do now. Printing does not require as much coordination as writing does. Skill in sports and athletics also requires coordination of muscles which develop as you grow older. Scientists have divided the stages of a person's growth as shown in the table.

Development of the Body		
Birth–2 years	Infancy	Learns to control muscles for body movements
2–10 years	Childhood	Develops coordination of large muscles for body activities
10–16 years	Adolescence	Develops coordination of smaller muscles for fine skills
16 years on	Adulthood	Develops judgment and control of activities

13-3 Is muscular coordination inborn or developed? Explain.

REVIEW

1. At what period in life does the height and weight of the body increase fastest?
2. How does the increase in weight of most organs in the body compare to the general growth of the body?
3. At what age period does the body go through the period of adolescence?
4. Which of the voluntary muscles do you learn to coordinate first?
5. Name the four main stages of growth in the body.

B/What Is Inheritance?

Children usually look like their parents. Living organisms look like all others of the same species in a general way. Would you have any trouble telling a dog and a cat apart? Dogs have certain characteristics, or *traits*, that are similar in all dogs. Similarly, cats have characteristics common in all cats. In the same way, humans have traits common in all humans. We walk on two legs, we have hands for grasping, and we have a highly developed nervous system. These traits are known as *species characteristics* (Fig. 13-4). All children *inherit*, or receive from their parents, the general characteristics of the human species.

But children also inherit *individual characteristics* which make them a little different from all other humans. These characteristics are inherited from their parents. Have you ever seen a child that looks very much like one of his parents? You can easily pick out the characteristics that are inherited. Perhaps the child does not look exactly like his parents. There are enough common characteristics present, however, to tell you that the child belongs in that family. Most children inherit characteristics from both parents. They may look like one parent in some ways and like the other parent in other ways (Fig. 13-5). The activity below illustrates inherited characteristics among your classmates.

13-4 What species characteristics are inherited?

INFER

To find out how children resemble their parents, have each student in the class report the color of the eyes and the hair of his mother and father. Compare the eye color and the hair color of each pupil with those of his parents. Do the students in the class usually resemble their parents in these two characteristics?

Perhaps you found that one parent has brown eyes and the other blue eyes. What color eyes might the child have? He could have either brown or blue eyes. Again, one parent might have dark, straight hair and the other parent might have light curly hair. The child may have dark curly hair or light, straight

13-5 What family characteristics are inherited?

hair. What determines the characteristics that a child will inherit?

Scientists know that some characteristics are *dominant* and others are *recessive*. That is, some characteristics usually appear more often in children than others. Therefore, if one parent has certain dominant characteristics, the chances are that the children will inherit them. The table below lists some dominant and recessive characteristics that are found in humans.

| INHERITED HUMAN CHARACTERISTICS ||
Dominant	Recessive
Black or brown eyes	Blue eyes
Large bones	Small bones
Curly hair	Straight hair
Dark hair color	Light hair color
Normal sight	Color-blindness
Normal skin color	Lack of skin color
Normal mentality	Mental defectiveness

Some inherited characteristics may be mixed, however. For example, one parent may have dark hair and the other parent may have light hair. The child may have a hair color that is between the dark and light colors. This is known as a *blending* of the two characteristics. Also, some characteristics may be inherited by only the females or only the males. If the father is bald, will the girls in the family also inherit the baldness? Probably not, since baldness is a trait usually found in males. The male children may inherit the bald characteristic and lose their hair when they become adults. It is interesting to note that even the shape of the bald spot seems to be inherited!

Children may be unlike their parents. Once in a while, children have a trait that cannot be found in the parents, the grandparents, or in any other members of the family. Such a trait is called a *mutation*. A mutation is a *new* trait that appears suddenly in an offspring. It then becomes a family trait for that individual and can be passed on to the next generation

CHAPTER 13 GROWTH OF THE BODY / 381

with all the other characteristics. For example, an animal or a baby may be born without any coloring in the body cells. The skin is very light, the hair is white, and the eyes are pink (Fig. 13-6). This type of animal or person is called an *albino*. It is caused by a mutation in the color cells in the body. White mice, with pink eyes, are examples of albino animals that pass this trait on to their offspring. Perhaps you can find mutations as the following activity suggests.

EXAMINE

Examine different plants and animals around the home to see if you can find mutations from their normal appearance or structure. For example, a four-leaf clover is a change from the normal three-leaf plant. Describe any mutations you find.

13-6 Why is an albino robin considered a mutation?

What caused mutations? Scientists believe that some mutations are caused by X rays and other radiations that affect the reproductive cells of the parents. Many early studies of mutations were performed with fruit flies.

Herman J. Muller (1890–1968), an American scientist, was the first to learn that mutations can be caused by radiations. For example, fruit flies with normal red eyes produced a generation of white-eyed flies when the eggs were exposed to X rays (Fig. 13-7). He found that radiations increased the rate of mutation in fruit flies about 150 times above normal. This important discovery was made in 1927. Other investigators, working independently, confirmed Muller's findings. Muller was awarded the Nobel Prize in 1946 for his discovery of this cause of mutations.

Characteristics may be studied in several generations. Inheritance is difficult to study in humans because it takes a long time to produce new generations. Also, because people move a great deal, it is difficult to keep accurate records. However, scientists have studied some families for several generations. The evidence shows that human inheritance follows the same patterns and laws that apply to all living organisms.

13-7 Why do some of the fruit flies have stunted wings?

13-8 Why do identical twins have similar characteristics?

13-9 In what ways does the environment affect a person's behavior?

382 / UNIT 4 MAN EXPLORES THE HUMAN BODY

The science of *heredity* (hih-RED-ih-*tee*) has been helped through the study of *identical twins* (Fig. 13-8). Heredity is the study of the way in which characteristics are passed from parents to offspring. For all practical purposes, identical twins are the same person, duplicated. They both develop from a single fertilized egg cell which somehow split into two parts. These two portions develop into two individuals. Can you see why identical twins have the same inherited characteristics? Identical twins look alike, develop in the same way, and usually behave the same. However, there are differences for even identical twins are not absolutely alike. Scientists believe that many of these differences are caused by the surroundings or *environment* in which the children grow up. Characteristics that are exactly alike are inherited, although different traits in the twins are probably caused by their environment.

Characteristics are affected by the environment. We know that a person's inherited characteristics may be influenced by the environment, but how much does the environment affect us? Body size is mostly controlled by heredity. Yet, in general, children are one or two inches taller than their parents. Better diets and modern medical care for babies and young children have produced bigger and stronger bodies. We see, then, that the environment can influence some inherited characteristics very much.

Is there such a person as a "born thief"? Probably not, because a person learns to become a thief, just as a person learns to be a good, honest citizen. What causes a person to develop desirable or undesirable characteristics? The environment in which he grows up and the kinds of people he chooses for his friends often determine the way he thinks and behaves (Fig. 13-9). What you learn in school helps to determine what you will do when you grow up. (In fact, studying different subjects in school also can lead to a discovery of a satisfying hobby.)

All the characteristics you inherit, plus those you acquire from your environment determine your *personality,* or the kind of person you become. You cannot change your inherited characteristics, but you can trade undesirable influences and habits for desirable ones.

CHAPTER 13 GROWTH OF THE BODY / 383

REVIEW

1. Name the two general groups of characteristics inherited by all children.
2. List several dominant characteristics in humans.
3. List several recessive characteristics in humans.
4. What causes mutations in living organisms?
5. What human traits may be affected by the environment?

C/What Are the Chemical Regulators?

Endocrine glands control body activities. A *gland* is an organ that produces powerful chemical substances which affect the body in some way. We have already learned that glands, like the digestive glands, pass on their chemical products through tubes or *ducts*. Other glands in the body, however, do not have ducts. Instead, their products are absorbed directly by the blood and carried in this way to affect the parts of the body. These ductless glands are called *endocrine* (EN-duh-krihn) glands.

Endocrine glands produce important chemical substances called *hormones* (HAWR-monz). Hormones are very active substances. They control the rate of growth and development of specific parts of the body and regulate the activity of all the body systems. In addition, the endocrine glands affect the activities of other glands. Figure 13-10 shows endocrine gland tissues.

Endocrine glands are located in different parts of the body. The various endocrine glands make up the *endocrine system* in the body. Figure 13-11 on page 384 will help you to locate the endocrine glands.

If you touch your throat just below the voice box, you can feel a mass of spongy tissue, the *thyroid* (THYE-roid) *gland*. It is located in the neck attached to the front part of the trachea (Fig. 13-12) on page 384. Other endocrine glands are not as easy to locate. The two *adrenal* (uh-DREE-nuhl) *glands* are attached to the top of the kidneys. The *pancreas* (PAN-

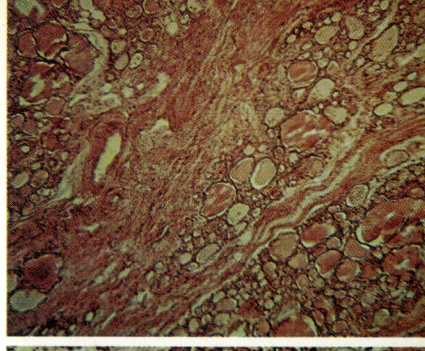

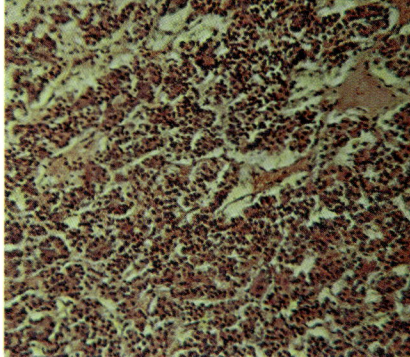

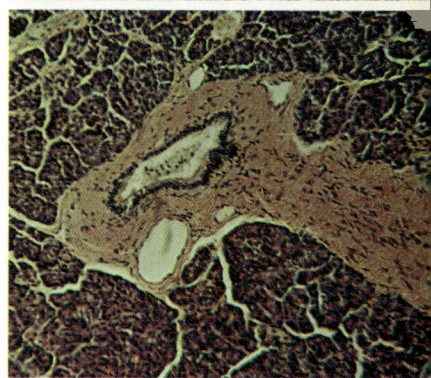

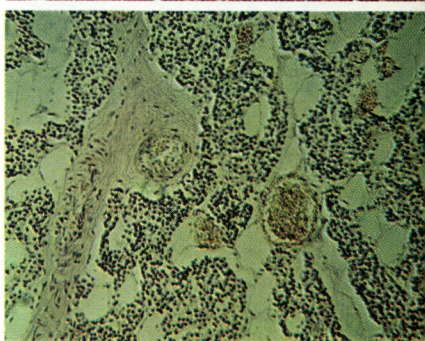

13-10 Examples of endocrine gland tissue (top to bottom): thyroid, pituitary, adrenal, parathyroid.

384 / UNIT 4 MAN EXPLORES THE HUMAN BODY

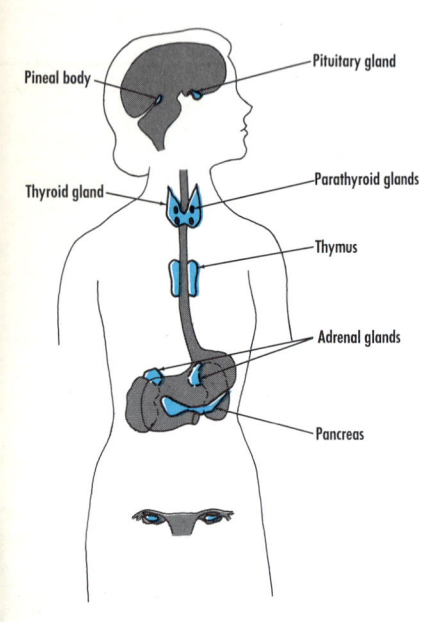

13-11 Locate the glands of the endocrine system.

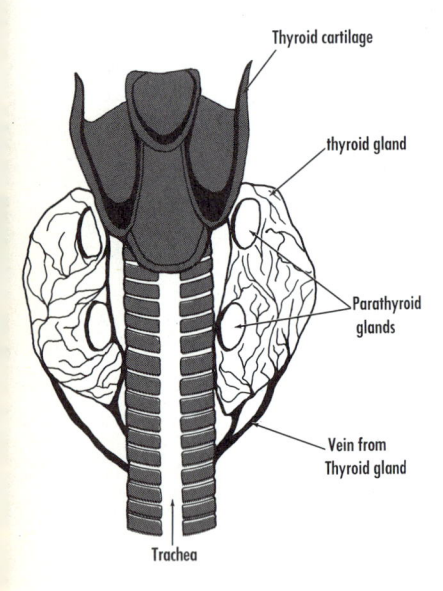

13-12 What are the functions of the thyroid and parathyroid glands?

kree-uhs) *gland* is located under the stomach. It produces hormones as well as digestive juices. These two functions of the gland are separate. The gland may be working normally in digestion and yet may not be producing hormones normally.

The thyroid tissue includes four other small glands called the *parathyroid glands*. The functions of these two glands in the body are quite separate. Biologists who first studied the thyroid gland had difficulties. They did not realize that the parathyroids are located in the same tissue.

Two endocrine glands are found in the brain. The *pituitary* (pih-TYOO-ih-*tay*-ree) *gland* is attached to the underside of the brain. If you put your finger in your mouth and touch the roof as far back as you can reach, you are pointing in the direction of this gland. The *pineal* (PIH-nee-uhl) *gland* is located in the middle of the brain.

The male and female reproductive glands are also endocrine glands. The main function of each in the body is to produce sperms and eggs for reproduction. They also produce hormones affecting the growth and development of the body, especially during the adolescent period.

Various hormones affect the body in different ways. The thyroid gland produces thyroxin (thye-RAHK-suhn) which controls the body's use of food and oxygen to release energy. If this gland is not functioning normally in a child, a condition, *cretinism* (KREE-tihn-*ihzm*), may result in which the child is physically stunted and mentally retarded.

The adrenal glands produce several hormones. The best known is *adrenalin* (uh-DREN-uh-lihn). This hormone is produced when you are angry or frightened. Have you ever seen a cat's fur rise when the cat is suddenly frightened? Adrenalin causes the muscles of the body to tighten in a cat. The tiny muscles attached to each hair also become tense, causing the hairs to rise. This hormone also causes the liver to release sugar into the blood. Sugar supplies the muscles with energy. This is a way of protecting the body by preparing you to fight or to run.

You may have heard of the disease called *diabetes* (dye-uh-BEET-eez). One form of diabetes is caused by lack of *insulin* (IHN-suh-luhn) normally produced in the pancreas. This hormone regulates the body's use and storage of sugar. If the pancreas does not produce enough insulin, digested sugar cannot be used or stored properly. As a result, the sugar level in the blood rises. The kidneys now remove some extra sugar from the blood. That is why a test of the urine to detect excess sugar is made when a person receives a complete physical examination.

When the parathyroid glands are removed from an animal, it develops *convulsions* (kuhn-VUHL-shuhnz). Convulsions are uncontrolled contractions of the muscles that eventually result in death. Scientists are able to show that *parathormone* (par-uh-THAWR-mone), the hormone produced by these glands, controls the use and storage of calcium and phosphorus in the body. When this hormone is not present, the level of calcium and phosphorus in the blood drop. This affects the control of all the muscles. The heart and breathing muscles also develop convulsions and, without aid, the animal dies.

The pituitary gland has a vital role in the endocrine system. Its hormones are needed to start other endocrine glands to secrete their hormones. *Growth hormones* produced by this gland also control the size of the body. Too much hormone produces a giant, and too little produces a midget (Fig. 13-13). Other hormones produced by the pituitary gland control the amount of water the kidneys remove from the blood and the contractions of muscles in different organs.

Male animals usually can be distinguished from females of the same species by their different outward appearances. These differences are *secondary sex characteristics*. For example, male birds usually have brighter colored feathers than females. Male deer have antlers but the females do not. Men have beards, but women's faces are smooth.

The hormones produced by the reproductive glands are responsible for the secondary sex characteristics in the body. The male glands produce *androgen* (AN-druh-juhn), and the female glands produce *estrogen* (ES-truh-juhn) as the main hormones. Sometimes the body does not produce the proper amounts of these hormones. Unfortunate secondary sex characteristics, like a bearded woman, may result. The table summarizes what we have learned about the endocrine system.

13-13 What are the causes of giantism and dwarfism?

THE ENDOCRINE SYSTEM

Name of gland	Hormones produced	Functions in the body
Thyroid	Thyroxin	Controls the use of food and oxygen
Adrenal	Adrenalin	Stimulates the body in case of emergency
Pancreas	Insulin	Regulates the use and storage of sugar
Parathyroid	Parathormone	Controls the use and storage of calcium and phosphorus
Pituitary	Growth hormones (others)	Controls growth and regulates action of other glands
Reproductive (male)	Androgen	Produces male secondary sex characteristics
Reproductive (female)	Estrogen	Produces female secondary sex characteristics

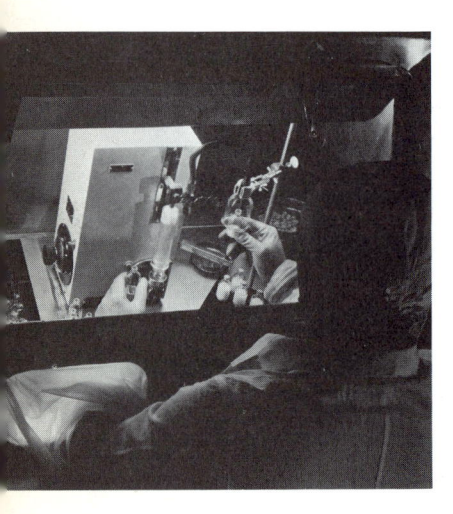

Scientists have discovered ways of obtaining hormones from certain animal glands for use as substitutes in the human body. Thus, if a person is suffering from diabetes, insulin obtained from sheep pancreas can be used to replace the insulin lacking in the body. Other hormones have been prepared in the laboratory (Fig. 13-14). Scientists constantly do experimental work with hormones to help discover how people can live longer and healthier lives.

13-14 How do scientists obtain artificial hormones?

REVIEW

1. What are the two main types of glands found in the human body?
2. How do the secretions produced by endocrine glands reach the parts of the body that they influence?
3. Name the main endocrine glands in the body.
4. What two endocrine glands are located in the neck region of a human being?
5. What two endocrine glands are located in the brain of a human being?

D/How Does the Body React?

The nervous system controls voluntary muscles. Voluntary actions are the actions of muscles that move when you want them to move. When you walk, run, climb a rope, or scratch your ear, you are performing voluntary actions. You control the actions of these muscles through your nervous system. Figure 13-15 helps you locate some of the main parts of the nervous system. Do the following activity, with care, on muscle coordination.

EXPLAIN

Try doing a stunt with your hands, like rubbing your abdomen with one hand and patting the top of your head with the other hand at the same time. Can you do this? Practice five minutes a day for a week. What does this show you about learning to control voluntary actions?

Some actions of voluntary muscles are involuntary and are called *reflex actions* (Fig. 13-16). Your nervous system controls these muscles without your awareness. For example, if something comes toward you suddenly, you blink your eyes. If a foreign object lodges in your nose or throat, you sneeze or cough to remove the irritating object. Most reflex actions are natural protections for the body. The next activity illustrates several reflex actions.

DISCOVER

To show different reflex actions, tickle the inside of a student's nostrils with a feather. What happens? Move your hand suddenly near the student's eyes. What happens? What kind of muscles are used in these reflex actions? What is the purpose of reflex actions in the body?

Other kinds of reflex actions are learned. We learn to swim, to write, and to play a musical instrument. At first, you have to practice each movement carefully. As learning progresses, the actions become automatic, and the conscious part of the

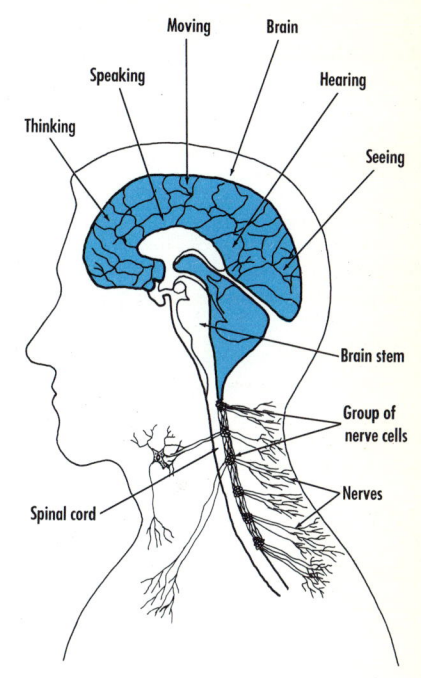

13-15 Locate the control centers in the brain.

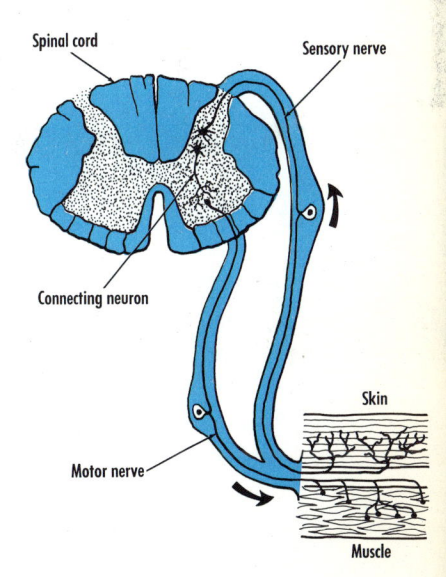

13-16 The path of a reflex action. Why are such actions important?

brain no longer has to think of each action. These learned activities become acquired reflexes or *habits.*

The nervous system controls the involuntary muscles. The important internal organs are composed of involuntary muscles which are controlled by certain parts of the brain. However, involuntary muscles are controlled without any conscious thinking on your part. The beating of the heart, the action of the breathing muscles, and the movement of the digestive organs are all examples. Just think of what it would be like if you had to consciously control all the actions of the involuntary muscles.

The nervous system controls the actions of glands. All the actions of the body are influenced by muscles, nerves, or glands. Did your eyes ever "water" when something got into them? *Tear glands,* located in each corner of the eyes, produce tears. This fluid washes over the surface of eyes. Why do tears form when you cry? The nervous system acts on these glands when you are emotionally upset.

Glands in the mouth and nose produce a thick, lubricating substance, called *mucus.* It coats the linings of the mouth and nose, protecting the tissues. We have already learned about the action of the digestive and endocrine glands.

Nerve fibers carry impulses to and from the brain. Whenever you see, hear, or feel anything, the brain receives a message from one or more of the *sense organs.* The main groups of sense organs in the body are those in the eyes, the ears, the skin, the nose, and the mouth. Any sensation in any of these sense organs produces *nerve impulses* in the nerves connecting the sense organs with the brain. Nerve fibers that are carrying impulses from the sense organs to the brain are known as *sensory nerves.*

Nerve fibers also carry impulses from the brain and spinal cord to muscles and glands. These are called *motor nerves* because the impulses they carry produce certain actions of the body. If a sensory nerve is injured, we lose all sense of feeling in that part of the body. If a motor nerve is injured, we lose the ability to control that part of the body. Compare the sense of sight of one of your classmates with his other senses in the next activity.

IDENTIFY

Blindfold a student and hand him an object he hasn't seen. Let him examine the object in any way, except by sight. Hide the object and remove the blindfold. Have the pupil make a drawing of the object and describe it to the class. Bring out the object and let him compare it with his description. Were the other senses as accurate as sight?

Scientists have developed special instruments to measure nerve impulses. Using a tiny probe, 1/25,000 of an inch in diameter, they can measure the *electrochemical current* produced in a single nerve cell! These impulses travel about 300 miles an hour. If you are driving an automobile and another automobile stopped suddenly in front of you, how long would it take you to stop? It takes time for the impulse to travel from your eyes to the brain warning the brain of the danger. It takes time, too, for the brain to send an impulse to the muscles of your leg and foot to apply the brakes. If you are driving fast, there may not be time for all this to happen, and you may have an accident.

Sense organs are affected by conditions around the body. When stimulated, the nerves from the sense organs carry impulses to the brain. For example, light, striking the eye, activates the nerves in the sense organs in the eye, sending impulses to the brain. When the brain receives these impulses, it interprets them as something you see. In the same way, sound waves picked up by the ear send impulses from the ear to the brain. Nerves from the sense organs of touch, taste, and smell, upon stimulation, also carry impulses to the brain.

Since all nerve impulses reaching the brain are electrochemical reactions, why don't you smell with your ears or hear with your eyes? The nerves from the different sense organs connect with different parts of the outer layer of the brain. An impulse coming in to that part of the brain is automatically interpreted as a sensation from that sense organ. For example, a sharp blow on the back of the head will cause you to see "stars." Actually, the eyes have not seen anything. However, the area connecting the nerves from the eyes is at the back of

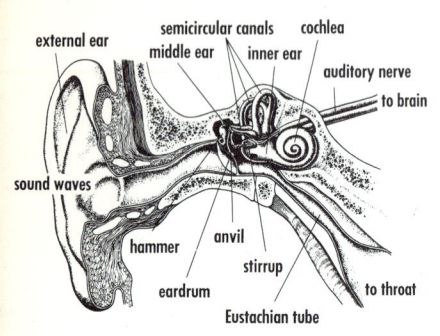

390 / UNIT 4   MAN EXPLORES THE HUMAN BODY

13-17 Locate the three main parts of the ear.

the brain. Hence, when this area is affected by a blow, the brain interprets this as an impulse from the eyes. You "see" flashes of light.

Sound waves must be present in order for us to hear. You may have learned that vibrating bodies set air in motion, creating sound waves. Unless there is a medium to carry the waves, a person cannot hear.

The *outer ear,* shaped like a funnel, directs the sound waves to the *eardrum,* a membrane which vibrates "in step" with the air vibrations. The eardrum separates the outer ear from the *middle ear.* As the membrane vibrates, a chain of three tiny bones, extending across the middle ear, also vibrate. These bones, the *hammer, anvil,* and *stirrup,* transmit vibrations to the *inner ear.*

The innermost section of the ear is composed of a snail-like spiral passage called the *cochlea* (KAHK-lee-uh). The cochlea is filled with fluid and lined with nerve fibers. As the vibrations from the bones of the middle ear reach the cochlea, its fluid vibrates. This causes the hairlike nerve fibers to be stimulated. These impulses are transmitted by the nerve fibers to the auditory nerve. Finally, they reach the part of the brain that interprets hearing (Fig. 13-17).

Anything that interferes with the vibrations in the ears causes a loss of hearing. For example, a punctured eardrum that does not vibrate as it should, causes a person to be partly deaf. Calcium deposits on the three little bones, attached to the drum, prevent these bones from vibrating. As a result, the person does not hear well. Any injury to the auditory nerve, preventing impulses from reaching the brain, results in deafness. A person should have his hearing checked from time to time (Fig. 13-18). You can learn more about sound with a classmate in the next activity.

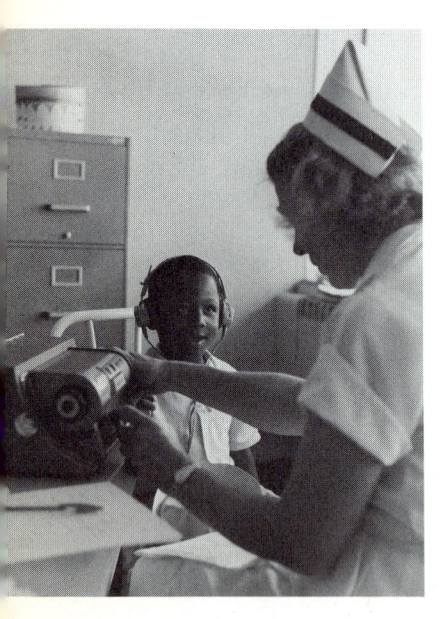

EXPLAIN

Blindfold a student. Have him sit in the center of the room with the rest of the class seated quietly around the walls of the room. Have him cover one ear. Move a loud-ticking alarm

13-18 Why should hearing be checked periodically?

CHAPTER 13 GROWTH OF THE BODY / 391

clock around the room. Can he determine the exact direction the clock is from his ear? Repeat, but this time keep both ears uncovered. Can a person determine the direction of sound better with one ear or with both ears? Can you explain what is the reason for this?

Seeing is a response to light waves. As you probably know, the human eye may be compared to a camera. Both are sensitive to light waves. The eye, however, is far better than any camera. It takes a picture, sends an impulse to the brain, erases the picture, and is ready for the next one in about one-tenth of a second. The control of the amount of light entering the eye, and the focusing of the lens are reflex actions. Figure 13-19 shows the structure of the eye.

We see when light waves enter the eye through an opening, called the *pupil.* Its size is regulated by the *iris* (EYE-ruhs), the colored part of the eye. A *lens* lies behind the pupil opening of the iris. The contraction of the muscles, supporting the lens, changes the shape of the lens. In this way, varying light waves are focused on the surface of the *retina* (RET-uh-nuh). The layers of cells making up the retina are stimulated by light waves. This stimulation sends impulses to the brain through the *optic nerve.* The inside of the eyeball is filled with a jelly-like liquid called *vitreous* (VIH-tree-uhs) *humor.*

What are some common defects of the eye that need correction by eye glasses? Slight differences in the shape of the eyeball might result in *nearsightedness* or *farsightedness.* As you can see by looking at Fig. 13-20, in the nearsighted eye, the eyeball is just a tiny bit too *long.* The defect causes the image to focus in front of the retina. Glasses, consisting of concave lenses, correct this condition. In the farsighted eye, the eyeball is just a little too *short,* and the image focuses behind the retina. Convex lenses are used to correct this condition. A third common defect of the eye is *astigmatism* (uh-STIHG-muh-*tihz*-uhm). It is due to irregularities in the shape of the lens or *cornea* (KAWR-nee-uh), the outer covering of the eye. Glasses which have been specially ground are able to correct this condition.

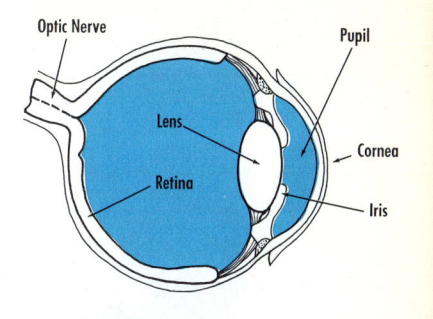

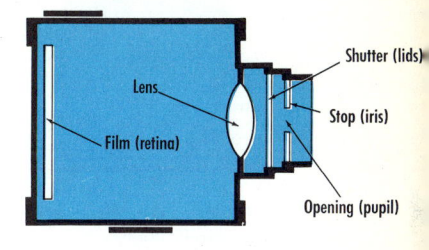

13-19 In what ways is our eye like a camera?

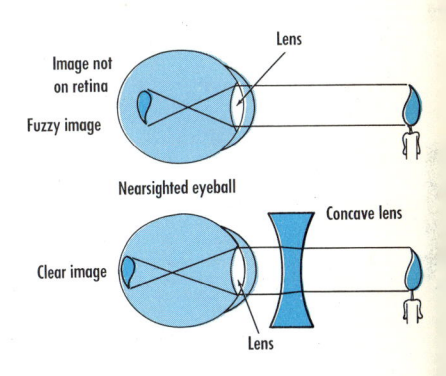

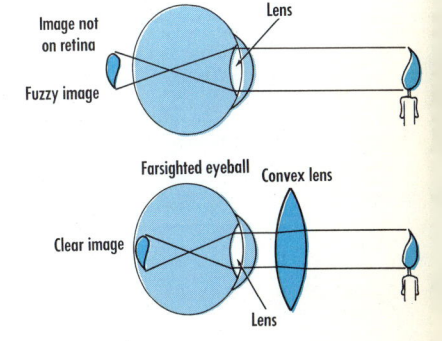

13-20 How can eyeglasses correct defects in sight?

392 / UNIT 4 MAN EXPLORES THE HUMAN BODY

Smelling and tasting are chemical responses. The simple act of smelling is an electrochemical process that takes place in the *olfactory* (ahl-FAK-to-ree) *area*. The process is activated when molecules are given off into the air by a substance like perfume or ammonia. The odor reaches the nerve endings of smell which are located in the moist lining of the nose.

The sense of smell is one of the most sensitive. A chemist with the finest instruments can detect in air a gas that is diluted one part in a million. The human nose can detect some odors present in the air as one part in a billion. This is 1000 times more sensitive than the best scientific instruments. Yet, the sense organs for smell in each nostril cover a space the size of a postage stamp! (Fig. 13-21). Find out the relationship between taste and smell in the following activity.

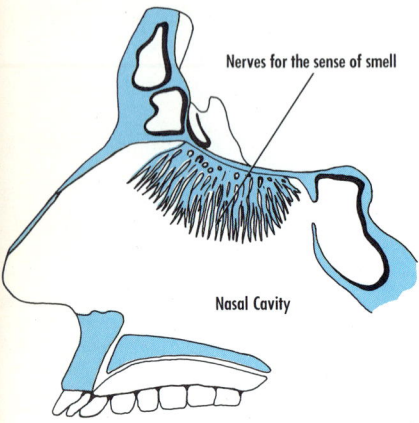

13-21 Where are the sense organs for smell located?

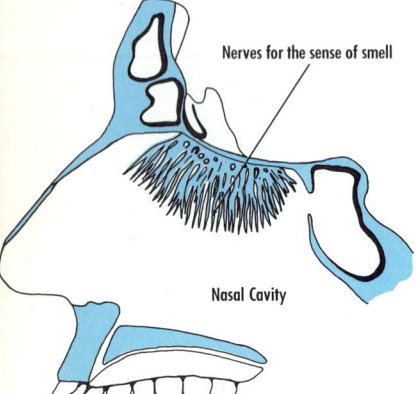

DISCOVER

Blindfold a student and close his nostrils carefully with a clothespin. Have him taste small pieces of apple, carrot, onion, potato, etc. How many can he identify? Now remove the clothespin and repeat the test. Explain how the sense of smell is used in tasting foods.

Recently, scientists developed a mechanical "sniffer." It is almost as sensitive as the nose in identifying certain odors. Most smells and tastes are results of reactions to different chemical solutions and gases. Therefore, a mechanical sniffer is useful in checking if every batch of a substance, like perfume or coffee, has the same composition. This is a complicated job. In coffee, for example, over 50 odors have been identified. The correct combination of these must be present to give the coffee its best flavor.

The sense of taste is located in the *taste buds* in the tongue (Fig. 13-22). These lie along the front, sides, and back of the tongue. As foods are mixed with saliva, the chemical reaction stimulates nerve endings in the taste buds, sending impulses to the brain. The sense of taste in man can distinguish sweet, sour, salty, and bitter. Many of the things we taste are really

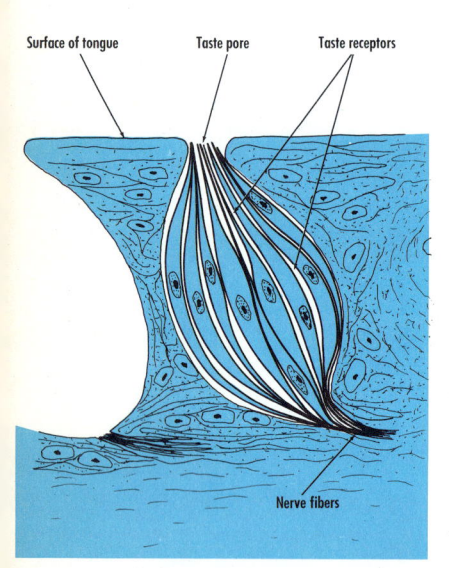

13-22 In which areas of the tongue are taste buds located?

the result of both the sense of taste and of smell. Try to determine the taste areas of the tongue by doing the activity below.

DISCOVER

Wipe a student's tongue dry with a clean towel. Using a medicine dropper, place drops of sugar solution on different parts of the tongue. Can you taste sugar better on some parts of the tongue than on others? Wipe the tongue dry. Repeat the test with salt solution and lemon juice. Make a drawing of the tongue and label the parts most sensitive to these three tastes.

REVIEW

1. Give some examples of natural reflex actions.
2. What are some involuntary muscle activities controlled by the nervous system?
3. In what direction are nerve impulses carried by: (*a*) sensory nerves, (*b*) motor nerves?
4. What type of stimulation produces impulses in the auditory nerves? in the optic nerves?
5. Name three common defects of the eye.

THINKING WITH SCIENCE

A. *On a separate sheet of paper, write the numbers 1 to 15. Some of the following statements are true and some are false. Rewrite the statements, changing the terms in italics if necessary, to make them all true.* **Do not write in your book.**

1. The most rapid growth of the body takes place between *three and four* years of age.

2. Glands that secrete substances directly into the blood are known as *duct glands*.

3. An involuntary movement of voluntary muscles to protect the body is called a *reflex action*.

4. The appearance and behavior of all animals in a group are known as *individual* characteristics.

5. A trait that is found in parents and inherited more often than other traits is a *recessive* trait.

6. *Sense organs* are the special structures in the body that detect sensations around us.

7. The chemical substances produced by the endocrine glands are called *hormones*.

8. A new trait that appears in the offspring and becomes a family characteristic is known as a *mutation*.

9. The beating of the heart and the action of the digestive organs are examples of *voluntary* muscle actions.

10. *Sensory nerves* carry impulses from the brain and spinal cord to muscles and glands.

11. An animal or plant that lacks all natural coloring is known as an *albino*.

CHAPTER 13 GROWTH OF THE BODY / 395

12. Light waves enter the eye and stimulate sense organs in the *cornea*.

13. *Insulin* controls the way the body uses and stores sugar.

14. Sound waves strike the *eardrum* inside the ear, causing it to vibrate.

15. The sense organs involved with the sense of taste are called the *vitreous humor*.

B. *Write the answers to the following in your notebook. Be sure to use complete sentences with correct spelling and grammar.*

1. Make a chart showing the rate of growth in height and weight up to the age of 18 years for an average boy and an average girl.

2. Describe some of the body changes that take place during adolescence.

3. Compare muscle coordination at each of the four main stages of growth in the body.

4. Make a list of species characteristics and individual characteristics which you believe you have inherited.

5. Describe the action of the pancreas as: (*a*) a duct gland, (*b*) an endocrine gland.

6. Name the main effects in the body of the hormones produced by the: (*a*) thyroid gland, (*b*) adrenal glands, (*c*) reproductive glands.

7. What is the difference between a natural and an acquired reflex action?

8. What happens to a particular part of the body when: (*a*) a sensory nerve is injured, (*b*) a motor nerve is injured?

9. Describe some of the common defects in seeing and tell how they can be corrected.

10. What is the relationship between the sense of taste and the sense of smell in tasting the flavor of a food?

RESEARCH IN SCIENCE

1. If you can observe a baby over a period of time, determine when he is able to coordinate different actions of the body.

2. Make a chart comparing some of your traits with those found in your parents, grandparents, or other relatives.

3. If you know a pair of identical twins, make a list of their traits that are the same and of those that are different. Explain any differences.

4. Make a report on how hormones are prepared by drug companies.

5. Make a display of empty bottles that contain hormone preparations and explain the use of each one in the body.

6. Use a loud-ticking watch to test the hearing of various members in the class; make a chart showing the results.

chapter 14

Health
of the Body

A/What Causes Disease?

Many diseases are caused by microscopic plants and animals. In your studies of living organisms, you learned about various kinds of microorganisms. It would be helpful to review your understanding of these organisms to better investigate their harmful effects on our health.

The single-celled plants, called *bacteria*, are tiny and can be seen only under the high power of a light microscope. One group of bacteria appears as little spheres, or balls, called *coccus* (KAH-kuhs) bacteria. Pneumonia and blood poisoning are caused by forms of these bacteria. A second group of bacteria looks like little short rods, the *bacillus* (buh-SIHL-uhs) bacteria. Tuberculosis and diphtheria are caused by this kind. The third group of bacteria is shaped somewhat like tiny corkscrews, the *spirillum* (spye-RIH-luhm) bacteria, as shown in Fig. 14-1 (page 398). Cholera is a disease caused by this type.

Bacteria perform life functions similar to other living cells. However, unlike the green plants, most bacteria cannot make their own food. They must use other plant or animal substances

398 / MAN EXPLORES THE HUMAN BODY

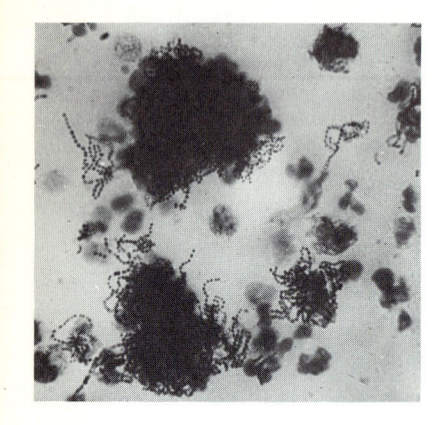

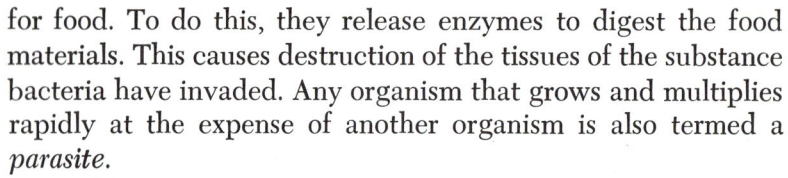

for food. To do this, they release enzymes to digest the food materials. This causes destruction of the tissues of the substance bacteria have invaded. Any organism that grows and multiplies rapidly at the expense of another organism is also termed a *parasite*.

In addition to producing enzymes to digest the food, bacteria also give off waste products, called *toxins*. Toxins are harmful substances which act as poisons. Toxins interfere with normal body functions. When this occurs, you have the disease which these bacteria cause. The following activity will help illustrate the differences in appearance of the three forms of bacteria you have just learned.

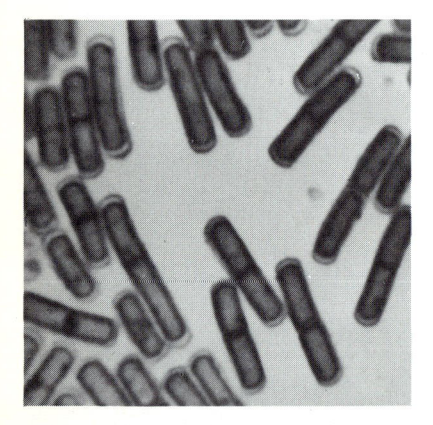

EXPLORE

Examine prepared slides showing the three types of bacteria with the high power of the microprojector or microscope. In addition, study photographic color slides showing these forms of bacteria. Compare the size and shape of the three types of bacteria discussed previously.

Describe the shape of each form of bacteria that you see. Which are the largest bacteria? Which are the smallest bacteria? Name some of the diseases in man that are caused by each kind of bacteria.

Viruses are smaller than bacteria. If you have ever had a cold, you were probably suffering from a *virus* disease. Viruses are the smallest living organisms. Most viruses are complex protein compounds. They appear to be nonliving; however, once they are in a living cell, viruses become active. The viruses now use the living substances of the cell to grow and multiply, causing disease. Viruses cannot be seen with ordinary microscopes. Figure 14-2 shows the appearance of a virus as seen through an *electron microscope*. An electron microscope can magnify objects hundreds of thousands of times larger than an ordinary microscope. Viruses are difficult to study because only a few can be grown outside of living cells. Some diseases caused by viruses are measles, smallpox, polio, and colds. You can grow tobacco mosaic virus on the leaves of tobacco plants in the next activity.

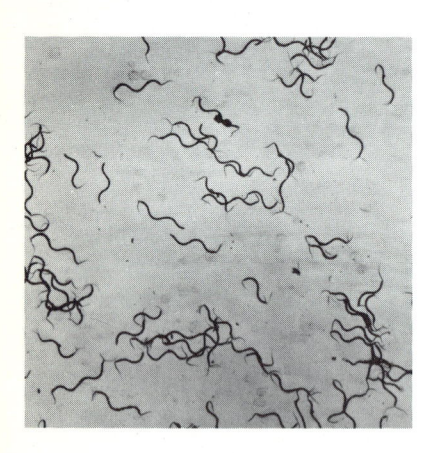

14-1 Identify these three types of bacteria.

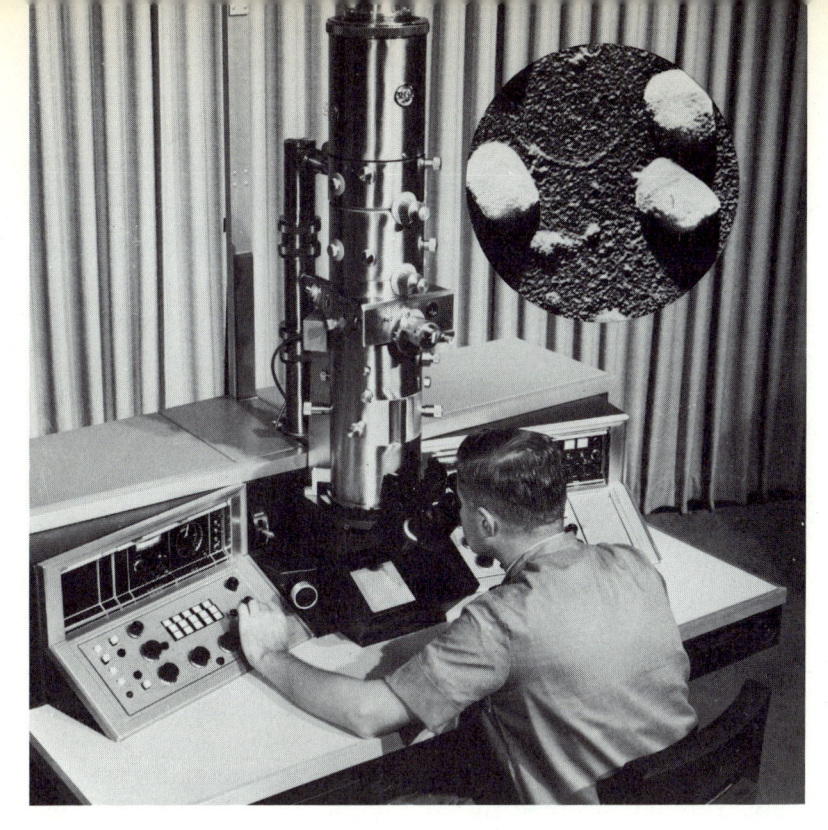

14-2 Why are electron microscopes used to study viruses?

EXPERIMENT

Tobacco mosaic virus can be grown on the leaves of a tobacco plant in the laboratory. Obtain some young tobacco plants and pot them separately. (These can be raised in a few weeks from tobacco seeds). Mash a small amount of fresh pipe tobacco with two thin pieces of wood or tongue depressors until the wood is wet with tobacco juice. (Most commercial forms of tobacco are infected with tobacco mosaic virus.) Discard the loose tobacco. Rub the leaf hard enough to bruise the tissue with the two pieces of wood, as shown. Using clean pieces of wood, treat a leaf of another plant the same way. What happens to the infected plant after a few days?

Infecting plant with tobacco mosaic virus

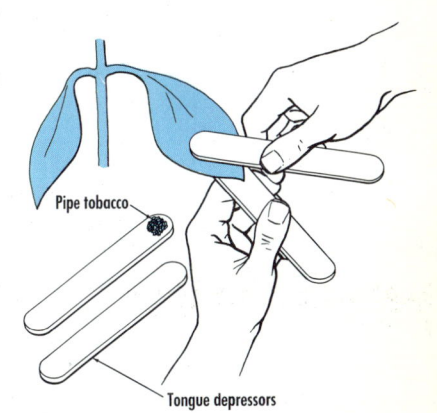

Pipe tobacco

Tongue depressors

Some fungi and protozoans are parasites. All disease-causing organisms are parasites in the body, but we usually apply the term only to those *fungi* and those *protozoans* which cause diseases in the body. Diseases caused by fungi are seldom fatal, but they are irritating and difficult to cure. Ringworm and athlete's foot are two skin diseases caused by fungi.

Protozoans are one-celled animals; many are quite harmless. However, certain protozoans cause severe diseases when they enter the body. Malaria and African sleeping sickness are two protozoan-caused diseases infecting hundreds of thousands of people every year. Many disease-causing protozoans are introduced into the body by the bites of insects.

Infectious diseases are caused by microorganisms. Bacteria, viruses, molds, and protozoans that cause diseases in the body are microscopic in size. Hence, these forms of life are called *microorganisms*. Diseases caused by some microorganisms are called *infectious* (ihn-FEK-shuhs) *diseases* because they usually spread from person to person or from animals to persons. Some infectious diseases are spread by water and food taken into the body or by the air we breathe. We shall learn more about the way diseases are spread in the next section.

Diseases may be caused by improper functioning of the body. You have learned that the normal growth and development of the body is controlled, in part, by endocrine glands. If these glands produce too little or too much of their hormones, *glandular diseases* may result.

Some diseases of the heart, blood vessels, and kidneys are known as *organic diseases*. Some condition in these organs prevents them from functioning normally. Severe damage to the body may result. Organic diseases cannot be treated like infectious diseases because microorganisms do not cause them.

Another common organic disease is *cancer,* a wild multiplication of a group of cells in the body. This results in destruction of normal body tissues and in the production of tissue poisons (Fig. 14-3). Unless cancerous growths are discovered early and treated by trained medical scientists, death is almost certain to follow.

Leukemia (lyoo-KEE-mee-uh) is a cancer of the blood. It is a disease in which the white blood cells multiply at a rapid rate. The red blood cells gradually decrease. Unfortunately, no permanent cure for cancer through the use of drugs has yet been found. However, scientists work constantly to find out why cancer cells reproduce rapidly in the body and how to prevent and cure this disease.

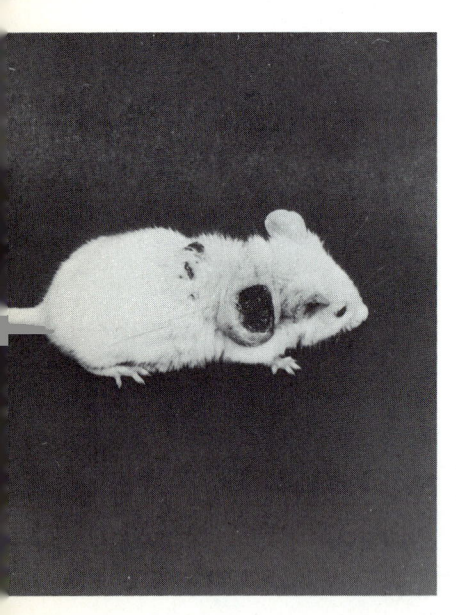

14-3 Locate the skin cancer on this rat.

Disease may be caused by radioactive materials. Radiations from radioactive substances also may cause disease. We have already learned that radiations affect the reproductive cells in living organisms, producing mutations. Radioactive materials, however, also affect the body directly.

Scientists, experimenting with radioactive substances, found that radiations burn the skin and cause the hair to fall out. They found that blood is damaged when a person is exposed to radiations because the ability of the bone marrow to form blood cells is destroyed. Without white blood cells, the body cannot fight infections. If large doses of radiation are received, the person dies.

Disease may be caused by allergies in the body. Do you know any of your friends who break out in a rash if they eat strawberries, or start to sneeze if they pet a cat? These people are sensitive to something that enters the body in the food they eat or the air they breathe. An *allergy* (AL-uhr-jee) is a disease caused by the body's unfavorable reaction to a foreign substance that is ordinarily harmless. Allergic reactions may take the form of skin irritations, sneezing and difficulty in breathing, reddening and watering of the eyes, digestive disturbances, and fever. Find out more about the causes of allergies as directed in the next activity.

RECORD

Look in health books and in encyclopedias to find out more about the causes of allergies. Make a chart showing the number of common substances to which different members of the class may possibly be allergic. Do you find that some members of the class are allergic to more than one substance?

The body may develop an allergy to almost any substance. Some people are allergic to certain foods. Others cannot wear wool clothes, and some are allergic to animal fur. Hay fever and asthma, caused by breathing certain plant pollens floating in the air, are two of the more common allergy diseases that afflict man (Fig. 14-4).

14-4 Why should the ragweed plant be destroyed?

Most allergies can be prevented by simply avoiding the offending substance. In some cases, relief has been obtained through the use of *antihistamines* (*an*-tih-HIHS-tuh-*meenz*). These are powerful drugs and should be taken only on a doctor's advice. In very severe cases, it may be necessary to treat the patient with a series of "shots," or *inoculations* (ihn-*ahk*-yoo-LAY-shuhnz) of the offending substance. Tiny doses of the substance are injected in the patient at intervals. The body slowly becomes accustomed to the substance, and the patient has relief.

MAIN CAUSES OF DISEASE

Type of Disease	Cause	Examples
Infectious diseases	Microorganisms	Pneumonia, polio, tetanus, tuberculosis
Glandular diseases	Incorrect amounts of certain hormones in body	Diabetes, cretinism
Organic diseases	Improper functioning of body	Heart disease, cancer
Allergic diseases	Reaction to foreign proteins, dust, animal hair	Hay fever, asthma
Radiation diseases	Exposure to radioactive materials	Skin injuries, bone cancers, leukemia

REVIEW

1. What are the three main forms of bacteria?
2. List some examples of diseases caused by: (*a*) bacteria; (*b*) viruses.
3. Name four general groups of disease-causing microorganisms.
4. Give two examples of organic disease.
5. Name the five main types of disease that can affect the body.

B/How Do Microorganisms Infect Us?

Disease-causing microorganisms may enter through the skin. Microorganisms are present around us all the time. Most are harmless to the healthy body, but some, as you have learned, produce infectious diseases. One way that microorganisms enter the body is through the skin on your hands. For example, "pink eye," is a disease that causes the eyes to become red. It may be contracted by using an infected towel or handling an object from a person with this disease. The microorganisms get on your hands and then into your eyes when you rub them. The way infection by contact takes place is shown in the activity below.

EXPLAIN

Obtain two apples, one of which is fresh and solid and one with a decay spot on it. Pass a needle through a flame to sterilize it. Stick the needle into the rotten part of the apple. Now stick the needle with some of the rotten material on it into the fresh apple. Discard the rotten apple and keep the other apple for several days. Did a rotten spot form where the needle was pushed in? What caused this to happen? What does this show you about the way in which an infection may be spread? How can infections be prevented?

Most skin diseases are spread by contact with infected materials. *Athlete's foot* and *ringworm* are two other diseases that can easily spread from one person to another by contact. To prevent such infections, you should not use towels, combs, drinking cups, or other articles which are used by other people.

Microorganisms may also enter the body through a cut or break in the skin. Even the smallest opening can admit millions of microorganisms into the blood. If some of these are disease-causers, an infection may result. One example of a disease caused by microorganisms entering this way is tetanus. This deadly disease causes convulsions of the muscles and may result in death (Fig. 14-5). Another disease is blood poisoning.

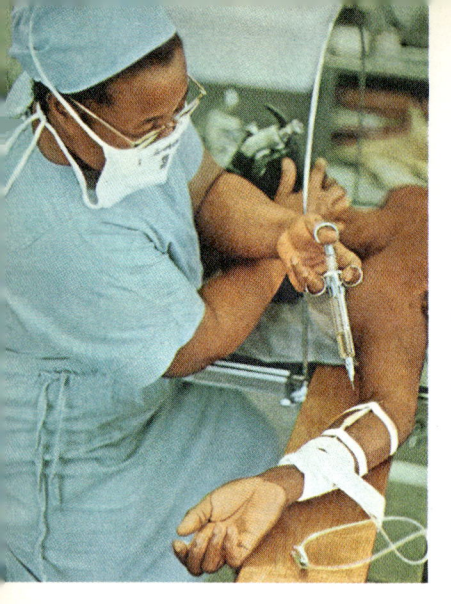

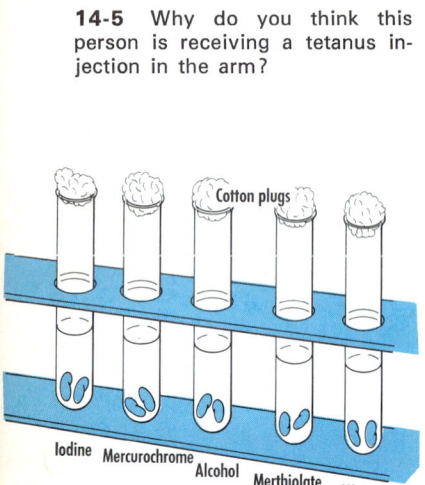

404 / MAN EXPLORES THE HUMAN BODY

The skin and mucous linings of our body are our most important defenses against serious diseases infecting the body by contact.

Any serious break in the skin should be treated by your doctor or at a hospital emergency room. Report any puncture wound made by a rusty nail immediately. These wounds may admit and seal in tetanus microorganisms.

Any small break in the skin should be treated promptly with one of the many *antiseptics* available. An antiseptic kills most microorganisms upon contact and helps to prevent infection. *Iodine, mercurochrome, alcohol, and merthiolate* are reliable antiseptics commonly found in the home. We can show how an antiseptic reduces the growth of bacteria in the following activity.

14-5 Why do you think this person is receiving a tetanus injection in the arm?

DISCOVER

Soak some beans in water for several hours. Half fill five test tubes with water. Drop two beans in each tube, as shown. Paste a label on each tube and mark the name of the antiseptic which will be added to it. To each of four test tubes, add about 10 milliliters of a different antiseptic. To the fifth tube, add the same amount of water. Why is this done to the fifth tube? Plug the tubes with cotton and leave them in a warm, dark place for several days. Smell each tube every day to find out which shows decay. Which antiseptics prevent decay the longest? What does an antiseptic do to bacteria that may grow in the tubes?

Scientists have better ways to study the growth and behavior of bacteria than that used in our experiment. To grow properly and form colonies that can be seen without a microscope, bacteria need a food supply, moisture, and a warm temperature. The first two conditions are supplied by preparing a *nutrient agar* solution for the bacteria to grow on. To learn how bacteria grow and spread, *Petri* (PEE-tree) *dishes*, containing an agar solution, may be prepared.

For convenience, many laboratories use commercially prepared nutrient agar. You can learn how to prepare an agar solution in the activity.

EXPLORE

Sterilize the Petri dishes in a pressure cooker at a temperature of about 150°C for 20 minutes. This temperature is reached when the pressure is 15 pounds per square inch. Dissolve a bouillon cube in 500 milliliters of hot water. Add 15 grams of commercial agar. Heat and stir until all the agar is dissolved. Put a piece of clean cloth over a funnel and filter the solution into a clean beaker. Sterilize again in the pressure cooker for 20 minutes at 150°C. While the solution is still hot, pour a little in the bottom half of each Petri dish. Put the cover on each dish immediately. Rotate the dish gently to spread the agar solution as a thin film. This amount of solution should be enough to prepare 25 to 30 Petri dishes.

Why were the Petri dishes sterilized before using? Why was the agar solution sterilized after filtering? Would you expect to find any live bacteria in the prepared Petri dishes? Can you explain the purpose of the agar film on the bottom of each dish? Why were the Petri dishes covered immediately, after they were prepared?

Microorganisms are present in the air. You have seen dust particles floating in the air when a strong beam of light shines through a window. You know, too, that air contains tiny droplets of water vapor. Microorganisms of all kinds are found on dust and in water. You can show this easily in the activity given below.

EXAMINE

Expose a prepared Petri dish to bacteria in the air by removing the cover for 15 minutes. You can try different places with a number of dishes, like the classroom, in the halls at passing time, on the playground, etc. Replace the cover and paste a label on the dish to mark the place where it was exposed. Use an unexposed dish as a control. Leave the dishes in a warm, dark place for several days. Do you see little colored spots forming on the agar similar to those in the photograph? How many can you count in each dish? Each spot is a colony usually started by one microorganism. Where did you find the greatest number of microorganisms in the air?

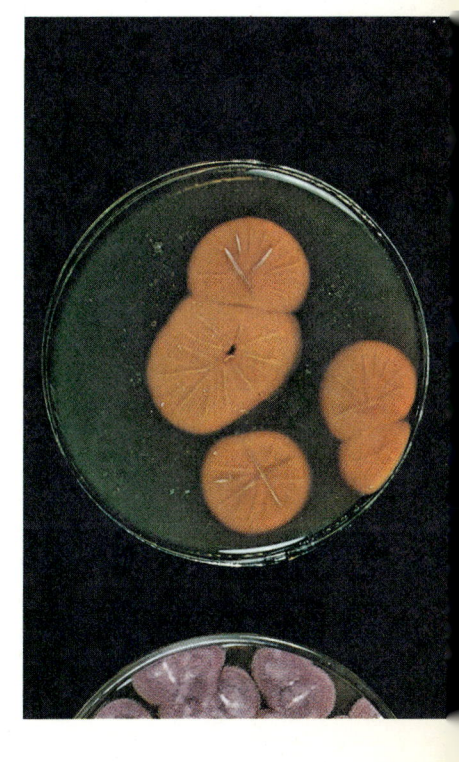

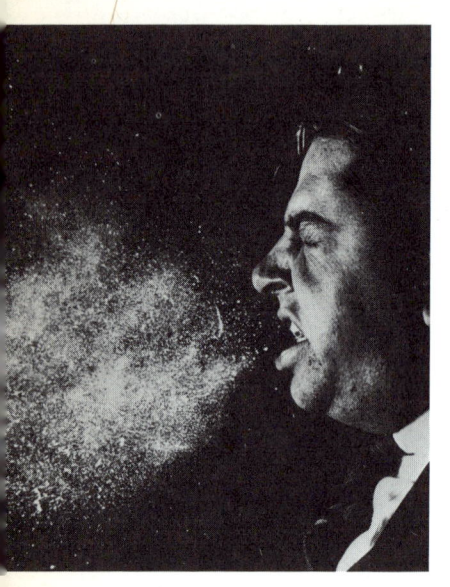

406 / MAN EXPLORES THE HUMAN BODY

Many common diseases are caused by air-borne microorganisms. Colds, measles, mumps, and chickenpox are a few examples. How do these microorganisms get into the air? A person with a cold may cough or sneeze without covering his mouth and nose with a handkerchief. As a result, he sprays millions of microorganisms into the air (Fig. 14-6). If you happen to breathe this air, you may become infected.

Water and food may carry microorganisms. The water supply for most cities comes from large rivers or lakes open to the air. If the water has been used before, billions of microorganisms may be in the water. For this reason, water is filtered and treated with chlorine to kill disease-causing microorganisms. You must remember that water taken from a stream for drinking purposes may contain disease-causing microorganisms, even though the water may look clean. The water should be boiled or chemically treated to make it safe for use. A dangerous disease that results from drinking unsafe water is typhoid fever.

Milk, fresh from a cow, may contain organisms causing tuberculosis. Most milk used in the home has been *pasteurized* (PAS-chuhr-eyezd) to destroy bacteria. You can learn how milk is kept from souring in the next activity.

14-6 Why is it proper to cover your nose before sneezing?

EXPLORE

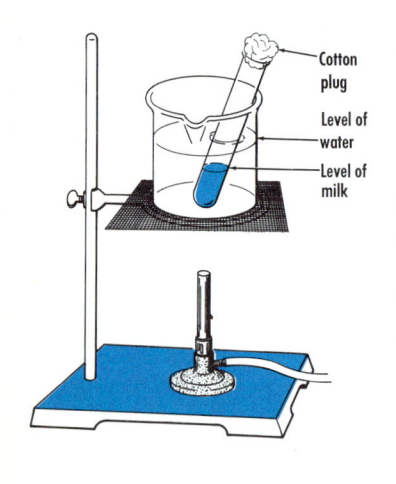

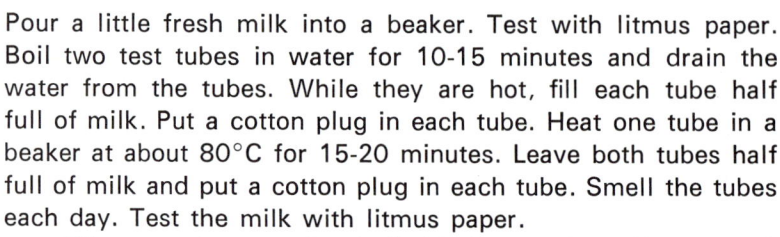

Pour a little fresh milk into a beaker. Test with litmus paper. Boil two test tubes in water for 10-15 minutes and drain the water from the tubes. While they are hot, fill each tube half full of milk. Put a cotton plug in each tube. Heat one tube in a beaker at about 80°C for 15-20 minutes. Leave both tubes half full of milk and put a cotton plug in each tube. Smell the tubes each day. Test the milk with litmus paper.

Why were the test tubes boiled before putting milk in them? What is the purpose of the cotton plugs? What did the litmus paper test show with fresh untreated milk? Describe the odor of each tube after several days. What does a strong odor indicate? What was the result of testing the milk in each tube with litmus paper after several days? What caused the milk in the unheated tube to sour more quickly?

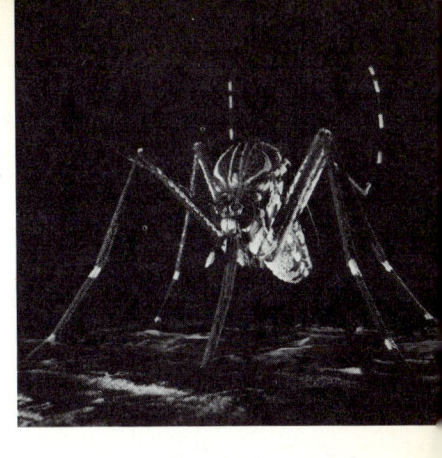

Since microorganisms are all around us, the food we eat and drink should be covered until we use it. This is important because dust in the air, and the careless handling of food by people, may introduce dangerous microorganisms. Modern food markets and restaurants take precautions to prevent this as much as possible. Uncooked food should still be washed carefully. Of course, all food that we eat should be served in a sanitary manner.

Insects may infect the body with microorganisms. Malaria and African sleeping sickness are caused by protozoans that enter the bloodstream. These diseases are transferred from one person to the next through an insect. When the infected insect bites a victim, the microorganisms are introduced into the blood. The *anopheles* (uh-NAHF-uh-leez) *mosquito* carries malaria, and *tsetse* (TSEE-TSEE) *flies* carry sleeping sickness. Other insects that transmit diseases are the *Aedes* (ay-EE-deez) *mosquito* (Fig. 14-7) which carries yellow fever, *rat fleas* which spread bubonic plague, and *body lice* which spread typhus fever.

The common *housefly* is one of the worst disease carriers. The fly carries many dangerous microorganisms on its hairy body and legs (Figure 14-8).

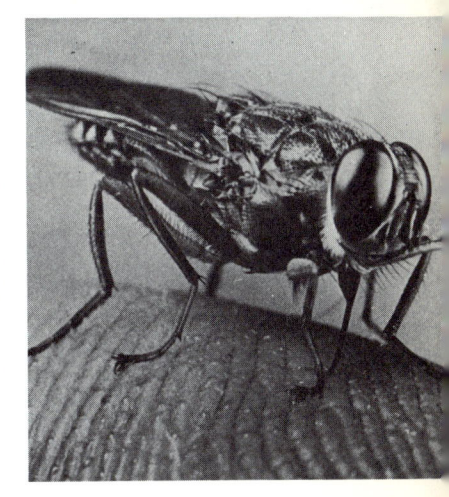

Flies feed on garbage and visit places where there are waste materials containing large quantities of bacteria (Fig. 14-9). Flies will fly directly from these sources to your table. As the fly walks over food, millions of bacteria drop off on the food. If some of these are disease-causing microorganisms, you may become ill. Flies are responsible for causing about twenty different diseases in humans and animals. Do the following activity to show how flies spread microorganisms.

14-7 What disease is carried by the Aedes mosquito?

14-8 How do flies transmit disease?

14-9 How can breeding places for flies be eliminated?

OBSERVE

To show how a fly spreads microorganisms, use a prepared Petri dish. Catch a live fly and slip it under the cover of the dish. **(Wash your hands with soap and water).** Let it walk around on the agar. Remove the cover to let the fly out, and cover the dish again. Leave the dish in a warm, dark place for several days.

What forms on the surface of the agar? Describe what you see. Where did these microorganisms come from? What does this show you about the way flies carry disease?

Microorganisms growing in the body cause disease. To cause disease, microorganisms must invade the body. Once inside the body, the microorganisms find excellent conditions for growth. There is enough food and moisture present, and the body temperature is just right! As the microorganisms grow and multiply rapidly in the body, they destroy cells. Toxins also are given off. As the body fights back at the invaders we notice *symptoms* of the disease. Symptoms reveal the presence of disease-causing microorganisms as well as the reaction of the body trying to overcome the infection. Some common symptoms of some diseases are fever, chills, a rash on the skin, or an upset stomach.

In the previous section, we divided diseases into five main types. Each disease has its own causes and symptoms, and each is different from other diseases.

Keeping free of disease is one of the body's most important functions because serious injury or death may result from infection. You may recover from a disease without treatment, but in many cases, treatment and drugs under a doctor's care can help greatly. When you recover from an infectious disease, the body has destroyed the disease-causing microorganisms and the harmful effects of their toxins.

REVIEW

1. Name three diseases of the skin contracted through direct contact.
2. How can microorganisms present in a break in the skin be destroyed?
3. Name three air-borne diseases.
4. Name two diseases that are able to enter the body from food or water.
5. List five insects that are known to transmit diseases from their bites.

C/What Are Our Natural Defenses?

The body has natural defenses against diseases. The best way of controlling infectious diseases is to keep the disease-causing microorganisms out of the body. The body is adapted to do this in several ways. The outer skin is a tough layer that prevents microorganisms from entering. Tears, saliva, and digestive juices are slightly antiseptic which kill many microorganisms upon contact. The cells making up the lining of the nose, throat, and air passages secrete a thick, sticky fluid, the *mucus*. Mucus traps and kills microorganisms we breathe in the air. Similarly, the wax in your ears and the hairs in your nose prevent many organisms from entering.

Suppose some disease-causing microorganisms pass this "first line" of defense and invade the body. *White blood cells* present in the blood throughout the body can pass through the walls of the capillaries into the tissues. They destroy many microorganisms by surrounding and ingesting them.

Another way in which the body fights microorganisms is by a rise in temperature. We say the body has a fever. It is a means of defense because microorganisms are weakened by higher temperatures. It is easier, then, for white cells to destroy them. Find out the conditions that promote the growth of microorganisms by doing the activity below.

DISCOVER

Use several prepared Petri dishes. Open each dish, press your fingers on the agar, and cover each dish immediately. Put one dish in bright sunlight, another in a warm, dark place, and a third in a refrigerator. Leave the dishes for several days. Then, examine them for growth of colonies. Which one showed the most growth? What are the best conditions for growth?

The body may also enclose certain, harmful bacteria with a growth of cells. Tuberculosis bacteria cause the body to react in this way (Fig. 14-10). Little masses of tissues, called

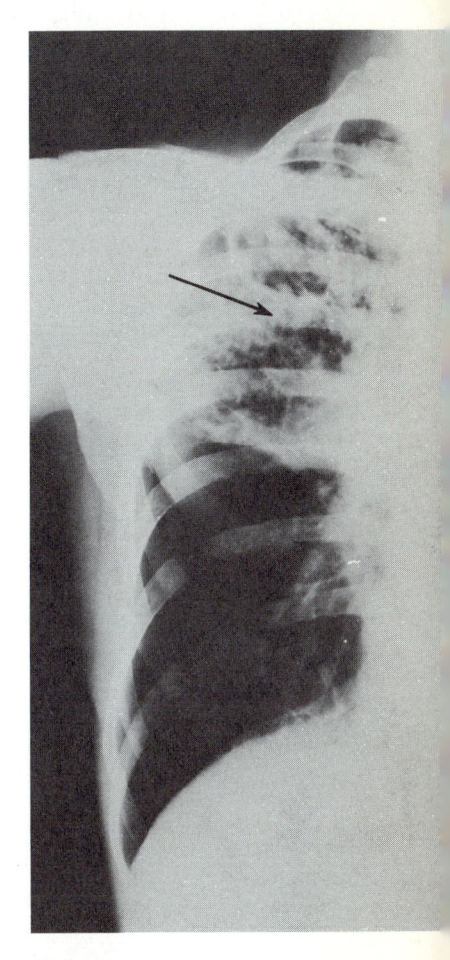

14-10 The light area (arrow) shows a tubercular infection. Why should a person have an annual Xray?

tubercles (TYOO-buhr-kuhlz), form in the lungs around the infection (Fig. 14-10). The bacteria are not destroyed, but they are kept from spreading and infecting other cells in the lungs. However, there is the danger that the bacteria will break out to start an active infection again. This is one reason why tuberculosis is a difficult disease to control.

Finally, chemical substances in the blood called *antibodies* destroy other dangerous microorganisms producing toxins. Certain antibodies called *antitoxins* neutralize toxins directly. Other antibodies called *lysins* dissolve bacteria. Still another group of antibodies cause certain bacteria to clump; they are known as *agglutinins* (uh-GLOO-tih-nihnz). Specific antibodies are produced by the body to fight specific infections. Thus, an antibody that is effective against typhoid fever does not have any effect on smallpox. Also, antibodies seem to be produced only in cases of bacterial and virus infections. Protozoans, molds, and some bacteria and viruses do not appear to influence the body to form antibodies.

The body can develop an immunity to some diseases. When the body produces antibodies, they usually remain in the blood for varying lengths of time after the disease has been overcome. This gives the body an *immunity* to that disease. For example, diseases like mumps, chickenpox, and smallpox produce antibodies in the person. He cannot contract the disease again for a number of years, and in some cases for life.

You can be made immune to some diseases without contracting the actual disease by means of a *vaccination*. Scientists have discovered that the body does not always know the difference between live, active microorganisms and weakened or dead ones. Therefore, disease-causing microorganisms are grown in the laboratory and then weakened or killed (Fig. 14-11). This material, known as a *vaccine* (vak-SEEN), is injected into the body. This stimulates the cells to produce antibodies, just as the body does in an actual infection. If active disease-causing microorganisms now enter, the antibodies are present to protect the body.

One of the newer vaccines developed by scientists to control diseases is the *polio vaccine*. Two forms of this vaccine are now

14-11 The influenza virus is grown in chick embryos and then is weakened or killed. Why?

CHAPTER 14 HEALTH OF THE BODY / 411

available. The Salk vaccine, developed by *Dr. Jonas Salk (1914–)* in 1953, uses killed virus and is injected into the body. The Sabin vaccine, developed by *Dr. Albert Sabin (1906–)* and released for general use in this country in 1961, uses weakened virus. It is taken by mouth.

There are several tests a doctor can use to find out if you are immune to certain diseases. For example, the *tuberculin test* shows whether tuberculosis bacteria are present in the body. If you have ever had the disease or now have an active infection when the test is made, a positive reaction is produced (Fig. 14-12). A positive tuberculin test is followed by a chest X ray to find out if there is an active case of tuberculosis. Since this disease can be arrested if discovered early, the tuberculin test saves many lives.

Suppose, in spite of proper precautions, you contract a serious disease. The body attempts to produce antibodies to overcome the infection, but the infection is too serious. The physician helps you by injecting *serum* (SEE-ruhm) containing antibodies previously produced in the body of another animal. For example, tetanus antitoxin and diphtheria antitoxin can both be produced in the blood of a horse purposely infected with the weakened bacteria. These antitoxins, introduced into

14-12 Tuberculin is injected under the skin (left). Describe the positive reaction to the test (right).

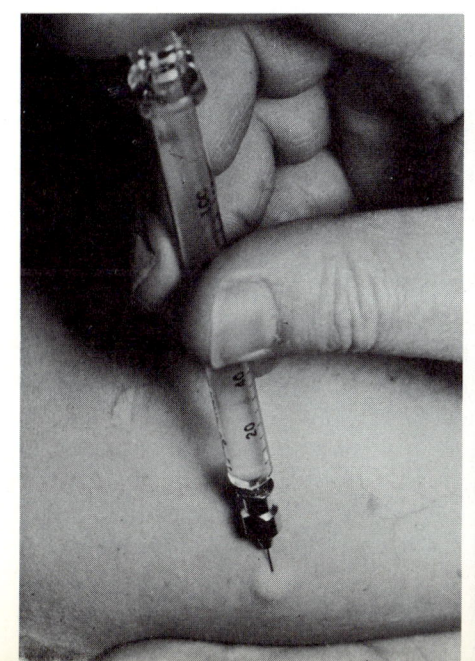

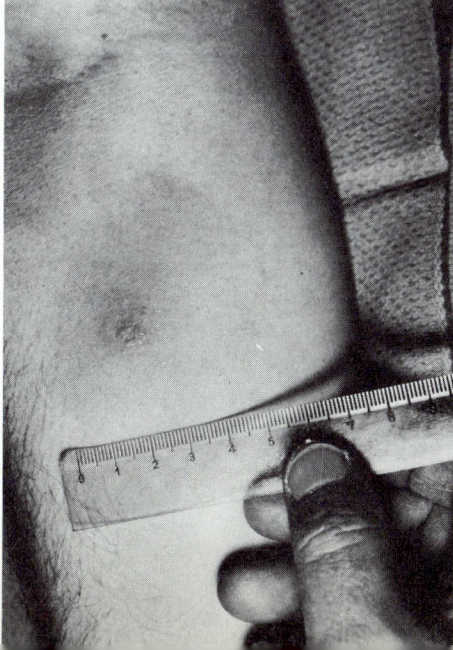

412 / MAN EXPLORES THE HUMAN BODY

the human body, act to overcome the effects of the toxins produced by the same bacteria in your body.

New drugs have been developed to treat diseases. In many cases, a severe infection cannot be overcome by the body alone, and drugs are used. One early group of drugs, the *sulfa drugs,* is effective against many disease-causing microorganisms, especially those that cause infections of the respiratory system. However, sulfa drugs should be used only under a physician's direction because they produce harmful side effects.

A group of newer drugs are the *antibiotics* (*an*-tih-bye-AH-tihks), substances made by microorganisms. Antibiotics are used against a variety of disease-producing agents. Since the discovery of *penicillin* (pen-ih-SIHL-ihn) in 1929 by *Sir Alexander Fleming (1881–1955),* a British scientist, many other antibiotics have been discovered (Fig. 14-13). Most of these drugs are produced during the growth of molds. Examine such a mold in the activity below.

14-13 What did Sir Alexander Fleming discover?

DESCRIBE

Place a ripe orange in a jar containing a moist piece of toweling paper. Cover the jar with a clean lid to keep the paper moist. In a week or so you will see a mold forming on the orange. Place a little of the mold on a clean slide. Examine it with a magnifier. Describe the appearance of the mold. What antibiotic drug is prepared from a similar kind of mold?

A number of antibiotics are available today to fight infections. One of these is *streptomycin* (*strep*-to-MYE-sihn), used in combating respiratory diseases and tuberculosis. Others are *aureomycin* (*aw*-ree-o-MYE-sihn) and *chloromycetin* (*klaw*-ro-mye-SEE-tihn), both effective against virus pneumonia. Newer antibiotics include *terramycin* (*ter*-uh-MYE-sihn), *erythromycin* (uh-*rihth*-ro-MYE-sihn), and *tetracycline* (*tet*-ruh-SYE-kleen). An ideal antibiotic is one that is effective against a broad range of harmful organisms. It must, in addition, not injure body tissues or disturb body functions. Consult your physician or druggist on the source of the antibiotics listed above. Find out from them the diseases these antibiotics are effective against. To see the action of an antibiotic do the next activity.

INFER

To show how an antibiotic prevents the growth of bacteria, cut a small disc of filter paper (use a paper punch). With tweezers, dip it in an antibiotic solution made by dissolving a *penicillin tablet* in boiled water that has been allowed to cool. (**Caution: Avoid contact with your skin.**) Place the soaked paper disc in the center of a prepared nutrient agar dish. Pour a few drops of bacteria culture on the nutrient agar. Tilt the dish back and forth to spread the bacteria culture and replace the cover. Leave the dish in a warm, dark place for several days. Then, examine the dish with a magnifier. Do you see a clear space around the disc? What has happened to the bacteria colonies near the soaked disc?

However, like all drugs, antibiotics are not cure-alls. They never should be used without the advice of a physician. Much harm can be done by an overdose of drugs and some people become allergic to some of them, especially penicillin. In addition, by using a drug for minor infection, microorganisms may become resistant to the drug. As a result, the drug's effectiveness is reduced when it is needed to fight a major infection.

REVIEW

1. What are some ways the body keeps microorganisms from entering?
2. How do white blood corpuscles destroy bacteria?
3. Name three main types of antibodies that are produced in the body.
4. Name a medical test for detecting tuberculosis.
5. List at least five antibiotic drugs that are used to control diseases.

D/What Are Organic Diseases?

Heart disease can be controlled by rest and drugs. *Heart disease* is one of the main causes of death in our country today. Many people suffer from high blood pressure, hardening of the arteries, or coronary attacks. These are all diseases of the circulatory system. High blood pressure may be caused by worry and fear or by working too hard under tension. Hardening of the arteries usually is a result of old age. Coronary attacks are caused by a blood clot plugging up a blood vessel furnishing blood directly to the heart. If the heart's blood supply is blocked off, the heart cannot function properly and a heart attack results.

Heart attacks can be cured partly by complete rest to give the heart a chance to repair itself. The degree of damage to the heart can be found by taking an *electrocardiogram* (ee-*lek*-tro-KAR-dee-uh-*gram*). Physicians may prescribe drugs to relax the heart and to help it function normally again. Medical scientists study diseases of the heart and circulatory system to discover how these heart attacks may be prevented. When more information is found, many people will live longer and stay in better health (Fig. 14-14).

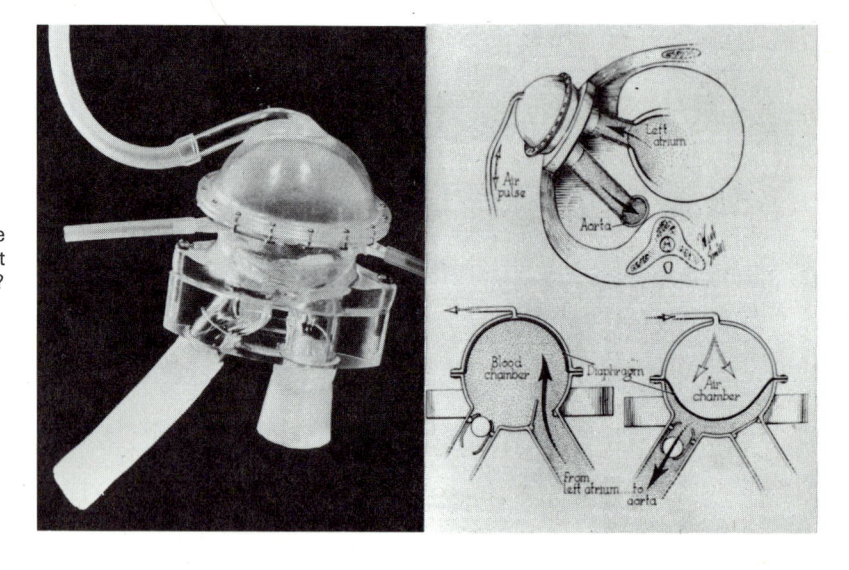

14-14 How do you think the invention of a mechanical heart saves lives in the operating room?

CHAPTER 14 HEALTH OF THE BODY / 415

Cancer can be controlled by early discovery and treatment. *Cancer* is a disease caused by a group of cells that have suddenly begun to divide rapidly in an uncontrolled manner. This uncontrolled growth destroys the normal cells around the cancer, producing poisonous products in the body. Although a cancer may start growing anywhere, it is most often found in the throat, lungs, stomach, and intestines.

The best way of controlling cancer is to discover it in its early stages. Everyone should have a physical examination at least once a year. If any of the common signs of cancer, as shown in the table, are noticed, a doctor should be consulted at once.

Common Signs of Cancer
Any sore that does not heal in a few weeks.
Continued bleeding from any body opening.
A lump on the body which changes shape or gets larger.
A great deal of indigestion.
A gradual change in the color of a wart or a mole.
Periods of diarrhea followed by constipation. A change in bladder habits.
Continued hoarseness or coughing.

What can be done if a cancer is found? Some ways in which cancers are treated are by surgery, X rays, or radiation. Surgery removes the growth from the body. If the cancer is found early enough, this usually cures the patient. X rays also are used to destroy cancer cells. In addition, radioactive substances produce radiations that kill cancerous cells. For example, certain cancers can be treated with radioactive cobalt (Fig. 14-15).

Radioactive materials are also used inside the body to treat certain cancers. Radioactive iodine, taken into the body, is carried by the blood to a cancerous thyroid gland. Here, the radiations destroy the cancer cells. Radioactive phosphorus, too, is absorbed by the bones. It is useful in treating cancer of bones and leukemia.

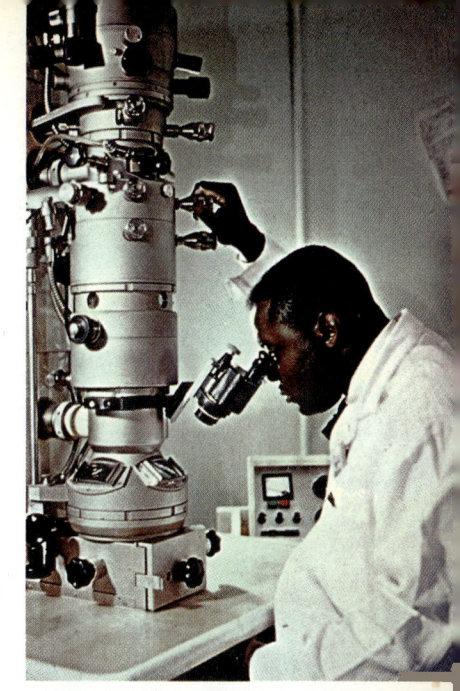

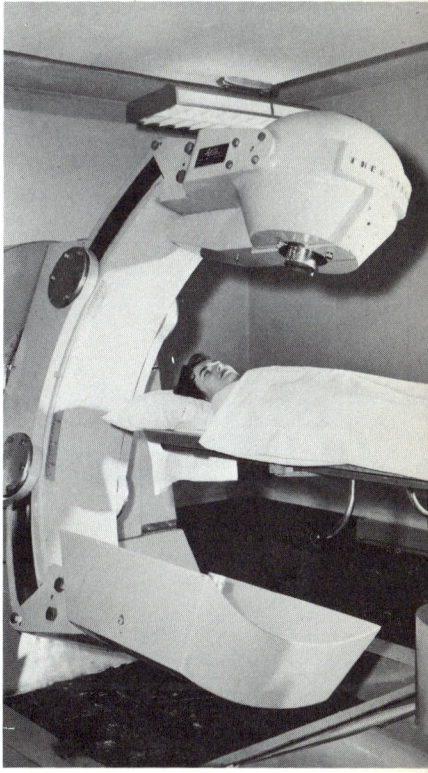

14-15 The electron microscope (above) and radioactive cobalt (below) are used to detect and treat what disease?

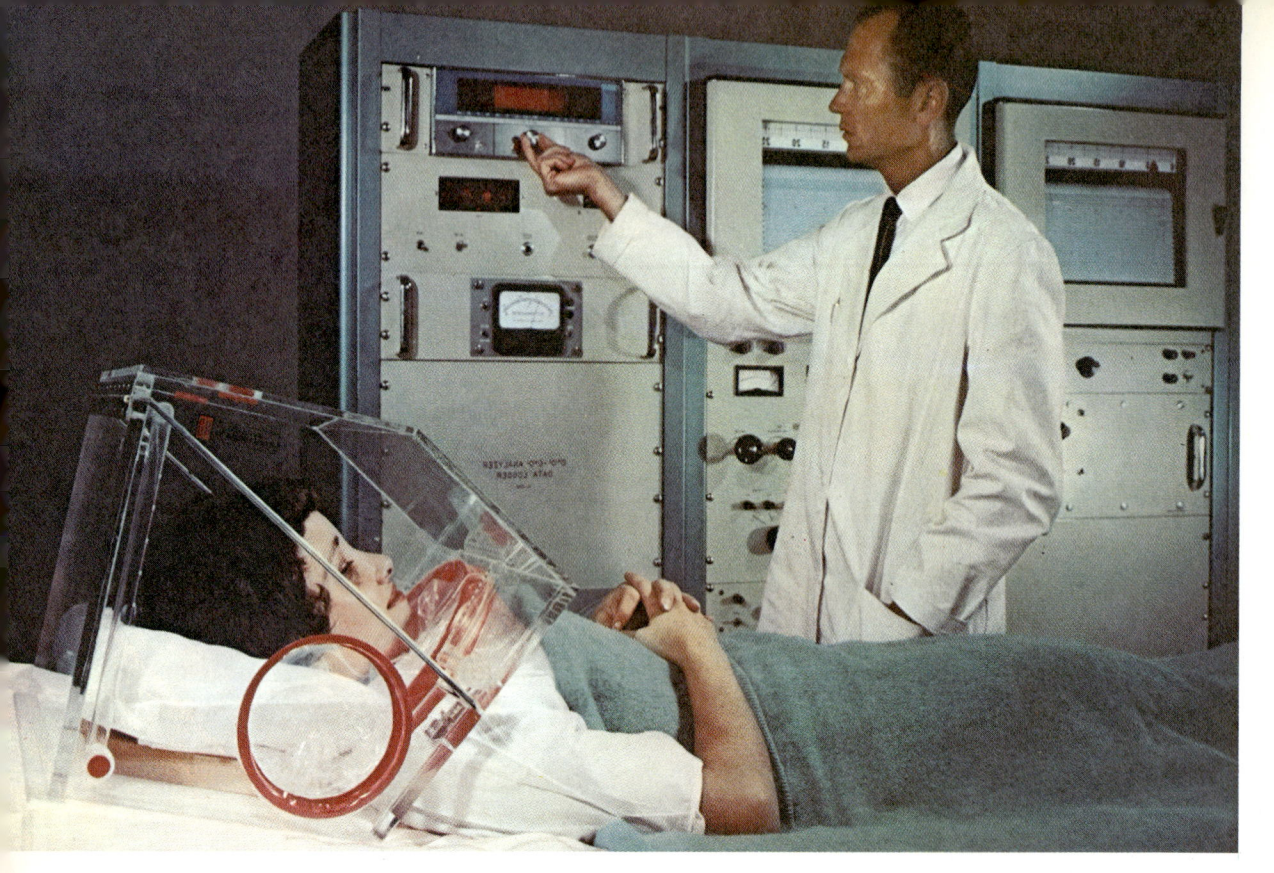

14-16 How are radioactive isotopes used to locate diseased organs?

Other radioactive substances may be placed directly in the cancerous tissue. These might be in the form of little pellets or as radioactive threads sewn into the tissue. Radioactive gold, cobalt, and iridium have been used in this way to treat cancer in the chest area. The radioactive dose is carefully measured. After a certain number of days or weeks the radioactivity disappears, and the material becomes harmless.

Radioisotopes are used to locate diseased organs. Many common substances can be made artifically radioactive by placing them in nuclear reactors. These materials then produce radiations, much like naturally radioactive material. They are called *radioisotopes* (*ray*-dee-o-EYE-so-tops). Their use has opened a whole new method of locating diseased organs (Fig. 14-16).

Suppose a brain tumor is suspected. A radioisotope tracer is injected into the blood. If there is a tumor in the brain, the radioactive material collects in the tumor. A detector, like a Geiger counter, is used to measure the radiation from the area.

Different radioisotopes collect in different parts of the body. For example, radioactive iodine collects in the thyroid gland, radioactive potassium collects in the nervous system, and radioactive calcium and magnesium collect in the muscles. Hence, by using radioisotopes, a suspected disease in a certain part of the body can be traced.

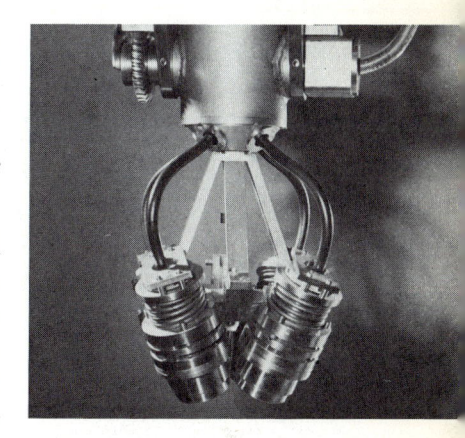

Mental illness can be controlled by proper treatment. Mental illness is a disease of the nervous system which can happen to anyone. Mental illness may be caused by strain under which some people live and work. Many other factors contribute to mental illness. If a person constantly feels that he is not liked by others, that everyone is "picking on him," all the time, or that he can never do well in school or his work, he may suffer a mental breakdown.

Since mental illness affects the nervous system, the way a person thinks and acts changes. Most people are cured and returned to normal life. Special medical help is usually needed, however. The person may have to be placed in a mental hospital with specially trained doctors.

New methods of brain surgery have recently been developed using high-intensity sound waves. Brain surgeons have found that in certain illnesses, it is necessary to cut away a tiny bit of diseased tissue from the brain. A new device now produces sound waves, above the range of human hearing, that can destroy a bit of tissue deep in the brain (Fig. 14-17). The sound waves can be accurately pinpointed, leaving the surrounding brain tissue undamaged.

14-17 High frequency sound waves are used instead of knives. How are they used in brain surgery?

REVIEW

1. Name some diseases of the circulatory system.
2. In what parts of the body do cancers most often grow?
3. What are the three main ways in which cancer cells can be destroyed in the body?
4. In what part of the human body does radioactive iodine collect?
5. What are some factors that may produce mental illness?

E/What Are Habit-Forming Drugs?

Alcohol affects the nervous system. Many people think that alcohol "peps up" the body. Actually, even small amounts of alcohol depress the nervous system. Although alcohol releases energy in the body cells, it is generally not considered a food because of its harmful effects. The reason why some people drink alcohol is to be "sociable." The deadening effect on the nervous system relaxes the body, making the person feel more at ease with people. Larger amounts of alcohol, however, seriously affect muscular control and judgment. Excessive amounts of alcohol can result in unconsciousness and even death.

Alcohol also affects the body by lowering the body's resistance to pneumonia, tuberculosis, and influenza. Persons who frequently drink large amounts of alcohol damage their kidneys, liver, and heart. Do the activity below to find out the effect of alcohol on muscle tissue.

EXPLAIN

Place two pieces of uncooked beefsteak, one in a beaker of water, the other in a beaker of ethyl alcohol. Observe 24 hours later. What has happened to the meat in the alcohol? Can you explain your findings? What may you conclude about the effect of alcohol on muscle tissue?

There is proof that excessive use of alcohol shortens a person's life. Alcohol not only damages the body, but it also affects the home and social life of many people. Some people become *alcoholics*. Alcoholics lose all desire to become or remain useful citizens.

A major problem in today's society is the high number of automobile accidents caused by people who drive under the influence of alcohol (Fig. 14-18). This does not refer only to people who are drunk, but includes those who have had only one or two drinks. Even a small intake of alcohol makes it difficult for a person to coordinate his actions and use good

14-18 Why is it dangerous to drive under the influence of alcohol?

judgment. Many authorities now think that most automobile accidents result from drivers under the influence of alcohol.

The time required for the driver of an automobile to recognize an emergency and press his foot on the brake is known as the *reaction time*. The distance a car travels during this reaction time and the braking distance necessary to stop the car for different speeds are shown in the chart (Fig. 14-19).

14-19 This chart shows the distances a car will travel at various speeds during the reaction time. What effect does alcohol have on reaction time?

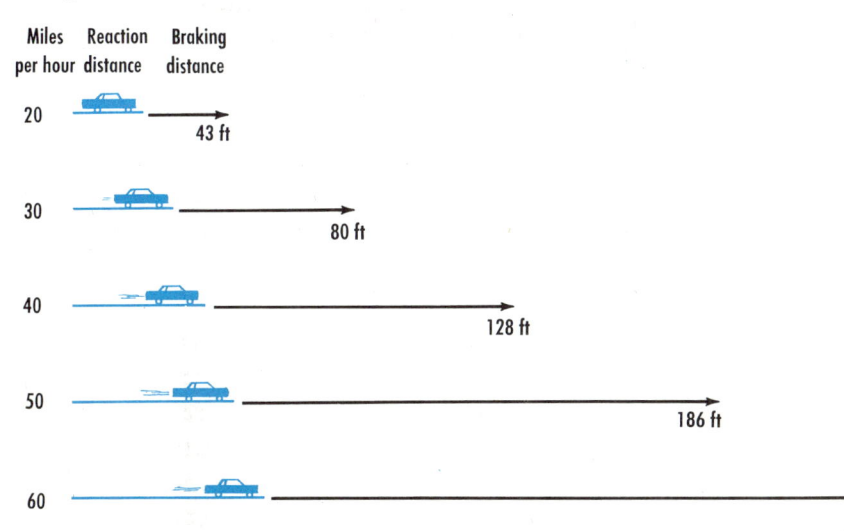

Miles per hour	Reaction distance	Braking distance
20		43 ft
30		80 ft
40		128 ft
50		186 ft
60		254 ft

Many people who drink think that their driving and thinking are not affected when they are behind the wheel of a car. Scientific tests show this is not true. One drink can almost double the reaction time of a person. Remember, that a car moving at 40 miles an hour goes about 35 feet before the alert driver can apply his foot on the brake. If the reaction time is doubled, the car travels 70 feet before the driver begins to apply the brakes! The extra reaction time might well result in an accident and death.

Tobacco affects the heartbeat, digestion, and blood pressure. Tobacco contains a deadly drug, called *nicotine*. In fact, one of the most effective insect sprays in use consists of a strong solution of nicotine. Fortunately, the burning of tobacco oxidizes most of the nicotine. People who inhale tobacco smoke, however, accumulate tars in the nose, throat, and lungs. Tars produce "smoker's cough" in people. It is thought by many scientists to be a contributing factor in cancer of the lungs in later life (Fig. 14-20).

Even the small amount of nicotine from one cigarette causes an immediate rise in blood pressure and puts a strain on the heart. A pulse count on a person right after smoking usually

14-20 What is the heavy smoker doing in this cartoon?

shows an increase in the rate of heartbeat. This is one reason why athletic coaches ask the boys on their teams not to smoke cigarettes. Nicotine also has a depressing effect on the appetite of boys and girls. This discourages their desire for the proper amounts and kinds of food that their bodies need. You will be able to see some of the harmful substances that are contained in tobacco smoke by doing the following activity.

OBSERVE

To show some of the products in burning tobacco, set up the equipment, as shown. The pumping actions draw air through the pipe with each upward stroke of the pump. The upward stroke represents a puff on the pipe. Light the tobacco in the pipe while you are drawing air through it with the pump. Keep the tobacco burning by pumping at about the same rate as a pipe-smoker puffs on his pipe. After five minutes, remove the glass jar and examine its water contents. How does it smell? What is the color of the water? Does the water have an oily feel? Taste a **small drop** of the liquid. Describe its taste.

Burning tobacco produces substances which irritate the body. The tars irritate the linings of the mouth, nose and throat. Nicotine and carbon monoxide, which could not be seen in the activity, have poisonous effects, too, in the body. It should be clear to thinking people that smoking helps to damage a healthy body.

Narcotic drugs are dangerous and habit-forming. We have been discussing the harmful effects of alcohol and tobacco on the body. However, a person usually can stop drinking or smoking if he has enough will power to do so. Alcohol and tobacco are habit-forming, but not in the same sense as a true narcotic drug.

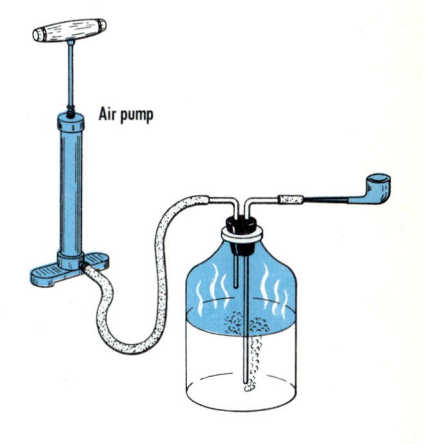

The use of narcotic drugs becomes a deadly habit. Once the person begins to use the drug, he becomes an addict and finds it almost impossible to stop. In addition, a person taking narcotic drugs must have larger and larger amounts of the drug to satisfy the craving of his body. If the use of the drug is stopped, the person becomes violently ill and mentally upset (Fig. 14-21). It takes months of treatments to break the narcotic drug habit. To obtain money in order to support their habit, narcotic addicts often turn to crime.

Small amounts of narcotic drugs may be used as part of a medical treatment, but only under a physician's direction. *Morphine* (MAWR-feen), *heroin* (HAYR-o-ihn), and *cocaine* (ko-KAYN) are the better known of the narcotic drugs. Remember that the use of narcotic drugs can become an uncontrollable and dangerous habit. Narcotic drugs of this type are so dangerous to human health that their sale is controlled by federal law. Therefore, a druggist must keep accurate records of all these drugs sold.

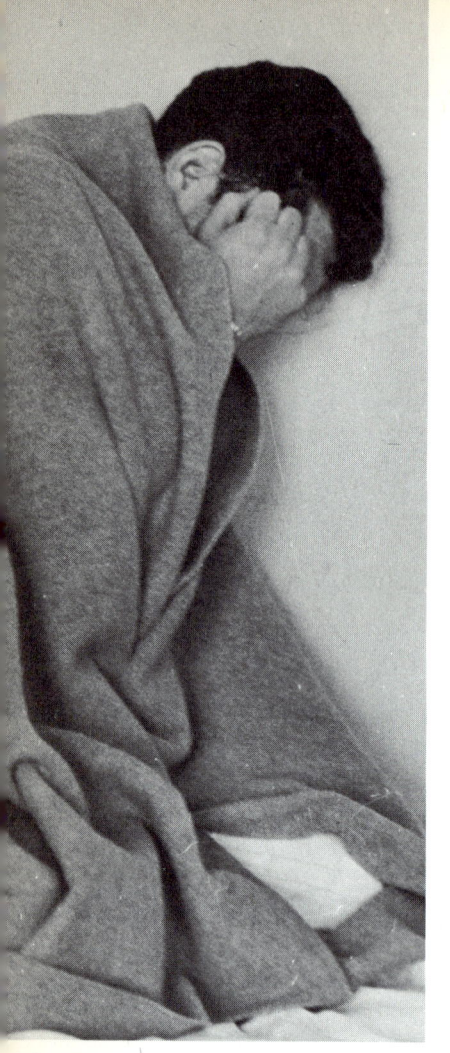

14-21 Why do narcotic addicts become violently ill when the drug is taken away.

REVIEW

1. What is the effect of alcohol on the nervous system?
2. What organs in the body may be injured through continued use of alcohol?
3. How does alcohol affect the reaction time of a person driving an automobile?
4. What are the main effects of tobacco in the body?
5. Name three well-known narcotic drugs which affect the body.

THINKING WITH SCIENCE

A. *On a separate sheet of paper, write the numbers 1 to 15. After the number of each question, write the term that correctly completes the statement.* **Do not write in your book.**

1. Diseases caused by microorganisms growing in the body are called _____.

2. A habit-forming drug that is injurious to the body is known as a(n) _____.

3. A chemical substance that prevents the growth of bacteria on contact is called a(n) _____.

4. A disease caused by the improper functioning of some part of the body is known as a(n) _____.

5. A radioactive material used in locating and treating certain diseases is called a(n) _____.

6. A complex protein substance that grows only in living cells to cause disease is a(n) _____.

7. A disease caused by the body's reaction to some foreign material is a(n) _____.

8. A substance produced by the cells in the body to fight a specific infection is a(n) _____.

9. An outward sign of a disease is known as a(n) _____.

10. The resistance of the body to a specific disease-causing organism is called a(n) _____.

11. A substance producing an immunity in the body to a specific disease when injected is called a(n) _____.

423

12. Milk that has been heated to destroy disease-causing organisms is said to be _____.

13. A substance containing antibodies and injected into the body to fight a specific disease is called a(n) _____.

14. A disease caused by body cells reproducing rapidly and wildly is called a(n) _____.

15. A drug found in tobacco that increases blood pressure when taken into the body is _____.

B. *Write the answers to the following in your notebook. Be sure to use complete sentences with correct spelling and grammar.*

1. In what ways do harmful bacteria affect the body to produce a disease?

2. What are some examples of human diseases caused by: (*a*) molds, (*b*) viruses, (*c*) protozoans?

3. What is the main difference between an infectious disease and an organic disease?

4. How does an allergy differ from an infectious disease?

5. What are the main conditions needed by bacteria for growth?

6. How is the body adapted to destroy microorganisms that invade the blood?

7. Explain the difference between a vaccine and a serum as used in the body.

8. What happens to the normal cells in a part of the body when cancerous cells begin to grow?

9. How are radioactive materials used to treat certain cancers?

10. What are some of the effects of alcohol taken into the body?

RESEARCH IN SCIENCE

1. Test the effects of different kinds of insecticides on houseflies.

2. Prepare a chart comparing the number of people who die each year in this country from different diseases.

3. Ask your doctor to show you X rays of diseased body organs and to explain how the diseases were discovered.

4. Prepare a demonstration to show the effects of various soaps as antiseptics.

5. Prepare a demonstration to show how water is purified for human use.

READINGS IN SCIENCE

Asimov, Issac, *The Human Body: Its Structure and Operation.* Houghton, 1963. Chapter by chapter information is given about the human skeleton, muscles, and the major systems: respiratory, circulatory, digestive, excretory, and reproductive.

Bacon, M. K., and Jones, M. B., *Teen-Age Drinking.* Thomas Y. Crowell, 1968. The book presents up-to-date facts and figures. Social drinking, customs and taboos are discussed.

Boettcher, Helmuth, *Wonder Drugs.* Lippincott, 1963. The book tells about many of the strange remedies that have been used through the ages in the treatment of disease. Shows how some of these still have value today.

Cain, Dr. Arthur H., *Young People and Smoking.* The John Day Co., 1964. Dr. Cain presents the facts about teen-age smoking and smokers. The book represents ten years of research.

Dolen, Edward Jr., *Adventure With a Microscope.* Dodd Mead, 1964. Deals primarily with Robert Koch's life and the discovery of the tuberculosis bacillus. Tells of other great medical discoveries based on Koch's germ theory of diseases.

Grant, Madeline, *Wonder World of Microbes.* McGraw-Hill, 1964. Delves into secrets of DNA and RNA as related to the study of viruses. Work of Salk and Sabin in their preparation of polio vaccine is described.

Groch, Judith, *You and Your Brain.* Harper and Row, 1964. This provides an accurate account of what is presently known about the human brain and nervous system. Presents a broad survey of the evolution, anatomy, and function of the brain. It discusses such topics as consciousness, sleep, learning, etc.

Harvey, Tad, *Exploring Biology.* Doubleday, 1963. The book traces the history of biology from crude beginnings to modern cell theory. It includes up-to-date information on organic evolution, molecular biology, and other basic ideas.

Hutchins, Carleen, *Life's Key: DNA.* Coward-McCann, 1961. This book explains the role of DNA in heredity. Many illustrations and directions for making a model of DNA are included.

McBain, W. H., and Johnson, R. C., *The Science of Ourselves.* Harper and Row, 1962. The book deals with the senses, with learning, and with memory and forgetting. The effects of environment on inherited traits are discussed. General methods of animal training will interest many students.

Poole, Lynn and Gray, *Electronics in Medicine.* Whittlesey House, 1964. This illustrated book describes how electronic instruments are being used in hospitals and laboratories for medical research, diagnosis, and treatment. Much of this is based on the work at the Johns Hopkins Medical School.

Reinfeld, Fred, *Miracle Drugs and the New Age of Medicine.* Sterling, 1962. This book highlights present research with viruses and cancer and the promise of future results. Included is a brief history of medicine, the discovery and establishment of the "germ theory" of disease, and an account of the discovery, production and uses of modern "wonder drugs."

Simon, Tony, *The Heart Explorers.* Basic Books, 1966. Opens with an account of the earliest heart surgery on record and carries the reader through to the latest work on heart transplants. Explains the use of heart-lung machines and "artificial hearts" in surgery.

White, Anne and Lietz, Gerald, *Man the Thinker.* Garrard, 1967. Deals with the brain of man and his nervous system. Discusses instinct, stimulus-response learning, language, and memory.

Words in Science

Key to Pronunciation

This key will help you to pronounce the science words in your text and in your WORDS IN SCIENCE list. In the parentheses you will find the word respelled in alphabet characters and divided according to pronunciation. The syllable having the main stress is printed in capital letters; the one having the lighter stress appears in italics in lower case. For example, the word "scientific" appears as (*sye*-uhn-TIHF-ihk).

Vowel Symbol	Pronunciation	Word
A, a	(MAT-uhr)	matter
AY, ay	(DAY-lee)	daily
AH, ah	(FAHR)	far
OW, ow	(TOW-uhl)	towel
E, e	(BET-uhr)	better
EE, ee	(EE-vuhn)	even
IH, ih	(AK-tihv)	active
YE, ye; eye; i(*)e	(HYE-uhr) (EYE-luhnd) (TUHR-mite)	higher; island; termite
O, o; o(*)e	(BO-nee) (SKOPE)	bony; scope
AW, aw	(SAW)	saw
OI, oi	(KOIN)	coin
OO, oo	(ROOL)	rule
UH, uh	(PUHL)	pull
	(NUHT)	nut
	(BET-uhr)	better
	(buh-NAN-uh)	banana
UH, uh	(*ahp*-uhr-AY-shuhn)	operation

* Consonant

Consonant Symbol	Pronunciation	Word
B, b	(BAY-bee)	baby
CH, ch	(NAY-chuhr)	nature
D, d	(DIHD)	did
F, f	(FIHF-tee)	fifty
G, g	(GIHFT)	gift
H, h	(uh-HED)	ahead
HW, hw	(HWAYL)	whale
J, j	(JUHJ)	judge
K, k	(KIHN)	kin
L, l	(LIHL-ee)	lily
M, m	(MUHR-muhr)	murmur
N, n	(O-nuhr)	owner
NG, ng	(HELP-ihng)	helping
P, p	(PEP-uhr)	pepper
R, r	(RAYR-lee)	rarely
S, s	(SAWRS)	source
SH, sh	(SHELF)	shelf
T, t	(uh-TAK)	attack
TH, th	(THROO; THAYR)	through; there
V, v	(GIHV)	give
W, w	(uh-WAY)	away
Y, y	(YAHRD; KYOO)	yard; cue
Z, z	(ZON)	zone
ZH, zh	(MEZH-uhr)	measure

absolute humidity (hyoo-MIH-dih-tee), the amount of water vapor per cubic foot of air.

acetylcholine (uh-*set*-ihl-KOH-leen), a chemical substance released by a nerve fiber when an impulse reaches the motor end plate.

adaptation (uh-*dap*-TAY-shuhn), the adjustment of organisms to their environment.

adolescence (ad-uh-LES-uhns), the teen years during which many body changes take place including the development of secondary sex characteristics.

aeration (ayr-AY-shuhn), the process of allowing water to mix with the air.

air mass, widespread body of air all of which has the same general temperature, humidity, and other weather conditions.

air sacs, the microscopic structures in the lungs in which the exchange of oxygen and carbon dioxide in the blood takes place.

albino (al-BYE-no), person or animal with unusually light skin, pink eyes, and nearly white hair all due to partial or complete lack of natural pigment.

algae (AL-jee), large group of primitive plants, ranging in size from microscopic to the visible as in seaweeds such as kelp, having chlorophyll but lacking true roots, flowers, stems, and leaves.

allergy (AL-uhr-gee), a disease, ordinarily harmless, caused by the body's unfavorable reaction to a foreign substance.

altimeter (al-TIH-muh-tuhr), radio or barometric instrument used to measure altitude.

ameba (uh-MEE-buh), microscopic water animal having only one cell and no definite shape. It represents one of the simplest forms of life.

amino (uh-MEE-no) **acid,** one of a group of compounds of which proteins, present in all living tissues, are composed.

amphibian (ah-FIH-bee-uhn), animal of a class including frogs, toads, and salamanders, which as larvae have gills and live in water and as adults have lungs and live on land.

anemia (uh-NEE-mee-uh), a condition of the blood where there may be insufficient oxygen carried to the cells because of a lack of hemoglobin.

anemometer (*an*-uh-MOM-uh-tuhr), instrument for measuring wind speed.

aneroid (AN-uhr-*oid*) **barometer,** a barometer made up of an airtight box containing a partial vacuum and having a flexible top to which a pointer is attached. Changes in atmospheric pressure cause the top to bend in or out, moving the pointer.

anthracite (AN-thruh-*site*) **coal,** a very hard coal that burns with little smoke or flame.

antibiotic (*an*-tih-bye-AH-tihk), a drug produced from certain microorganisms that overcomes disease-causing organisms in the body.

antibody (AN-tih-*bah*-dee), chemical substance produced by the body which helps destroy microorganisms and their products in the body.

appendage (uh-PEN-dihj), outgrowth such as a limb, tail, horn, or other secondary part attached to the head or body of an animal.

arachnid (uh-RAK-nihd), any of a large

group of insect-like animals including spiders, scorpions, mites, and ticks. Arachnids differ from insects in having four pairs of legs and lacking wings and antennae.

artery (AHR-tuhr-ee), a blood vessel in which the blood flows away from the heart.

arthropod (AHR-thro-*pohd*), member of the largest phylum in the animal world, which includes spiders, crabs, and insects. Arthropods have a hard outside skin or shell and jointed legs.

asexual reproduction, reproduction involving one parent.

astigmatism (uh-STIHG-muh-*tihz*-uhm), a defect in vision caused by an irregularly curved cornea or lens in the eye.

atmosphere (AT-muh-*sfeer*), the layer of gases that surrounds the earth.

atom, the smallest particle of an element that has all the properties of the element.

aqueducts (AK-wah-duhkts), closed pipes or open ditches that carry water to areas where it is not plentiful.

bacillus (buh-SIHL-uhs), bacterium that is rodlike in shape.

bacteria (bak-TEE-ree-uh), large class of microscopic, one-celled organisms considered to be plants, but also having animal characteristics. Their action is the cause of a number of diseases and of many processes, including decay, fermentation, and soil enrichment.

barometer (buh-RAH-muh-tuhr), instrument used to measure air pressure.

basal metabolism (BAY-suhl me-TA-bo-*lihzm*), the rate at which the body uses up energy while at rest, indicating general metabolism.

bends (BENDZ), the collection of nitrogen bubbles in the joints, nerve tissue, brain, muscles, and other body tissues producing severe pain.

beriberi (BE-ree-BE-ree), a vitamin deficiency disease caused by lack of thiamin.

biome (BYE-ome), the interrelationships that exist among plants, animals, and climate in a given area.

biotic (bye-AH-tihk) **community,** an interdependent group of plants and animals living in a particular area.

bituminous (bye-TYOO-mih-nuhs) **coal,** a soft coal that burns with a smoky flame.

block mountain, a mountain which is formed by the raising of a large area of the earth's crust.

budding, a form of asexual reproduction in which a cell divides into two unequal cells.

buoyant (BAW-yuhnt) **force,** the upward force which a liquid or gas exerts on a body placed in it.

calorie (KAL-uh-ree), the unit of heat equal to the amount of heat needed to raise the temperature of one gram of water one degree Celsius. The calorie is used to measure the amount of heat energy in foods.

calorimeter (*kal*-uh-RIHM-uh-tuhr), a device used to measure the amount of heat produced by a substance.

cambium (KAM-bee-uhn) **layer,** the active growing layer of one type of plant stem.

cancer (KAN-suhr), an organic disease identified by a wild growth and multiplication of body cells.

capillary (KAP-uhl-eh-ree), a tiny, thin-walled blood vessel.

capillary (KAP-uhl-eh-ree) **action,** the way water rises to the surface of the ground.

carbohydrate (*kahr*-bo-HYE-drayt), any of a class of compounds of carbon, hydrogen, and oxygen, such as sugars, starches, and cellulose, which are manufactured by plants and are the ultimate source of animal food.

carbon dioxide, a common gas found in the atmosphere given off as a waste product of human respiration and taken in by plants for use in the process of photosynthesis.

carnivore (KAHR-nih-*vawr*), meat-eating animal, such as the dog or tiger.

cartilage (KAHR-tih-lihj), tough, rubbery tissue that forms much of the skeleton of young animals and infants and usually develops into bone; gristle.

cell membrane, a thin living membrane which surrounds the cytoplasm of a cell.

chemosphere (KEM-uh-*sfeer*), layer of air which overlaps the stratosphere, mesosphere, and thermosphere and where chemical processes take place.

chlorophyll (KLOR-roh-fihl), the green pigment found in plants, capable of absorbing energy from the sun.

chromatin (KRO-muh-*tihn*), the substance in a cell's nucleus that contains the genes.

chromosome (KRO-muh-*som*), thread-shaped structure in every cell nucleus which help determine inherited characteristics.

cilia (SIH-lee-uh), short, threadlike extensions of the cytoplasm that are developed by some cells.

cirrus (SIHR-ihs) **clouds,** light, feathery clouds usually seen between seven and ten miles above the earth.

climate (KLIH-muht), an average of weather conditions over a long period of time.

coagulation (ko-*ag*-yool-LAY-shuhn), the use of chemical compounds to separate foreign particles from water.

coccus (KAH-kuhs), bacterium that is spherical in shape.

coelenterate (sih-LEN-tur-ayt), any of a phylum of invertebrate sea animals, such as jelly fish, corals, sea anemones, and hydras, with thin-walled bodies, a large body cavity, and only one opening, the mouth.

community, a group of plant and animal populations living within some sort of natural boundaries, such as in a pond.

condensation (*kahn*-den-SAY-shuhn), the process by which a gas or vapor changes into a liquid.

conduction (kuhn-DUHK-shuhn), the transfer of heat energy by collision of molecules.

conjugation (*kahn*-joo-GAY-shuhn), a simple form of sexual reproduction usually seen in one-celled plants or animals.

conservation, the wise use of the earth's natural resources.

convection (kuhn-VEK-shuhn), circular currents within a liquid or gas due to differences in density between hotter and cooler portions and resulting in the transfer of heat.

core (kor), 1. of the sun, the very hot gaseous center. 2. of the earth, the very dense innermost portion.

corpuscle (KOR-puhs-uhl), a red or white blood cell.

crust (KRUHST), the outermost layer of the earth.

crustacean (kruhs-TAY-shuhn), any of a class of tough-shelled animals, most of which live in water, such as crabs, lobsters, and shrimp.

cumulus (KYOO-myoo-luhs) cloud, dome-shaped, billowy clouds that may pile up to reach great heights.

cyclone (SYE-klone), weather condition in which winds whirl around a center of low air pressure.

cytoplasm (SYE-to-*plaz*-uhm), the living substance in a cell surrounding the nucleus.

deficiency (dee-FIHSH-uhn-see) disease, disease caused by a lack of particular vitamins in the diet.

delta (DEL-tuh), the roughly triangle-shaped piling up of earth and sand at the mouth of a river.

depletion (dee-PLEE-shuhn), of the soil, the using up of minerals necessary for plant growth.

desert (de-zuhrt), an area where there is a continual lack of water.

diabetes (dye-uh-BEET-eez), disease due to insufficient amount of insulin in the blood stream.

diaphragm (DYE-uh-fram), 1. a thin, movable shutter for regulating the intensity of light. 2. the muscular wall between the chest and abdomen used in breathing.

diatom (DYE-uh-tahm), one of a large group of very tiny one-celled algae living in water and having beautifully marked, glass-like shells.

diffusion (dih-FYOO-zhuhn), the passage of food through the villi of the small intestine into the bloodstream.

digestion (dye-JES-chuhn), the process in which food is changed to soluble forms for use in the body.

dihybrid (DYE-hye-brihd), a plant or animal which is the result of crossing the separate characteristics.

distillation *(dihs*-tih-LAY-shuhn), the evaporation of water and the condensation of vapor back to water.

DNA, deoxyribonucleic acid, a super molecule found in the nucleus of a cell, which transmits heredity information and controls cell activities.

dominant (DAHM-ih-nuhnt) characteristic, trait which is inherited and appears more often than others in the offspring.

dunes (DYOONZ), hills of sand caused by the blowing of wind over bare soil.

earthquake, sudden trembling or shaking of the ground, usually caused by a shifting of rock layers along a fault or fissure under the earth's surface.

echinoderm (ih-KYE-no-duhrm), one of a large group of sea animals, including starfish and sea urchins. Most echinoderms have a tough spiny skin and are made up of several similar segments extending outward from a common center.

ecology (ee-KAHL-uh-jee), branch of biology that deals with the relationships of organisms to each other and to their environment.

electrocardiogram (ee-lek-tro-KAR-dee-uh-*gram*), an instrument used to detect heart damage.

element (EL-uh-ment), the simplest form of matter that cannot be broken down into other substances by ordinary means.

embryo (EM-bree-o), early form of animal during development from a fertilized egg and before being hatched or born.

embryology (*em*-bree-AHL-o-jee), branch of biology that deals with the structure and development of embryos.

endocrine (EN-duh-krihn) **glands**, glands which secrete hormones directly into the bloodstream.

energy (EN-uhr-jee), 1. the ability or capacity to do work. 2. force which causes motion or produces changes in matter.

environment, all the forces, influences, and conditions of the surrounding world which act on an organism.

enzyme (EN-zy-em), protein that starts or speeds up a chemical action in other substances without undergoing any permanent change itself. Enzymes are produced by all living organisms.

equator (ih-KWAYT-uhr), an imaginary circle around the earth that is equally distant from the North and South Poles and divides the earth into Northern and Southern Hemispheres.

era (IHR-ruh), one of the major divisions of geologic time of the earth's history.

erosion (ee-RO-zhun), wearing away of land by action of water, wind, and ice.

estivation (*es*-tih-VAY-shuhn), a quiet resting period during the summer. Such animals as the frog estivate during the summer.

Euglena (yoo-GLEE-nuh), in zoology, any of a large genus of green, one-celled, aquatic protozoans with one or more flagella. Euglenas move about like animals, but contain granules of chlorophyll with which they synthesize food like plants.

evaporation (ee-*vap*-uh-RAY-shuhn), the change in state from a solid or liquid into vapor.

excretion (eks-KREE-shuhn), the process of getting rid of waste material by a living organism.

exosphere (EKS-o-*sfeer*), the layer of air beyond the ionosphere where the composition of the atmosphere changes from helium to hydrogen.

fault, a deep break in the earth's crust caused by the movement of crustal blocks.

fermentation (*fuhr*-men-TAY-shuhn), the breakdown of sugar into carbon dioxide and alcohol.

fertilization (*fuhr*-tih-lih-ZAY-shuhn), the union of the sperm nucleus with the egg nucleus.

fibrin (FY-brin), an insoluble protein substance that forms a network of threads in a blood clot.

filtration (fihl-TRAY-shuhn), removal of sediment from water by passing it through rocks, gravel, and sand.

fission (FISH-uhn), 1. the breakdown or desintegration of the nucleus of an atom, with the release of large quantities of energy. 2. the process of cell division in most one-celled organisms.

flotation (flo-TAY-shuhn), process by which impurities are separated from the minerals in ores.

fluoridation (*flyoor*-ih-DAY-shuhn), the addition of fluoride salts to the drink-

ing water to help reduce tooth decay.
folded mountains, mountains formed by the uplifting of the earth's crust resulting from pressure within the earth.
fossil (FAH-sihl), the remains of an ancient plant or animal which have been preserved in rock layers or hardened resins of trees; also, a preserved trace of an ancient organism, such as a dinosaur footprint in mud that has turned to rock.
front, the irregular but definite boundary between warm and cold air masses.
fungus (FUHNG-guhs), one of a group of plants, without leaves or green color, which feeds upon plants or animal matter. Bacteria, molds, mushrooms, toadstools, and mildews are kinds of fungi.

gametes (GAM-eets), either of the two reproductive cells (egg and sperm) that can unite to form a new individual.
gas, the state of matter that has no definite shape and no definite volume.
gem, a mineral prized for its color, luster, and hardness.
gene (JEEN), unit in the chromosomes which carries hereditary characteristics.
geology (jee-AH-luh-jee), the study of the history, structure, and composition of the earth.
geyser (GYE-zuhr), a hot spring that throws a jet of steam and hot water into the air, often at regular intervals.
glacier (GLAY-shuhr), a large slow moving mass of snow and ice.
glucose, sugar produced in living organisms from starch and other sugars. Glucose is the major source of energy for metabolism, and in plants it plays an important part in photosynthesis.

habitat, the natural home of a plant or animal.
half-life, the period of time in which half the atoms of a radioactive element break down into other elements.
hemoglobin (HEE-mo-*glo*-bihn), a red protein substance in the red blood cells needed to carry oxygen.
herbivore (HUHR-bih-*vawr*), plant-eating animals, such as cows and horses.
heredity (huh-RED-ih-tee), 1. the passing on of traits from parent to offspring by means of genes. 2. all the traits passed on in this way, such as eye color or blood group.
hibernation (hye-buhr-NAY-shuhn), the passing of winter in a state resembling sleep, as do bears and raccoons.
hormones (HAWR-monz), chemical regulators of certain body functions secreted by the endocrine glands.
hurricane, vast and destructive tropical storm with very heavy rains and violent winds exceeding 73 miles an hour, that spiral around a calm center of low atmospheric pressure.
hybrid (HYE-brihd), an organism with both a dominant and a recessive gene for the same characteristic.
hydrosphere (HYE-druh-sfeer), the water layers covering the surface of the earth.
hygrometer (hye-GRAH-muh-tuhr), an instrument used to measure the relative humidity in the air.

igneous (IHG-nee-uhs) rock, rock formed by the solidification of molten material.
infectious (ihn-FEK-shuhs) disease, disease caused by the entrance, growth, and reproduction of micro-organisms

435

in the body and their effects on the system. Disease is communicated from one living organism to another.

ingestion (ihn-JES-chuhn), the taking of food into the body.

interdependence (ihn-tuhr-dee-PEN-dens), of living things, the influence or dependence on each other or one another.

inversion (ihn-vuhr-zhuhn), when the air near the surface of the earth is colder than the upper air.

invertebrate (ihn-VUHR-tuh-brayt), any of a group of animals having no back bones.

ionosphere (eye-AHN-uh-*sfeer*), layer of air that overlaps the chemosphere and reflects radio waves. It has several layers and runs generally from about 35 to 600 miles above the earth.

isotherm (EYE-so-thuhrm), line on a weather map which connects places having the same temperature at a particular time.

jet-streams, enormous currents of air having wind speeds that may reach 200 miles per hour.

law of segregation (se-grah-GAY-shuhn), a law of heredity which states that when two hybrids are crossed, one fourth of the offspring will be pure dominant, one half will be hybrid though resembling the pure dominant, and one fourth will be recessive.

leaching (LEE-ching), loss of soil fertility.

lichen (LYE-ken), a small plant without flowers or true leaves that grows flat on rocks or tree. Lichen are made up of fungi; and algae growing cooperatively together.

lithosphere (LITH-uh-sfeer), the solid crust of the globe, as distinct from layers of air and water.

littoral (LIHT-o-ruhl) zone, that portion of the sea which extends from the shoreline to where the water is approximately 250 meters deep.

magma (MAG-muh), the extremely hot rock material that occurs deep within the earth's crust; and from which igneous rocks and lava are formed.

mantle (MAN-tuhl), the earth layer found between the crust and the core.

meiosis (my-O-sihs), process of sex cell division in which the number of chromosomes is reduced to half the original number.

mesosphere (MEZ-uh-*sfeer*), the layer of air above the stratosphere.

metamorphic (*met*-uh-MOR-fihk) rock, rocks changed by heat and pressure.

metamorphosis (*met*-uh-MAWR-fuh-sihs), the bodily changes which organisms pass through from the egg to the adult stage.

metazoan (*met*-uh-ZO-uhn), any of the group of animals whose cells are differentiated into tissues and organs.

migration (mye-GRAY-shuhn), the seasonal movement of organisms from one region to another.

mimicry (MIHM-ih-cree), a form of protective coloration in which an animal closely resembles another kind of animal or an object in its environment.

mineral (MIHN-uhr-uhl), 1. nutrients that supply materials for growth and regulation of the body. 2. chemical compounds or uncombined elements found

in rocks or soil.
mitochondria (mye-to-KAHN-dree-uh), tiny granular bodies found in the cytoplasm of all cells having a nucleus. They contain enzymes necessary for the cell's processes.
mitosis (mye-TO-sihs), cell division in which the chromosomes are equally divided between the two daughter cells.
molecule (MAHL-ih-*kyool*), the smallest particle of a substance that can exist and still show the properties of the substance.
mollusks (MAHL-uhsk), any of a large phylum of animals having soft bodies usually enclosed in a hard shell.
moraine (maw-RAYN), material deposited by glaciers, usually consisting of large rocks.
mutation (myoo-TAY-shuhn), sudden variation in a plant or animal, which results from a change in the genes and chromosomes and can therefore be inherited.

natural resources, valuable materials found in and on the earth.
nucleus (NYOO-klee-uhs), 1. the positively charged dense central part of an atom which contains almost all of its mass. 2. the dense round or oval body present in living cells and controlling most of the activities of the cell.
nutrient (NYOO-tree-uhnt), essential food substances which are needed for proper growth and health of the body.

oceanographer (*o*-shuhn-AH-gruh-fuhr), scientist that studies the oceans and seas, including their currents, and chemical composition.
omnivore (AHM-nih-vawr), animal that eats both meats and vegetables.
ore (AWR), rocks or minerals containing enough of one or more elements, usually metals, to make the mining of it profitable.
organ (AWR-guhn), a part of an animal or plant, such as the heart, the eye, or a leaf, that has a particular structure and function.
organism (OAR-guh-*nihzm*), a complete living thing composed of one cell or many cells.
osmosis (ahz-MO-sihs), the process in which water and dissolved substances pass through a semipermeable membrane.
oxidation (ahk-sih-DAY-shuhn), the combining of a substance with oxygen which releases energy.
oxide (AHK-syed), the combination of oxygen with other elements.

paleontologist (*pay*-lee-ahn-TAHL-uh-jihst), a scientist who studies fossils and prehistoric forms of life.
paleontology (*pay*-lee-ahn-TAHL-o-jee), study of fossils and prehistoric forms of life.
paramecium (*par*-uh-MEE-see-uhm), microscopic one-celled protozoan shaped like a slipper and covered with cilia with which it swims.
parasite (PAR-uh-*site*), an organism which grows and lives only on or in another organism.
pasteurized (PASS-tyoor-ized), heated enough to kill dangerous germs, but not enough to alter its original nature to any great extent.

penicillin (*pen*-ih-SIHL-ihn), any of a group of powerful germ-killing drugs, made from molds and used in healing infectious diseases.

peristalsis (*per*-ih-STAHL-sihs), a wavelike contraction in the walls of digestive organs which moves food along.

photosynthesis (fo-to-SIHN-thuh-sihs), the complex process by which green plants manufactures sugars from water and carbon dioxide in the presence of sunlight; artificial light may replace sunlight.

plankton (PLAYNK-tuhn), the microscopic plants and animals which make up the primary source of food in oceans.

plastid (PLAS-tid), living bodies in the cytoplasm of plant cell which store or form substances important in the life processes of the cell.

pollen (PAHL-uhn), male sex products produced by the anthers of flowers.

pollination (*pahl*-ih-NAY-shuhn), the transfer of pollen from the stamen to the pistil of a flower.

pollutants (puh-LYOO-tuhnts), particles that result from burning and rise into the air.

pollution (puh-LYOO-shuhn), the contamination of air and water by the discharge of harmful or poisonous substances.

population (POP-yoo-*lay*-shuhn), the number of all species of a plant or animal found in a specific area at a specific time.

precipitation (pree-*sih*-pih-TAY-shuhn), any form of water, liquid or solid, which condenses and falls from the air.

protoplasm (PRO-to-*plaz*-uhm), the living material of a cell.

protozoan (*pro*-tuh-ZO-uhn), microscopic, one-celled animal.

pseudopodium (*soo*-do-PO-dee-um), pl. (*soo*-do-PO-dee-uh), a temporary extension of the protoplasm of a living cell.

radioactivity (*ray*-dee-o-ak-TIH-vih-tee), the spontaneous breaking down of the nuclei of certain atoms which gives off high-energy particles and rays.

radioisotope (*ray*-dee-o-EYE-so-tope), isotopes that are radioactive, used in medicine as tracers to locate and treat diseases in the body.

radiosonde (RAY-dee-o-*sahnd*), an instrument containing recording instruments that is used to aid in atmospheric study.

recessive (ree-SES-ihv), characteristic, inherited trait that does not appear when a dominant trait is present.

reflex action, an involuntary movement or function, such as the contraction of a muscle or the secretion of a gland.

regeneration (ree-*jen*-uhr-AY-shuhn), the ability of an organism to grow new parts to replace missing ones.

relative humidity (hyoo-MIH-dih-tee), ratio, expresses as a percentage, of the amount of water vapor in the air to the total amount the air could hold at the same temperature and pressure.

reproduction (*ree*-pro-DUHK-shuhn), the producing, by living things, of further individuals of their own biological species.

respiration (*res*-pih-RAY-shuhn), the process of taking in oxygen and giving off carbon dioxide by the cells of the body.

rhizome (RYE-zome), a thickened underground stem which contains stored food, from which new plants appear.

ribosomes (RYE-bo-sohmz), tiny bodies found in the cytoplasm where protiens are built from amino acids.

rickets (RIHK-itz), deficiency disease due to lack of vitamin D.

saprophyte (SAP-ro-*fite*), an organism that lives on non living organic matter.

scurvy (SKUHR-vee), deficiency disease caused primarily by lack of vitamin C.

seamount (SEE-mount), an isolated undersea mountain that rises sharply from the sea bottom 500 or more fathoms, but does not reach sea level.

sedimentary (*sed*-ih-MEN-tuh-ree) rock, rock formed by the settling and cementing together of materials under water, these rocks may be formed in layers from materials deposited by water, wind, ice, or other agents.

seismograph (SYEZ-mo-graf), an instrument used to record earthquake waves on the earth; this instrument can be used to record the time, duration, and intensity of an earthquake any place on earth.

smelting (SMEL-teeng), the process by which metal is obtained from the chemical compounds which make up the ore.

solute (SAHL-yoot), the dissolved substance in a solution.

solution (so-LYOO-shuhn), a mixing of two or more substances so that single molecules of one are evenly spread through the others.

solvent (SAHL-vent), the dissolving medium in a solution.

spirogyra (SPYE-ruh-JYE-ruh), green, fresh water algae containing chloroplasts with a spiral shape.

spontaneous (spahn-TAY-nee-uhs) generation, disproved belief that certain nonliving or dead materials could be transformed into living organisms.

spore (SPAWR), a reproductive cell surrounded by a hard, protective coat.

spirillum (spye-RIHL-uhm), bacteria that have a spiral form.

stalactite (stuh-LAK-tite), icicle-shaped mineral deposits found on the roofs of caves resulting from the dripping of ground water.

stalagmite (stuh-LAG-mite), deposits found on the floor of caves growing upward; also resulting from dripping water.

stomata (STO-muh-tuh), opening usually on the underside of a leaf, which enable gases and liquids to enter and leave the leaf.

storm, a major disturbance in the atmosphere, especially when marked by high winds, and rain, hail, or snow.

stratosphere (STRAT-uh-*sfeer*), the layer of air above the troposhere.

symbiont (SIHM-bye-ahnt), organisms which live on or in other organisms, but with both benefiting from each other.

system (SIS-tuhm), a group of organs performing a related function in the body.

thermometer (thuhr-MOM-uh-tuhr), instrument used to measure temperature, usually by means of the expansion and contraction of a material according to temperature.

thermosphere (THUHR-muh-sfeer), the layer of air above the mesophere where temperatures may be very high.

tissue (TIHSH-oo), a group of similar cell performing a similar function in an organism.

topsoil, soil at the surface containing both mineral and organic matter.

tornado (tawr-NAY-do), violent, whirling wind that travels rapidly in a narrow path, usually over land, and is seen as a twisting, dark cloud shaped like a funnel.

toxin (TAHK-sihn), a poison of plants, animals, or bacterial origin, especially one that stimulates the body to produce a counteracting antitoxin.

trachea (TRAY-kee-uh), in man, a passageway for air going to and from the lungs; connects the larynx with the bronchi.

transfusion (trans-FYOO-zhuhn), the taking of blood from one person and injecting it into the circulatory system of another.

trichocyst (TRIHK-o-shist), any of the minute stinging organs on the bodies of protozoans.

trilobite (TRYE-luh-bite), the earliest shell covered animal which appeared in the Paleozoic Era.

troposphere (TRO-puh-*sfeer*), the atmospheric layer closest to the earth.

vaccine (vak-SEEN), preparation containing killed or weakened germs of a disease, introduced into the body to make it resistant to attacks of that disease.

vacuole (VAK-yoo-ole), spaces scattered through the cytoplasm of a cell and containing fluid.

Van Allen radiation belt, a layer of charged particles in space around the earth.

vegetative (VE-juh-*tay*-tihv) **reproduction,** a common method of asexual reproduction in higher plants whereby pieces of the plant tissue are capable of growing into a complete organism.

vein (VAYN), blood vessel in which blood flows toward the heart.

vertebrate (VUHR-tuh-brate), any group of animals having a back bone.

villi (VILE-eye), fingerlike projections on the lining of the human small intestine.

vitamin (VYE-tuh-mihn), organic substance that is needed in the normal growth and regulation of the body.

volcano (vahl-KAY-no), an opening in the earth's crust from which molten rock, ashes, and steam pour out, usually forming a hill or mountain.

volume (VAHL-yoom), the measure of the amount of space which matter occupies.

water cycle, the process of water movement from the atmosphere to the ground by precipitation and from the ground to the atmosphere by evaporation.

water table, the level at which the ground is saturated with water.

weathering, the wearing away and breaking down of rocks.

zygote (ZYE-gote), the new cell resulting from the union of a sperm cell with an egg cell.

Index

abyssal plain, 122
acetylcholine, 367
adolescence, changes during, 376-377
ADP (adenosine diphosphate) 338-339
adrenal gland, 383
adrenalin, 384
Aedes mosquito, 407
aeration, of water, 63
agar, nutrient, 404
agglutinins, 410
air, composition of (table), 95; movement of, 134-135; pollution of, 99-102, 362-363; pressure of, 79-80, 89-90; properties of, 89-95
air masses, 145-146
air pollution, control of, 101-102
air pressure, 79-80, 89-90; measurement of, 148-149
air sacs, of birds, 275; of lungs, 360
albino, 297, 381
albumin, 326
alcohol, use of, 418-420
alcoholics, 418
algae, 231, 242-244
alligator, 274
allergy, 401-402

alloy, 68
altimeter, 91
alum, and water purification, 63
ameba, 251-252; reproduction by, 252-253
amino acids, 85, 325, 337, 348
ammonite(s), 32
amphibians, 268-271
androgen, 385
anemia, 328
anemometer, 135-136
aneroid barometer, 90
animal(s), classes of, 251; classification of, 193-195; cold-blooded, 225; as food consumers, 185; food habits of, 197-199; pond, 226-229; reproduction of, 186-187, 191-192; sea, 222-239; and seed scattering, 250
annual rings, 247
Anopheles mosquito, 407
Antarctic Ocean, 233
anthracite coal, 14
antibiotics, 412-413
antibodies, 410
antiseptics, 404
antitoxins, 410

anvil, of ear, 390
aorta, 357
Appalachian Mountains, formation of, 35-36
aquaculture, 130
aqueduct, 62
arachnids, 261-262
Archeozoic Era, 29, 31
Archimedes, 109-110
Arctic Ocean, 7
argon, 26, 99
artery, 357; pulmonary, 357
arthropods, 260-265
asexual reproduction, 187
astigmatism, 391
athlete's foot, 403
Atlantic Ocean, 7
atmosphere, 7, 79; layers of, 82-84; measurement of, 81-82
ATP (adenosine triphosphate), 338-339, 374
Audubon Society, 307
auditory nerve, 390
auricle, 355

bacillus, 239, 397
bacteria, 229, 238-240; and disease, 397-398; kinds of, 239, 397; reproduction by, 239-240

441

balance, of life, 300
balanced diet, 323
barograph, 90
barometer, 81, aneroid, 90
basal metabolism, 320
basic-oxygen process, 67
bathyscaphe, 107
bathysphere, 107
Beaufort wind scale, 136; table, 137
Beebe, William, 107
behavior, patterns of, 378
bends, 80, 112
beriberi, 330
biceps, 366
bile, 335
bile ducts, 335
biome(s), 211-212
biotic community, 210
birds, 274-276; feathers of, 275-276; marine, 233; observation of, 274-275
bituminous coal, 14
bladder, urinary, 364
blending, of characteristics, 380
block cutting, 304
block mountains, 36
blood, composition of, 352-353; deoxygenated, 362; oxygenated, 362; pressure of, 358-359; transfusion of, 354-355; typing of, 355
blood banks, 355
blowby system, 102
body, development of (table), 318
bone(s), growth of, 369; number of, 368
borax, 12
Borazon, 70
boron, 70
Boyle, Robert, 93
bromthymol blue, 204
bronchi, 360
budding, 187, 190
bulbs, 189
bunsen burner, 14
buoyant force, 109-110

caddis worm, 228
caissons, 94
calcite, 10
calcium phosphate, 369
calories, average daily needs (table), 319; defined, 87-88, 319; necessary, 319-321; table of, 321
calorimeter, 88
cambium layer, 247
cancer, 400, 415-417; lung, 362; signs of (table), 415
canopy, forest, 213
capillary, 357
capillary action, and soil water, 61
carbohydrate(s), 318, 324, test for, 324-325
carbon, 14, 27-28, 34
carbonates, 12
carbon dating, 27
carbon dioxide, 7, 98-99; as a waste product, 364
Carlsbad Caverns, 44
carnivores, 197, 277
cartilage, 266, 369
casein, 326
cell(s), blood, 352; defined, 175; egg, 284; and heredity, 282; numbers of, 348; parts of, 178; part of (table), 179; reproduction of, 283-284; respiration in, 338-339; sex, 284, 285; sperm, 284; study of, 345-346
cell division, 283-284; and ameba, 252-253
cell membrane, 178, 252, 346
cell theory, 175
Cenozoic Era, 28, 32
Celsius, Anders, 78
characteristics, blending of, 380; dominant, 287, 380; individual, 379; inherited (table), 380; recessive, 287-288, 380; of species, 379; transmission of, 281-294
Charles, Jacques, 93
chemosphere, 83-84

chlorella, 243-244
chlorine, 63
chloromycetin, 412
chlorophyll, 179
cholinesterase, 367
chromatin, 283
chromosomes, 282, 284, 285; number of, 284; sex, 286-287
cilia, 253
circulation, 357
cirro-cumulus clouds, 142
cirro-stratus clouds, 142
cirrus clouds, 142
clam, observation of, 259
classification, 193-195; table, 194
clay, 10
climate, changes in, 163-164; determination of, 157-164; and mountain ranges, 159-160; and wind direction, 159
climax community, 210
club mosses, 245
clotting, of blood, 354
clouds, ceiling of, 153; formation of, 142; types of, 142-143
coagulation, and water purification, 63
coal, 14, 69; anthracite, 14; bituminous, 14; pollutants from (table), 100
coal-tar, 68
coccus, 239, 397
cochlea, 390
coelenterates, 256
coenzymes, 348
coke, and coal, 69
cold-blooded organisms, 225
colonies, 239
communities, 210; biotic, 210; climax, 210
condensation, 125
conduction, 86
conductors, 86
conglomerate, 10
coniferous forest, 219

conifers, 245
conjugation, 243; of paramecia, 254
connective tissue, 349
conservation, defined, 52; forest, 303-304; mineral, 64-70; soil, 52-59; water, 60-63; wildlife, 303-304
constrictors, 273
consumers, 204
continental shelf, 122
contour planting, 56
contractile vacuole, 251-252, 254
convection currents, 87
convulsions, 385
coordination, muscle, 365-366, 377-378
copepods, 230
coral reefs, 232
cornea, of eye, 391
coronary attacks, 414
corpuscles, 352
counter-shading, 208
cover crops, 56
crayfish, 261-262
creation, of world, 296
critical balance, 196
Cromwell Current, 118
crop rotation, 58
crosses, genetic, 289-291
cross pollination, 298-299
crustaceans, 261
crystals, structure of, 17
cultures, 239
cumulonimbus clouds, 143
cumulus clouds, 142-143
current(s), ocean, 118-120
Cuvier, George, 2
cyclone, 150
cytoplasm, 178

Dana, James, 2
Darwin, Charles, 296
deciduous forests, 211, 214
deficiency diseases, 330
delta, 46
depletion, of soil, 53
desalinization, 127-128; methods of, 128-129
deserts, 217-218
De Vries, Hugo, 297
dew points, 125
diabetes, 384
dialog, 384
diamond, 12, 18
diaphragm, 360
diastolic pressure, 359
diet, balanced, 323
diffusion and cell membrane, 346; of food, 336-337
digestion, 174, 333-337; chemical, 334; chemical (table), 337 in mouth, 334; physical, 334; in small intestine, 335-336; in stomach, 335
digestive glands, 334
digestive juices, 333
dihybrid, 292
dinoflagellates, 230
disease(s), causes of (table), 402
disease(s), and bacteria, 397-398; cancer, 415; causes of (table), 402; defenses against, 409-413; and fungi, 399; glandular, 400; of heart, 414; infectious, 400; and insects, 407; and micro-organisms, 403-408; organic, 400, 414-417; and protozoans, 399-400, and radioactive materials, 401; virus, 398
distillation, and water purification, 62-63, 128
DNA (deoxyribonucleic acid, 283
dome-shaped mountains, 36
dominance, incomplete, 291, law of, 287
dominant characteristics, 287, 380
drugs, 412; narcotic, 421-422; sulfa, 412
dunes, 44
dust storms, 59

ear, parts of, 390
earth, age of, 26-27; composition of, 3-7; and external forces, 42-48; heating of, 85-89; history of, 24-35; inner core of, 415; interior of, 5-6; and internal forces, 35-41; makeup of, 4-5; shape of, 3; size of, 3; theories of formation, 24-25; tremors of, 38
earthquakes, 38-41
earthworm, 258
eating habits, 322
echinoderms, 260
ecology, 195, 212
egg cells, 191, 284
electric furnace, 68
electrocardiogram, 414
electrochemical current, 389
electrochemical impulses, 366
electron microscope, 398
elements, in earth's crust, 11-12; table, 11
Elodea, 182, 204
embryo, 192; plant, 248-249
embryology, 286
endocrine glands, 383-386
endocrine system, 383; table, 386
energy, 97; and carbohydrates, 324; release of, 338-339
environment, 382; natural, 209
enzymes, 334, 336; and ATP formation, 339-341
Eohippus, 294
epidermis, of leaf, 247, 248
epithelial tissue, 349
era(s), of geologic time table, 28-29, 31-33
erosion, defined, 8; gully, 54; till, 54; sheet, 54; soil, 54-59; wind, 58-59
esophagus, 335
estivation, of frogs, 271
estrogen, 385
Euglena, 195
evaporation, 123-124
excretion, 174, 364

443

exercise, and heart action, 359
exosphere, 84
extinction, 197
eye, parts of, 391

Fahrenheit, Gabriel, 78
farsightedness, 391
fats, 326-327; and pancreatic juice, 336; saturated, 326; test for, 326; unsaturated, 326
fatty acids, 185, 337
faults, 36
Federal Aviation Agency, 153
Fehling's solution, 324
fermentation, 240
ferns, 245
fertilization, 191, 286
fuels, 14
fibrin, 354
filament, of spirogyra, 242
filtration, 62; of water, 63
fish, kinds of, 266-270
fission, 187, 239-240; and ameba, 252-253
fissure, 6
flame test, color identification by, 20; table, 20
Fleming, Sir Alexander, 412
flotation process, 66
flower(s), 248; parts of, 248-249
flowering plants, 245-246
fluorides, 63-64
fluoridation, of water, 63
folded mountains, 35-36
food(s), as energy source, 317-318; uses of, 318
food chain, 197-198, 214; ocean, 230-231
food producers, 204
food vacuole, 251
forest(s), canopy of, 213; coniferous, 219; conservation of, 303-304; deciduous, 211, 214; destruction of, 300-302; diseases of, 302; rain, 211; tropical, 213-214
fossils, 10, 163, 294, 296;

defined, 26-27
frog, life history of, 270-271
front, cold, 146; defined, 145; stationary, 147; warm, 145
fruit, 249-250
fruit flies, 381
fungi, 240; and disease, 399

Galapagos tortoise, 274
Galen, Claudius, 316
Galileo, 78
gall bladder, 335
game management, 306-307
game preserves, 305-306
gametes, 191
gases, compression of, 93
gastric glands, 335
gems, 12
genes, 282-283; changes in, 297
geographical distribution, 295-296
geologic timetable, 28-29
Geology, defined, 2
geysers, 6
gills, 259
gizzard, of birds, 275
glaciers, 46-47; continental, 47; mountain, 47
glands, digestive, 334; gastric, 335; intestinal, 335; reproductive, 384, 385; salivary, 334; tear, 387
glandular diseases, 400
glucose, 180-181, 185, 337, 324
gluten, 326
gneiss, 10
grafting, 190
Grand Canyon, history of, 31
granite, 9
graphite, 12
grasshoppers, classification of (table), 193
grasslands, 211, 216-217
ground water, 43
growth, rate of, 374-375
growth hormones, 385
guard cells, 181, 248

Gulf Stream, 118
gully erosion, 54
gypsum, 12

habitat, 203; woodland, 216
half-life, 26
hammer, of ear, 390
Harvey, William, 316
hatcheries, fish, 307-308
heart, beat of, 358; of fish, 268; functions of, 355-356, 358
heartbeat, 358
heart disease, 414
heart-lung machine, 360
helium, 99
hemoglobin, 325, 352-353
herbivores, 197
heredity, defined, 281; laws of, 287; science of, 382
Herodotus, 1
heroin, 422
hibernation, of frogs, 270-271
highs, in weather, 149
Holy Scriptures, 171, 296
hormones, 383; growth, 385
housefly, 407
humidity, 150; absolute, 124; relative, 124
humus, 52
hurricane, 151
Hutton, James, 2
hybrid, 289
hybridization, 298-299
hydra, 187, 256-257
hydrochloric acid, 335
hydrosphere, 6-7
hydrophones, 112
hydrometer, 125
hyphae, 242

Ice Ages, 32, 34, 122
identical twins, 286; and heredity, 382
igneous rock, 8-9
immunity, 410
impulses, nerve, 388
incomplete dominance, 291
independent assortment, law

444

of, 292
Indian Ocean, 7
inert gases, 99
infections, and white blood cells, 353
infectious diseases, 400
ingestion, 173
inland waters, 221-228
inner core, of earth, 5
insect(s), 262-265; control of, 265; and disease, 407; and man, 265
insulin, 384-385
intestine(s), small, 335
intestinal glands, 335
interdependence, 204
International Geographic Year, (IGY), 121
invertebrates, 255
involuntary muscles, 388
ionosphere, 84
iris, of eye, 391
isobars, 149
isotherms, 159

jet-streams, 82
joints, kinds of, 370; of skeleton, 369

kidneys, 364; diseases of, 364; mechanical, 364

lakes, organisms found in, 224-225
land breeze, 140
large intestine, 337
Lassen Peak (volcano), 37
lava, 37
Lavoisier, Antoine, 78
law of dominance, 287
law of independent assortment, 292
law of segregation, 288-289
leaching, of soil, 53
leaves, 247-248
leeches, 258
legumin, 326
lens, of eye, 391
leukemia, 400

lichen, 198
life processes, 173; table, 174
lift pump, 95
ligaments, 369
limestone, 10, 12
Linnaeus, Carolus, 193
lithosphere, defined, 4-5
littoral zone, of ocean, 231
liver, 335
liverworts, 244
loam, 53
lows, in weather, 149
lungs, cancer of, 362; function of, 360-361

magma, 9
Malpighi, Marcello, 316
mammals, 192, 276-277; characteristics of, 276
mantle, of earth, 5
marble, 10
meiosis, 284-285
membranes, cell, 346; permeable, 346; semipermeable, 347
Mendel, Gregor, 287
mesosphere, 83
Mesozoic Era, 29, 32
metabolism, basal, 320
metallic minerals, 65-70
metamorphic rock, 10
metamorphosis, of insects, 263-264
metazoan, 251
microorganisms, and disease, 403-408
microprojector, 175
microscope, 175-177; electron, 398
Mid-Atlantic Ridge, 121
migration, 275
Miller, Stanley, 172
millibars, 148
mimicry, 209
mineralogy, 17
minerals, 10-12; chemical properties of, 19; color of common (table), 16-17; conservation of, 64-70; in diet, 328-329; table, 328; hardness of, 18-19; scale of (table), 18; metallic, 65-70; in ores (table), 65; physical properties of, 15-16; and soil, 58; structure of, 14; useful (table), 64
mitochondria, 179, 347-348
mitosis, 283-284
molds, 242
molecules, 172
mollusks, 259-260
molten rock, 25
molting, 260
moraine, 48; terminal, 48
morphine, 422
mosquito, Aedes, 407; Anopheles, 407
mosses, 244
motor end plate, 367
mountains, 219-220; block, 36; dome-shaped, 36; folded, 35-36
mucus, 388, 409
mulch, and soil, 61
Muller, Herman J., 381
muscular coordination, 377-378
muscles, contraction of, 366-367; coordination of, 365-366; involuntary, 365, 388; skeletal, 366; voluntary, 365, 366, 387
muscle tissue, 348, 349
mutations, 297, 380-381
mutation theory, 297

naming, scientific, 193
Nansen bottles, 114
narcotic drugs, 421-422
natural resources, 52
natural selection, 297
nearsightedness, 391
neon, 99
nerves, auditory, 390; impulses, 366-387, 388; olfactory, 392; optic, 391; sensory, 388
nerve tissue, 348

445

nervous system, 387-389
nicotine, 420-421
night blindness, 330
nitrates, 184
nitrogen cycle, 97-98
nitrogen-fixing bacteria, 184
nuclear membrane, 283
nucleus, 178
nutrients, classes of, 323; in common foods (table), 327; defined, 322; soil, 58-59

oceans, bottom of, 122-123; composition of, 116-117; currents of, 118-120; depth of, 112-113; of earth, 6-7; food from, 130; food chain in, 230-231; formation of, 115-116; life in, 229-233; pressure in, 108-109; salt in, 117-118; sampling, 113-114
oceanography, 106-107; physical, 230
oil pollution, 303
Old Faithful, 6
olfactory, nerve, 392
omnivores, 197
open-hearth furnace, 67
optic nerve, 391
ore, 11, 65; roasting, 66; smelting, 66
organs, 180, 349
organic diseases, 400, 414-417
organisms, change in, 294-299
osmosis, 346
outer core, of earth, 5
ovary, of flower, 248
ovules, 248
oxidation, 96
oxides, 12, 96
oxygen, 7, 96
oxygen-carbon dioxide cycle, 185

Pacific Equatorial Current, 118
Pacific Ocean, 6-7
paleontologist, 27

Paleontology, defined, 2
Paleozoic Era, 29, 31-32
palisade layer, of leaf, 248
pancreas, 335, 383-384
pancreatic juice, and fats, 336
paramecia, 253-254; conjugation by, 254; parts of, 254
parasite, 198, 398, 399-400
parathormone, 385
parathyroid glands, 384
Paricutin (volcano), 37
Pascal, Blaise, 90
Pasteur, Louis, 172, 187
pasteurization, 406
pea plants, characteristic of (table), 288
penicillin, 242, 412
permeable membrane, 346
peristalsis, 333-334
petiole, 247
Petri dishes, 404
petroleum, 14; products, 69
photochemical reaction, 363
photosynthesis, 181-185; and respiration compared, 185
Piccard, Auguste, 107
pig iron, 67
pineal gland, 384
pistil, 248
pituitary gland, 384
planaria, 191, 257-258
plankton, 230
plant(s), embryos, 248-249; and food, 198; as food consumers, 185; green, 180-185; groups of, 238-250; higher, 245-250; lower, 238-244; main groups of (table), 241; and photosynthesis, 180-183; reproduction of, 188-190, 248; seed-producing, 245-246
plasma, 352
plastid, 178
platelets, 352; and clotting, 354
pneumatic appliances, 94
polio vaccine, 410-411
pollen, 163, 248

pollination, 248; artificial, 287
pollutants, 99-100; air, 362
pollution, air, 99-100, 362-363; oil, 303
Polyceram, 69
pond, life in, 225-229
population density, 206
potassium, radioactive, 26
prairies, 211, 216
precipitation, forms of, 144
pressure, blood, 358-359; diastolic, 359; ocean, 108-109; systolic, 359; water, 109-112
Priestley, Joseph, 78
primates, 277
protective coloration, 208
protective resemblance, 209
proteins, 325-326; building of, 348; tests for, 326
Proterozoic Era, 29, 31
protoplasm, 172, 175; composition of, 345
protozoan, 251; and disease, 399
pseudopodium, 251
ptyalin, 334
pulmonary artery, 357
pupil, of eye, 391

quartzite, 10

radiation, of sun, 85
radioactive carbon, 183
radioactive material, and cancer treatment, 416-417; and disease, 401
radioactivity, 26
radioisotopes, 416-417
radiosonde, 81, 153
rain forests, 211
rain gauge, 144
reaction time, 419-420
recessive characteristics, 287-288, 380
red blood cells, 352
Redi, Francesco, 171, 186-187
reflex action, 387-388
regeneration, 190

446

reproduction, 186-192; by ameba, 252-253; asexual, 187; of bacteria, 239-240; cellular, 283-284; plant, 248; sexual, 191, 284; of spirogyra, 242-243; vegetative, 188
reptiles, 271-274; marine, 233
respiration, 174, 184-185, 360-362; cellular, 338-339; and photosynthesis compared, 185
retina, of eye, 391
rhizomes, 188
ribosomes, 348
rickets, 331
rill erosion, 54
ringworm, 403
rock, igneous, 8-9; molten, 25
Rocky Mountains, formation of, 36
roots, 246
root tip, study of, 284

Sabin, Albert, 411
salivary glands, 334
Salk, Jonas, 411
salt, 117
San Andreas Fault, 38
sandworms, 258
saphrophyte, 198, 241
satellites, weather, 155
saturated fats, 326
scales, of fish, 267
scavenger, 224
scuba gear, 108, 110, 111, 112
scurvy, 331
sea anemone, 231
sea breeze, 139
seamounts, 121-122
secondary sex characteristics, 385
sediment, ocean, 122-123
sedimentary rock, 9-10; formation of, 10
seeds, 249; scattering of, 250
segmented worms, 258
segregation, law of, 288-289

seismic soundings, 112
seismograph, 41
selection, 298; mass, 298, 299
selective cutting, 304
semipermeable membrane, 347
sense organs, 388, 389-393
sensory nerves, 388
sex(es), ratio of, 291
sex cells, 191, 284-285
sex chromosomes, 286
shale, 10
sharks, 265-266
sheet erosion, 54
shock waves, 39; primary, 39; secondary, 40; shear, 39; speed of, 40-41; surface, 39
siphon, 94, 259
skeletal system, 368
slag, 67
slate, 10
small intestine, 335
smell, sense of, 392
smelting, 66
smog, 363
snail, observation of, 259
snakes, types of, 272-273
soil, 52-53; conservation of, 52-59; depletion of, 53; erosion, 54-59; formation, 42; humus, 525; leaching, 53; loam, 52; nutrients, 58-59; topsoil, 53; and water movement, 127
solute(s), 116
solutions, 116; concentrated, 116; dilute, 116; saturated, 116
solvent, 116
species, dominant, 230
sphygmomanometer, 359
spirillum, 239, 397
spirogyra, 242-243; reproduction by, 242-243
sponges, 255
spongy layer, of leaf, 248
spontaneous generation, 187
spores, 188, 239-240
spore cases, 188, 241, 245

sucrose, 324
sulfa drugs, 412
sulfides, 12
stalactites, 44
stalagmites, 44
stamen, 248
starfish, 260
steel, production of, 67-68
stem, 246
steppes, 216
stethoscope, 358
stirrup, of ear, 390
stomach, 335
stomata, 181, 248
storms, 145-148; movement of, 147-148
stratosphere, 83
stratus clouds, 142-143
strip cropping, 56
swim bladder, 267-268
symbiont, 198
system(s), 180; body (table), 349
systolic pressure, 359

table(s), average daily calorie needs, 319; Beaufort wind scale, 137; body development, 318; body systems, 349; calories, 321; causes of disease, 402; characteristics of pea plants, 288; chemical digestion, 337; classification, 194; classification of grasshopper, 193; color of common minerals, 15, 16, 17 comparison of weather, 162; composition of air; 95; causes of disease 402; elements in earth's crust, 11; endocrine system, 386; flame test, 20; geologic time-table, 28-29; hardness scale of minerals, 18; inherited characteristics, 380; life processes, 174; main groups of plants, 241; minerals in diet, 328; minerals in ores, 65; parts of cells,

179; nutrients in common foods, 327; pollutants from coal, 100; signs of cancer, 415; useful minerals, 64; vitamins needed by the body, 332; wind barometer, 155-156
tadpole, 270
taiga, 219
taste, sense of, 392-393
taste buds, 392-393
temperature, 150; and climate, 157, 159; and inland waters, 222-224; maintenance of, 318; and weathering, 42
tendons, 366
tentacles, of Hydra, 257
terramycin, 242, 412
tetracycline, 412
theodolite, 153
theory of spontaneous generation, 171-172
thermometer(s), 81; wet and dry bulb (table), 126
thermosphere, 83
thunderstorm, 151-152
thyroid gland, 383
thyroxin, 384
Tiros I, 77
tissue, 180; connective, 349; epithelial, 349; kinds of, 348-349; make-up of, 348-349; muscle, 348, 349; nerve, 348, 349
tobacco, use of, 420-421
tobacco mosaic virus, 398-399
topsoil, 53
Torricelli, Evangelista, 78, 89, 90
toxins, 240, 398
tracers, 183
trachea, 360
tracheae, of insects, 262
traits, hereditary, 379
tree line, 196

tremors, earth, 38
triceps, 365
Trieste, 107
tropical forest, 213-214
troposphere, 82
tsetse fly, 407
tuberculin test, 411
tuberculosis, 410
tubers, 188-189
tundra, 218-219
twins, identical, 286
typhoid fever, 410

United States Bureau of Biological Survey, 307
United States Forest Service, 305
unsaturated fats, 326
uranium, 26; decay method, 26
urethra, 364
Urey, Harold, 172
urine, 364

vaccination, 410
vaccine, 410-411
vacuole, 178, contractile, 251; food, 251, 254
valves, of heart, 357
Van Allen radiation belt, 84
Van Leeuwenhoek, Anton, 175
variability, 281
vegetative reproduction, 188
vein, 357; pulmonary, 357
ventricles, of heart, 355
vertebrates, 255; structure of, 266
villi, 336-337
virus, 398-399; tobacco mosaic, 398-399
vitamin(s), 329-332; A, 330; B, 330; B_1, 330, B_2, 331; C, 331, test for, 329-330; D, 331; K, 332; needed by the body (table), 332
volcano, 37-38; observatory, 37

voluntary muscles, 387

waste products, of body, 364
water, aeration of, 63; conservation of, 60-63; distillation of, 62-63; and earth's surface, 6-7; filtration of, 63; fluoridation of, 63; and movement through soil, 127; need for, 332; pressure, 109-112; purification, 62-64
water cycle, 60-61
waterspout, 151
water table, 60
water vapor, 7, 124
weather, comparison of (table), 162; forecast of, 152-155
weathering, 42-44; and chemical reactions, 42-43
weather maps, 148-150, 153-154; prediction of, 154
weather satellites, 77, 155
weather stations, 152
weather, types of, 144-150
weight, increase in, 375-376
white blood cells, 352, 353, 354, 409
wildlife, conservation of, 303-304; decline of, 302-303; protection of, 305-306
wind, definition, 92; direction of, 150; erosion by, 58-59; measurements, 135-136; production of, 136-139; and surface features, 140-141
wind-barometer table, 155-156
worms, segmented, 258; types of, 257-258

X chromosome, 286

Y chromosome, 286
yeast, 240

zygote, 284

Acknowledgments

For some of the science definitions in the Words in Science list acknowledgment is made to the *Holt Intermediate Dictionary of American English.* The authors gratefully acknowledge the courtesy and cooperation of the following persons and organizations who have been kind enough to supply the photographs used in this book.

Alinari Art Reference Bureau: Unit 2 — Torricelli, Unit 3 — Redi
American Cancer Society: Fig. 14-16, 14-21
American Cyanamid: Fig. 14-1B
American Heart Association: Fig. 12-9B
American Museum of Natural History: Fig. 2-3B, 2-3C, 6-10, 6-28, 7-22A, 7-22B, 7-22C, 8-15B, 8-28, 9-25, 9-28, 10-9, 14-9
American Red Cross: Fig. 12-6
Australian News and Information Bureau: Fig. 9-37

Bausch and Lomb, Inc.: Fig. 14-1A, 14-1C
Bettman Archive: Unit 4 — Vesalius
Black Star: Fig. 14-19 (Burnett)
Brokaw, Dennis: Fig. 7-21, 7-25, 8-5, 9-6
Brookhaven National Laboratory: Fig. 7-7, 13-7
Buffalo Museum of Science: Fig. 2-6A, 2-6E, 2-6F

Camera Hawaii: Fig. 2-1, 5-11 (Werner Stoy)
Carolina Biological Supply Company: Fig. 9-7
Caterpillar Tractor Company: Fig. 3-3

Chicago Natural History Museum: Fig. 2-3D
Cities Service Co.: Fig. 3-17
Coast and Geodetic Survey: Fig. 2-18

Degginger, E. R.: Fig. 2-23, 5-13, 7-17, 7-18A, 7-18C, 7-20, 7-24A, 8-6, 8-8, 8-9C, 8-9D, 8-9F, 8-11, 8-15A, 8-18, 8-21, 8-22A, 8-24, 8-29A, 9-18B, 9-18C, 9-18D, 9-31A, 9-31B, 9-31C, 9-31D, 932-A, 9-32B, 9-32C, 9-32D, 9-33, 10-17, 10-18.
Department of Water Resources: Fig. 5-22
D.P.I.: Fig. 8-11

Ebony Magazine: Fig. 12-6B, 12-9B
Echlin: Fig. 9-5C
Esso Photo: Fig. 6-2, 6-17, 6-21

Fairchild Aerial Surveys: Fig. 2-14
Fish and Wildlife Service: Fig. 8-15D
Florida State News Bureau: Fig. 5-6
Ford Foundation: Fig. 14-9
Funk Bros. Seed Co.: Fig. 10A, 10B

General Biological Supply House: Fig. 7-16, 8-16A 8-16B, 8-16C, 8-16D
General Electric Research Laboratory: Fig. A-1, B-1
Grace, W. R. and Co.: 1-15
Granger Collection: Unit 2 — Lavoisier
Greenberg, Jerry: Fig. 4-14, Unit 3 — Unit opening

Hahn, Jan: Fig. 5-14
Harrity, Jerry: Fig. 14-10
Hoppock Associates: Fig. 7-4, 7-11, 7-13, 9-5A, 9-18A
Icelandic Air Lines: Fig. 5-16
Indiana Limestone Co.: Fig. 3-12
Lederle Laboratories: Fig. 14-11
Life: Fig. B-6 (Albert Fenn)
Lockheed: Fig. A-3
Los Angeles Air Pollution Control District: Fig. 4-17A, 4-17B
Luray Caverns: Fig. 2-20, 2-21
Marine Biological Laboratory: Fig. 5-9
Menschenfreund, Joan: Fig. 2-5
Merck, Sharpe and Dohme: Fig. 11-11, 13-14
Monkmeyer: Fig. 11-1A (Zimbel), 11-1B, 13-2 (Forsyth), 13-3 bottom (Hugh Rogers), 13-4 (Aigner), 13-5 (Fujihari), 13-8 (Strickler), 13-18 (Shelton)
Muench, Josef: Fig. 8-12, 8-15E
NASA: Fig. 4-3
National Audubon Society: Fig. 8-14B (Leonard Lee Rice), 10-10 (Richard C. Finke)
National Bureau of Standards: Fig. B-2, C-2
National Center for Air Pollution Control: Fig. 14-3
National Foundation March of Dimes: Fig. 10-1B
National Geographic Society: Fig. 5-2
National Medical Audiovisual Centre: Fig. 12-11, 14-9
National Park Service: Fig. 1-2
National Tuberculosis Association: Fig. 14-12A, 14-12B
New York Public Library: Unit 4—Galen, Harvey
Ostman: Fig. 8-9B (Holmaser), 8-13A (Bertil Hagert), 8-13C (Reppen), 8-17, 8-22B, 8-22C, 8-27 (Roberts)
P. D.: Unit 2 — Unit Opening Fig. 3-2, 6-7, 8-9E, 8-10, 9-36, 9-38B
Pfizer and Co.: Fig. B-4, 14-2, 14-15
Photo Researchers: Fig. 4-2, 10-16A (Joe Monroe), 6-8 (Richard Hoit), 7-2A, 7-9B, 7-18B, 8-4, 9-14B, 10-1A, 10-16B (Russ Kinne), 7-2B, 9-5B (Eric Gravel), 8-5A (Kenbrath), 8-5B (Norman Light), 9-21 (Peter David), 9-27 (Ray), 10-14 (Carl Koford), 10-14B (Ron Austing), 10-14C (Victor Englebert), 13-3 top (Van Bucher), 14-4 (Stephen Collins), 14-5 (Elizabeth Wilcox), 14-C, E (P. W. Grace)
Radio Times Hulton: Unit 1 — Cuvier
Ramsey: Fig. 5-1
Rapho Guillumette Pictures: Fig. C-1 (Hella Hammond)
Roberts, Allan: Fig. 8-2, 8-3, 8-13B, 8-29B, 8-12A, 8-12B, 13-6
Roy, Maurice: Unit 3 — Urey
San Franciso Water Department: Fig. 3-11
Shaub, B. M.: Fig. 1-5, 1-6, 1-7, 1-8, 1-9, 1-10, 1-12, 1-13
Shostal: Fig. B-5, 1-2, 11-3
Smithsonian Institute: Fig. 2-3A
Soil Conservation Service: Fig. 3-5
Spencer, Hugh: Fig. 8-26A, 8-26B
Squibb, E. R. Company: Fig. 11-6A, 11-6B
Standard Oil Company (New Jersey): Fig. 4-11
Taylor Instrument Companies: Fig. 6-16
Time: Unit 4 — Unit opening (Fritz Goro)
Triboro Bridge and Tunnel Authority: Fig. 14-3
UNESCO: Fig. 5-17, 6-6, 6-9
Union Carbide: Fig. 3-19
Union Pacific Railroad: Fig. 1-4, 2-22
United Air Lines: Fig. 14-10
United States Air Force: Fig. A-3 (left), 4-1
United States Department of Agriculture: Fig. 3-1, 3-4, 3-6, 3-7, 3-8, 5-18, 9-8, 11-4A, 11-4B, 11-4C, 11-4D, 11-5, 11-7A, 11-7B, 11-8A, 11-8B
United States Department of Commerce: Fig. 6-11, 6-18
United States Fish and Wildlife Service: Fig. 9-38A
United States Forest Service: Fig. 8-14A
United States Geological Survey: Fig. 2-11
United States Navy: Fig. 2-24, 4-16
United States Steel Co.: Fig. 3-14, 3-15, 3-18
University of Illinois: Fig. 14-18
UPI: Fig. 13-3, 14-13, 14-14
Walcott, Charles: Fig. 9-22A, 9-22B, 9-22C
Welsh Co.: Fig. 4-8
WHO: Fig. 14-8
Wide World Photo: Fig. 2-16, 2-19, 5-23, 6-19, 12-8B
Woods Hole Oceanographic Institute: Fig. 5-8B, 5-10
Yerkes Observatory: Unit 4 — Galileo
Zeiss, Carl: Fig. 10-3A, 10-3B, 10-3C, 10-3D